Multicriteria Decision-Making under Conditions of Uncertainty

Multicriteria Decision-Making under Conditions of Uncertainty

A Fuzzy Set Perspective

Petr Ekel

Graduate Program in Electrical Engineering
Pontifical Catholic University of Minas Gerais, Belo Horizonte, MG, Brazil
Graduate Program in Electrical Engineering
Federal University of Minas Gerais, Belo Horizonte, MG, Brazil
ASOTECH – Advanced System Optimization Technologies Ltda.
Belo Horizonte, MG, Brazil

Witold Pedrycz

Department of Electrical & Computer Engineering
University of Alberta, Edmonton, AB, Canada
Systems Research Institute, Polish Academy of Sciences
Warsaw, Poland

Joel Pereira, Jr.

ASOTECH – Advanced System Optimization Technologies Ltda.
Belo Horizonte, MG, Brazil

Registered Office(s)
John Wiley & Sons, Inc., 111 River Street, Hoboken, NJ 07030, USA

John Wiley & Sons Ltd, The Atrium, Southern Gate, Chichester, West Sussex, PO19 8SQ, UK

Editorial Office
111 River Street, Hoboken, NJ 07030, USA

For details of our global editorial offices, customer services, and more information about Wiley products visit us at www.wiley.com.

Wiley also publishes its books in a variety of electronic formats and by print-on-demand. Some content that appears in standard print versions of this book may not be available in other formats.

Library of Congress Cataloging-in-Publication data is applied for

Hardback: 978-1-119-53492-1

Cover image: Wiley
Cover Image: © Jorg Greuel/Getty Images

Set in 10/12pt Warnock by SPi Global, Pondicherry, India

Printed in the United States of America

V10014832_101819

Contents

Preface

The present book provides a comprehensive and prudently founded fuzzy-set-based framework for a challenging and extremely important area of decision-making. It reflects ways of representing and handling diverse manifestations of the uncertainty factor and the multicriteria nature of problems arising in system design, planning, operation, and control. The models and methods for multiobjective and multiattribute decision-making are presented along with their applications to a wide range of real-world problems coming from different areas. The book comes with a wealth of detailed appealing examples and carefully selected real-world case studies. As such, it stresses the hands-on nature of the exposition of the overall material.

Owing to the coverage of the material, we hope that this book will appeal to the communities active in various areas, in which decision-making becomes of paramount relevance: operations research, systems analysis, engineering, management, economics, and administration. Given the way in which the material is structured, the book can also serve as a useful reference material for graduate and senior undergraduate students on courses related to the areas indicated here, as well as courses on decision-making, risk management, numerical methods, and knowledge-based systems. The book could be also of interest to system analysts and researchers in areas where decision-making technologies are of paramount relevance.

The book covers the following fundamental topics spread across seven chapters:

- general questions of decision-making in problems of system design, planning, operation, and control, including optimization and decision-making problems, the uncertainty factor and its consideration, multicriteria decision-making, and the role of fuzzy sets in problems of decision-making;
- notions and fundamental concepts of fuzzy sets, including their interpretation, information granularity and fuzzy sets, fuzzy numbers, linguistic variables, operations on fuzzy sets and fuzzy numbers, fuzzy relations, and operations on fuzzy relations;

- design and processing aspects of fuzzy sets, including construction of fuzzy sets, aggregation operations, and fuzzy set transformations;
- models of multiobjective decision-making (<*X*, *F*> models) and their analysis, including the concept of Pareto-optimal solutions, approaches to incorporate additional information, methods of multiobjective decision-making, the Bellman–Zadeh approach to decision-making in a fuzzy environment as applied to multiobjective decision-making, multiobjective allocation of resources, and practical examples of solving multiobjective problems;
- models of multiattribute decision-making and their analysis based on fuzzy preference modeling (analysis of <*X*, *R*> models), including construction of fuzzy preference relations, preference formats, and transformation functions, optimization problems with fuzzy coefficients and their analysis, <*X*, *R*> models and techniques for their analysis, and practical examples of solving multiattribute problems;
- the classic approach to dealing with uncertainty of information, including notions of payoff matrices and characteristic estimates, choice criteria, and construction of states of nature;
- generalization of the classic approach to dealing with uncertainty, including its application to multicriteria decision-making under conditions of uncertainty, consideration of choice criteria as objective functions within the framework of <*X*, *F*> models, construction of objectives and elaboration of states of nature using qualitative information, general scheme of multicriteria decision-making under conditions of uncertainty, and examples of applying the general scheme.

The present book arises as a natural development and extension of the book *Fuzzy Multicriteria Decision-Making: Models, Methods, and Application* (co-authored by W. Pedrycz, P. Ekel, and R. Parreiras), published by John Wiley & Sons in 2011. It augments and deepens many of the topics covered in the previous book. Also, this book includes completely new, far reaching, and original results. For instance, we cover an approach involving the consideration of choice criteria of the classical approach to dealing with uncertainty of information as objective functions in multiobjective decision-making under uncertainty. The use of this approach helps the decision maker to overcome contradictions appearing in the analysis of multiobjective problems under conditions of uncertainty on the basis of aggregating payoff matrices.

In addition, currently, we encounter more and more problems whose essence requires the consideration of the objectives (for instance, investment attractiveness, political effect, maintenance flexibility, etc.) formed with the use of qualitative information (based on the knowledge, experience, and intuition of involved experts) at all stages of the ongoing decision process. Taking this into account, the results of the book are aimed at generating multiobjective solutions within the framework of the possibilistic approach, including multicriteria robust solutions, by constructing representative combinations of initial data, states of nature, or scenarios with direct usage of qualitative information presented along with quantitative information. It permits one to realize a process of information fusion within the multiobjective models. The described results provide the possibility for experts to apply diverse preference formats processed by transformation functions.

We would like express thanks our colleagues and friends: R.C. Berredo, A.F. Bondarenko, E.A. Galperin, A.C. Lisboa, R.M. Palhares, R.O. Parreiras, A.V. Prakhovnik (in memoriam), J.C.B. Queiroz, G.L. Soares, R. Schinzinger (in memoriam), D.A.G. Vieira, and V.V. Zorin for thorough discussions, encouragement, and support.

We would like to thank our graduate students: T.M.M. Coelho, L.R. Figueiredo, M.F.D. Junges, R.B. Pereira, V.F.D. Ramalho (the results of his M.Sc. dissertation have been helpful in writing Chapter 7 of the book), S.P. Rocha, J.N. Silva, L.M.L. Silva, and V. Tkachenko for their dedication and hard work.

We are also grateful to the team of professionals at Wiley, including Brett Kurzman, Victoria Bradshaw, Karthiga Mani, and Lynette Woodward, for providing expert advice and encouragement, and assistance during the tenure of the project.

1

Decision-Making in Problems of System Design, Planning, Operation, and Control

Motivation, Objectives, and Basic Notions

The main objective of this chapter is to offer the reader a broad perspective on the fundamentals of decision-making problems, provide their general taxonomy in terms of criteria, objectives, and attributes involved, emphasize the objectivity and relevance of the uncertainty factor, classify the types of uncertainty, discuss ways of considering the uncertainty factor, and highlight the aspects of rationality of decision-making processes. This chapter also highlights the fundamental differences between optimization and decision-making problems, between the concepts of optimal solutions and robust solutions as well as between approaches to their construction. The main objectives and characteristics of group decision-making are discussed. The role of fuzzy sets is stressed in the general framework of decision-making processes. The most important advantages of their application to individual as well as group decision-making are discussed. The chapter also clarifies necessary notations and terminology (for instance, $<X, F>$ models and $<X, R>$ models) used throughout the book.

1.1 Decision-Making and Its Support

The life of each person is filled with alternatives. From the moment of conscious thought to a venerable age, from morning awakening to nightly sleeping, a person is faced with the need to make certain decisions. This need is associated with the fact that any situation may have two or more mutually exclusive alternatives and it is necessary to choose one among them. The decision-making process, in the majority of cases, consists of the evaluation of alternatives and the choice of the most preferable from them.

Pospelov and Pushkin (1972) indicate that making the "correct" decision means choosing such an alternative from a possible set of alternatives, in which, by considering all the diversified factors and contradictory requirements, an overall value will be optimized. That is, it will be favorable in achieving the goal sought to the maximal possible degree.

Multicriteria Decision-Making under Conditions of Uncertainty: A Fuzzy Set Perspective,
First Edition. Petr Ekel, Witold Pedrycz, and Joel Pereira, Jr.

If the diverse alternatives met by a person are considered as a set, then this set usually includes at least three intersecting subsets of alternatives related to personal life, social life, and professional life. As possible examples, we can indicate, for instance, deciding where to study, where to work, how to spend time on a vacation, who to elect, and many others.

At the same time, if we speak about any organization, it faces different goals and achieves them through the use of diverse types of resources (material, energy, financial, human, etc.), and the performance of managerial functions such as organizing, planning, operating, controlling, and so on (Lu et al. 2007). To fulfill these functions, managers need to participate in the continuous decision-making process. Since each decision supposes a reasonable and justified choice realized among different alternatives, the manager can be called a decision-maker (DM). DMs can be managers at various levels, from a technological process manager to a chief executive officer of a large company, and their decision problems can vary in nature. Besides, decisions can be made by individuals or groups (individual decisions are usually made at lower managerial levels and in small organizations and group decisions are usually made at high managerial levels and in large organizations). As possible examples, we can indicate, for instance, deciding what to buy, where to buy, when to begin a production process, whom to employ, and many others. These problems can concern logistics management, customer relationship management, production planning, and so on.

A person makes simple, habitual decisions easily and frequently in an automatic and subconscious way, without too much intensive thinking. However, in many cases, alternatives are related to complex situations that are characterized by a contradiction of requirements and multiple criteria, ambiguity in evaluating situations, errors in the choice of priorities, and so on. All these factors substantially complicate a way in which decisions are being made.

Furthermore, various facets of uncertainty are commonly encountered in a wide range of problems of an optimization character, which are inherently present in the design, planning, operation, and control of complex systems (engineering, economical, ecological, etc.). In particular, diverse manifestations of the uncertainty factor are associated, for instance, with (Ekel 1999; Pedrycz et al. 2011):

- the impossibility or inexpediency of obtaining sufficient amounts of information with the necessary degree of reliability;
- the lack of reliable predictions of the characteristics, properties, and behavior of complex systems that reflect their responses to external and internal actions;
- poorly defined goals and constraints in the design, planning, operation, and control tasks;
- the infeasibility of formalizing a number of factors and criteria and the need to take into account qualitative (semantic) information.

Considering the essence of the manifestations of the uncertainty factor listed here, more concisely, it is possible (Stewart 2005; Durbach and Stewart 2012) to talk about internal uncertainties (related to DM values and judgments) and external uncertainties (defined by environmental conditions lying beyond the control of a DM).

The described situation with the uncertainty involved is to be considered as natural and unavoidable in the context of problems of complex systems. In principle, it is impossible to reduce these problems to exact and well-formulated mathematical problems; to do this, it is necessary, in one way or another, to "discard" the uncertainty and accept some hypothesis (Pedrycz et al. 2011). However, the construction of hypotheses is a prerogative of the substantial analysis; in reality, this is the formalization of informal situations. One of the ways to address the problem is the formation of subjective estimates based on knowledge, experience, and intuition of involved experts, managers, and DMs in general, and the definition of the corresponding preferences.

Thus, DMs are forced to rely on their own subjective ideas of the efficiency of possible alternatives and importance of diverse criteria. Sometimes, this subjective estimation is the only possible basis for combining the heterogeneous physical parameters of a problem to be solved into a unique model, which permits decision alternatives to be evaluated (Larichev 1987). At the same time, there is nothing unusual and unacceptable in the subjectivity itself. For instance, experienced managers perceive, in a broad and well-informed manner, how many personal and subjective considerations they have to bring into the decision-making process. On the other hand, successes and failures of the majority of decisions can be judged by people on the basis of their subjective preferences.

However, the most complicated aspect is associated with the fact that the essence of problems solved by humans in diverse areas has been changed in recent decades (Trachtengerts 1998). New, more complicated and unusual problems have emerged. For many centuries, people made decisions by considering one or two main factors, while ignoring others that were perceived to be marginal to the essence of the problem. They lived in a world where changes in the surroundings were few and new phenomena arose "in turn" but not simultaneously.

Presently, this situation has changed. A considerable number of problems, or probably the majority of them, are multicriteria in nature where it is necessary to take into account many factors. In these problems, a DM has to evaluate a set of influences, interests, and consequences that characterize decision alternatives. For instance, in decision-making dealing with the creation of enterprises, it is necessary to consider not only the expected profits and necessary investments, but also market dynamics, the actions of competitors, and ecological, political, social factors, and so on.

Considering all the aspects listed here, it is necessary to stress that recognition of the factor of subjectivity of a DM in the process of decision-making conflicts with the fundamental methodological principle of operational research: the search for an objectively optimal solution. Recognition of the right of a DM in the subjectivity of decisions is a sign of the appearance of a new paradigm of multicriteria decision-making (Kuhn 1962). However, in problems with multiple criteria, an objective component always exists. Usually, this component includes diverse types of constraints imposed by the environment on possible decisions (availability of resources, temporal constraints, ecological requirements, social situations, etc.).

A large number of investigations of the psychological character demonstrate that DMs, not being provided with additional analytical support, use simplified and sometimes contradictory decision rules (Slovic et al. 1977).

Besides, Lu et al. (2007) share the opinion given here (Trachtengerts 1998) and indicate that decision-making in the activities of organizations is more complicated and difficult because the number of available alternatives is much larger today than ever before. Due to the availability of information technologies and communication systems, especially the Internet and its search engines, it is possible to find more information quickly and therefore more alternatives can be generated. Second, the cost of making errors can be big enough because of the complexity of operations, automation, and the chain reaction that an error can cause in many parts, in both vertical and horizontal levels, of organizations. Third, there are continuous changes in the fluctuating environment and more uncertainties in the impacting elements, including information sources and information itself. It is also very important that the rapid change of the decision environment requires decisions to be made quickly. These reasons mean organizational DMs require increasing technical support to help make high-quality decisions. A high-quality decision related, for example, to bank management, is expected to bring greater profitability, lower costs, shorter distribution times, and increased shareholder value, attracting more new customers or resulting in a certain percentage of customers responding positively to a direct mail campaign.

Decision support consists of assisting a DM in the process of decision-making. For instance, this support may include (Trachtengerts 1998):

- assisting a DM in the analysis of an objective component; that is, in the understanding and evaluation of the existing situation and constraints imposed by the surroundings;
- revealing DM preferences; that is, revealing and ranking priorities, considering the uncertainty in DM estimates, and shaping the corresponding preferences;
- generating possible solutions; that is, shaping a list of available alternatives;
- evaluating possible alternatives, considering DM preferences and constraints imposed by the environment;
- analyzing the consequences of decision-making;
- choosing the best alternative from the DM's point of view.

Generally, computerized decision support is based on the formalization of methods for obtaining initial and intermediate estimates given by a DM and on the algorithms for a proper decision process. The formalization of methods for generating alternatives, their evaluation, comparison, choice, prioritization, and/or ordering is extremely complicated. One of the main complexities is associated with the fact that a DM, as a rule, is not ready to provide quantitative estimates in the decision process, is not accustomed to the evaluation of proper decisions on the basis of applying formal mathematical methods, and analyzes the consequences of decisions with significant difficulties.

In fact, decision support systems have existed for a long time; for example, one can refer here to councils of war, ministry boards, various meetings, analytical centers, and so on (Trachtengerts 1998). Although they were never called decision support systems, they executed the functions of such systems, at least partially.

The term "decision support system" appeared at the beginning of the 1970s (Eom 1995). There are several definitions of this concept, such as that given in Larichev and Moshkovich (1996): "Decision support systems are man–machine objects, which permit a DM to use data, knowledge, objective and subjective models for the analysis and solution of semi-structured or unstructured problems."

Taking into account this given definition, it is necessary to indicate that one of the important features of decision-making problems is associated with their structures. In particular, it is possible to distinguish among structured, semi-structured, and unstructured problems of decision-making (Simon 1977; Larichev and Moshkovich 1996; Lu et al. 2007). The last two types of decision-making problems are also called ill-structured.

In *structured* problems (quantitatively formulated problems), essential relationships are established so convincingly that they can be expressed in numbers or symbols that receive, ultimately, numerical estimates. Such problems can be described by existing "traditional" mathematical models. Their analysis becomes possible by applying standard methods.

Unstructured problems (qualitatively expressed problems) include only a description of the most important resources, indicators, and characteristics. Quantitative relationships between them are not known. These problems cannot be described by existing traditional mathematical models and cannot be analyzed by applying standard methods yielding "traditional" solutions.

Finally, *semi-structured* problems (or mixed problems) include quantitative as well as qualitative elements. As these problems are analyzed, qualitative, little-known, poorly explored, uncertain parameters have a tendency to dominate. These problems are positioned between structured and unstructured problems, having both structured and unstructured elements. The solution of these problems involves a combination of both standard solution procedures and active DM participation.

Taking into account this classification, typical problems in operational research can be called *structured*. This class of problem is widely used in the design, planning, operation, and control of engineering systems. For example, it is possible to talk about the design of forms of an aircraft hull, planning of water supply systems, control of power systems, and so on.

The distinctive characteristics of unstructured problems can be outlined as follows (Larichev and Moshkovich 1996).

- uniqueness of choice in the sense that, at any time, the problem is a new one for a DM or it has new properties in comparison to a similar problem solved in the past;
- uncertainty in the evaluation of alternative solutions;
- the qualitative character of the evaluations of problem solutions, most often formulated in verbal form;
- the evaluation of alternatives obtained only on the basis of the subjective preferences of a DM;
- the estimates of criteria obtained only from experts.

Typical unstructured problems are associated, for example, with planning new services, hiring executives, selecting a locale for a new branch, choosing a set of research and development projects, and others.

If we talk about semi-structured problems, their solutions are based on applying traditional analytical models as well as models based on DM preferences. As an example, one can point at the problem related to liquidation of the consequences of extraordinary situations associated with radioactive contamination (Trachtengerts 1998). In forming its solution, analytical models can be applied to define the degree and character of radioactive contamination for given temporal intervals. At the same time, models based on DM preferences can be applied in the choice of measures for liquidation of the consequences of radioactive contamination. It is also possible to qualify many problems associated with economic and political decisions, medical diagnostics, and so on as semi-structured problems.

Returning to the issue of computerized decision support, it should be noted that, due to the large number of components (variables, functions, and parameters) involved in many decisions, this has become a basic requirement to assist DMs in considering and analyzing the implications of various courses of decision-making (Lu et al. 2007). Besides, the impact of computer technologies, particularly Internet- and Intranet-based, on organizational management is increasing rapidly. Here, the main tendency is associated with the fact that computer applications in organizations are moving from transaction processing and monitoring activities to problem analysis and finding solutions (Lu et al. 2007). Internet- or intranet-based online analytical processing and real-time decision support are becoming the cornerstones of modern management, in particular within the elaboration of e-commerce, e-business, and e-government.

There is a trend toward providing managers with information systems that can assist them directly with their most important task, namely making decisions.

A detailed description of the advantages generated by applying computerized decision support systems for individual as well group decision-making is given in Lu et al. (2007). At the same time, the authors of Lu et al. (2007) indicate that the important issue is that, with computerized decision support technologies, many complex decision-making problems can now be handled effectively. However, these technologies can be better used in analyzing structured problems rather than semi-structured and unstructured problems. In an unstructured problem, only part of the problem can be supported by advanced tools such as intelligent decision support systems. For semi-structured problems, the computerized decision support technologies can improve the quality of information on which the decision is based by providing not just a single, unique solution, but a range of alternative solutions from the decision uncertainty regions. Their occurrence and their essence will be discussed in the next section.

1.2 Problems of Optimization and Decision-Making

Is there any difference between the notions of "optimization" and "decision-making?" Are these notions synonymous or not? Partial answers to these questions have been given in the previous section. However, deeper and more detailed considerations are required.

A traditional optimization problem is associated with the search for an extremum (minimum or maximum, in accordance with the essence of the problem) of a certain objective function, which reflects interests of a DM, when observing diverse types of constraints (related to allowable resources, physical laws, standards, industrial norms, etc.). Formally, it is possible to represent an optimization problem as follows:

$$F(x) \rightarrow \underset{x \in L}{\text{extr}} \tag{1.1}$$

where L is a set of feasible solutions in $\mathbf{R}^n$ defined by the considered constraints.

To solve the problem Eq. (1.1), it is necessary to find x^0 such that

$$x^0 = \arg \underset{x \in L}{\text{extr}}\, F(x) \tag{1.2}$$

If numerical details to Eq. (1.1) have been provided and we can obtain a unique solution without any guidance or assistance of a DM, then Eq. (1.1) forms an optimization problem.

Generally, an optimization problem may be complicated from the mathematical point of view, and a large amount of time might be required to generate a solution. Can human participation in the search for a solution be useful? Without any doubt, such participation could be useful, because, for instance, the

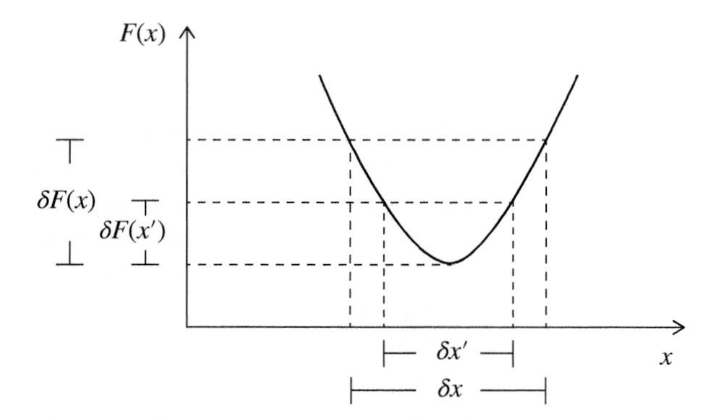

Figure 1.1 Decision uncertainty region and its reduction through the reduction of the uncertainty of information.

indication or change of initial points for a search or the introduction of heuristics can reduce the time necessary to obtain an optimal solution. However, this cannot change the solution and, in principle, a unique solution to the problem can be obtained without human participation.

At the same time, the presence of any type of uncertainty can require human participation in order to generate a unique solution to the problem.

For instance, the uncertainty of information gives rise to some decision uncertainty regions. As an example, Figure 1.1 (Pedrycz et al. 2011) demonstrates that the uncertainty of information $\delta F(x)$ in the estimation of an objective function $F(x)$ leads to a situation where formally the solutions coming from a region δx cannot be distinguished, thus generating a decision uncertainty region. Taking this into consideration, the formal formulation Eq. (1.1) can be transformed to the following form:

$$F(x,\theta) \to \underset{x \in L(\theta)}{\text{extr}} \tag{1.3}$$

where θ is a vector of uncertain parameters, whose existence changes the essence of Eq. (1.1). In particular, we can say that the solution Eq. (1.2) is an optimal one for a concrete realization of θ (a concrete hypothesis); however, for some other realization (another hypothesis), it is no longer optimal.

What are the ways to reduce this uncertainty region? The first way is to elicit information (let us not forget that any information has some cost), for example, by acquiring additional measurements or examining experts to reduce the level of uncertainty. As shown in Figure 1.1, the reduction of the uncertainty $\delta F(x)$ to $\delta F(x')$ permits one to obtain a reduced decision uncertainty region with $\delta x' < \delta x$.

However, if there is no possibility of reducing the uncertainty of information, it is possible to resort to another approach. It is associated with introducing

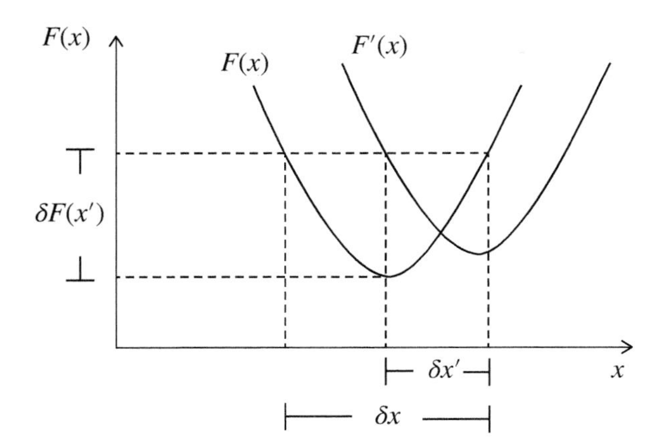

Figure 1.2 Decision uncertainty region and its reduction through the introduction of additional criteria.

additional criteria to try to reduce the decision uncertainty regions. As demonstrated in Figure 1.2, the introduction of the objective function $F'(x)$ allows one to reduce the decision uncertainty region as well, providing $\delta x' < \delta x$.

On the other hand, the existence of more than one objective function may be considered as uncertainty as well (Pedrycz et al. 2011): uncertainty of goals because this situation may be characterized as "we do not know what we want." Although the nature of this type of uncertainty is not the same as the uncertainty of available information, it also leads to the generation of decision uncertainty regions.

In particular, let us consider the simple problem of minimizing two objective functions $F_1(x) = F_1(x_1, x_2)$ and $F_2(x) = F_2(x_1, x_2)$, considering a set of feasible solutions L. We can transform L from the decision space to a certain region L_F of the space of objective functions $F_1(x)$ and $F_2(x)$ (or, simply, the objective space). In Figure 1.3, it is possible to observe that point a corresponds to the best solution ($\min_{x \in L} F_1(x)$) from the point of view of the first objective function. On the other hand, point b corresponds to the best solution ($\min_{x \in L} F_2(x)$) considered from the viewpoint of the second objective function.

Is point c a solution to the problem? Yes, it is. Is it possible to improve this solution? Yes, we can do it by passing to point d. Can we improve this solution? Yes, this is possible by passing to point e. Can we improve this solution? This is possible by passing to point f. Is it improving? No, we cannot advance here. It is possible to pass to point g but this step does not make the resulting solution any better: we can improve it from the point of view of $F_1(x)$, but deteriorate its quality from the point of view of $F_2(x)$. In a similar way, by passing to point h, we can improve the solution from the point of view of but deteriorate it from the point of view of $F_1(x)$.

Thus, formally, the solution to the problem presented in the objective space is a boundary Ω_L^P of L_F located between points a and b. The set $\Omega^P \subseteq L$

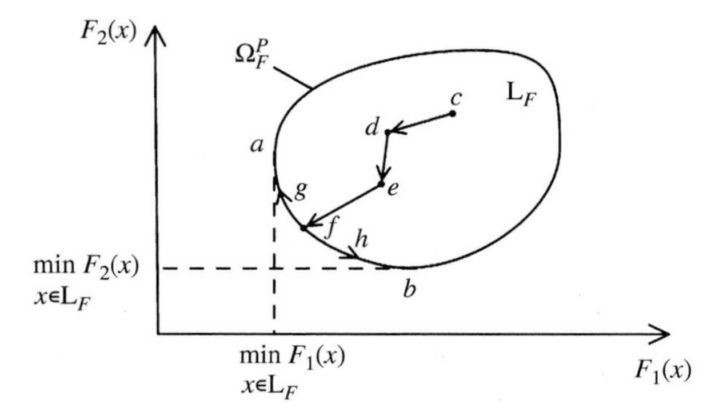

Figure 1.3 The concept of Pareto-optimal solutions.

corresponding to Ω_F^P is the problem solution, which is called a Pareto-optimal solution set. This concept of optimality was proposed by Edgeworth (1881) and was further generalized by Pareto (1886). Although we say that Ω^P is the problem solution, from a formal point of view, this is not a solution that can be implemented. In reality, it is the decision uncertainty region. The choice of a particular Pareto-optimal solution is based on the DM's involvement.

The more difficult situations are associated with problems where uncertainty of information exists as well as the uncertainty of goals. Some features of their analysis are discussed in the next section.

The problems of an optimization character, which include the uncertainty of information and/or the uncertainty of goals and demand human participation in their solution, are inherent problems in decision-making. Taking this into consideration, it is necessary to make some additional observations.

One of the most important criteria (Larichev 1984) for classifying decision-making problems is the existence or lack of an objective model established for the corresponding problem. Note here that it is not uncommon to encounter situations where it is impossible to talk about the existence of objective functions in decision-making problems. The models that can be used for analyzing these problems reflect "a point of view" and, in a more general sense, the "world outlook" of a DM. In these cases, an obvious question is how to choose actions that correspond, in the best way, to the preferences of a DM (Keeney and Raifa 1976; Pedrycz et al. 2011) and are based on his/her knowledge, experience, and intuition. Taking this into account, semi-structured and unstructured problems, classified in the previous section, are subjects of decision-making.

In conclusion, the following general tendency is visible. If we intend to solve an optimization problem, we generally look for the best solution. If we talk about a decision-making problem, the methodology used to solve it is quite distinct: we do not look for the best solution, but apply information arriving from different sources and try to eliminate some alternatives, which are dominated by other alternatives, in order to reduce the decision uncertainty regions.

1.3 Uncertainty Factor and Its Consideration

The incorporation of the uncertainty factor in constructing mathematical models, related to problems of an optimization character, serves as a means for increasing their adequacy and, as a result, the credibility and factual efficiency of decisions based on their analysis.

Durbach and Stewart (2012) classify the following five formats of uncertainty:

- probabilities;
- decision weight;
- explicit risk measures;
- fuzzy numbers;
- scenarios.

However, it is possible to distinguish two general approaches to taking into account the uncertainty of initial information: probabilistic and possibilistic. The first one is based on the preferential use of statistical data. Sometimes, this approach can be useful. However, generally, it is difficult or impossible to speak about the future, applying only information of the past. Taking this into account, we try to utilize the possibilistic approach, which consists of constructing and analyzing diverse representative combinations of initial data, states of nature, or scenarios, on the basis of applying information of the past as well as guesses, assumptions, and so on, based on knowledge, experience, and intuition of the involved experts.

The presence of several representative combinations of initial data, states of nature, or scenarios makes the process of decision-making under conditions of uncertainty completely different from traditional ways of solving problems by traditional methods of operational research. The majority of approaches, methods, and techniques of operational research is based on the conception of the existence of an "optimal solution." Considering this, most solution search strategies of operational research are directed at obtaining this "optimal solution." Under conditions of uncertainty, the notion of the "optimal solution" does not work: the "optimal solution" obtained for one scenario is not the "optimal solution" for another scenario. Taking this into account, it is necessary to talk about generating so-called "robust solutions": solutions that are intended to satisfy all possible scenarios in a maximal degree. However, if attempts at constructing these solutions for monocriteria problems are well reflected in the literature (for instance, Luce and Raiffa 1957; Raiffa 1968; Belyaev 1977; Webster 2003), the questions of their generation for multicriteria problems are not covered in the literature. Considering this, one of the main objectives of this work is to fill this gap, in particular, on the basis on combining two branches of mathematics of uncertainty such as the elements of game theory and fuzzy set theory.

Taking this into account, it becomes necessary to indicate that the general methodology of analyzing diverse decision-making problems that will pass through the book is different from the methodology of operational research: as was emphasized in the previous section, we do not search for a unique problem solution, but search for alternatives dominated by others ones, reducing, step-by-step, the number of non-dominated solution alternatives and, therefore, reducing the regions of solution uncertainty.

1.4 Multicriteria Decision-Making: Multiobjective and Multiattribute Problems

The uncertainty of goals in decision-making is an important manifestation of uncertainty that relates to a multicriteria character of many problems encountered in the project, planning, operation, and control of complex systems of different nature. Some professionals in the field of decision-making and systems analysis (for example, Lyapunov 1972) affirm that, from a general point of view, this type of uncertainty is the most difficult to overcome and handle because "we simply do not know what we want." In reality, this type of uncertainty cannot be effectively captured only on the basis of applying formal models and methods, as sometimes the unique information sources are the individuals who make decisions.

Multicriteria decision-making is related to making decisions in the presence of multiple and conflicting criteria. Multicriteria decision-making problems may range from our daily life decision problems, such as a purchase of a car, to those affecting entire nations, as in the judicious use of money for the preservation of national security (Lu et al. 2007).

However, even with the existing diversity, all multicriteria decision-making problems share the following common characteristics (Hwang and Yoon 1981):

- multiple criteria: each problem has multiple criteria, which can be objectives or attributes;
- conflicting criteria: multiple criteria conflict with each other;
- incommensurable units: criteria may have different units of measurement;
- design/selection: solutions to multicriteria decision-making problems are either to design the best alternative(s) or to select the best one among previously specified finite alternatives.

Taking this into account, two types of criterion should be distinguished: objectives and attributes. In such a manner, multicriteria decision-making problems can be classified into two wide classes (Hwang and Masud 1979; Hwang and Yoon 1981):

- multiobjective decision-making;
- multiattribute decision-making.

Although the main difference between these two classes, as the authors of Hwang and Masud (1979) and Hwang and Yoon (1981) indicate, is that the first concentrates on continuous decision spaces and the second focuses on problems with discrete decision spaces, the first class may include problems with integer, Boolean, and discrete variables, for instance Villareal and Karwan (1982) and Ramesh et al. (1986).

To proceed with the next steps, some basic concepts and terminology are given next based on Hwang and Masud (1979); Hwang and Yoon (1981); and Lu et al. (2007).

Criteria are the standard of judgment or rules to test acceptability. In the multicriteria decision-making literature, they indicate objectives and/or attributes.

Objectives are the reflection of the desire of DMs and indicate the direction in which DMs want to concentrate. Multiobjective decision-making problems, as a result, involve the design of alternatives that optimize or at least satisfy the objectives of DMs.

Goals are entities desired by DMs and expressed in terms of a specific state in space and time. Thus, while objectives give the desired direction, goals give a desired (or target) level to achieve.

Attributes are the characteristics, qualities, or performance parameters of alternatives. Multiattribute decision-making problems involve the selection of the "best" alternative from a pool of preselected alternatives described in terms of their attributes.

Generally, multiobjective decision-making is known as the continuous type of multicriteria decision-making and its main characteristics are that the DM needs to achieve multiple objectives while these objectives are non-commensurable and conflict with each other. A multiobjective decision-making model includes a vector of decision variables, objective functions that describe the objectives, and constraints. The DM attempts to maximize or minimize the objective functions.

At the same time, multiattribute decision-making is related to making preference decision (that is, comparison, choice, prioritization, and/or ordering) over the available alternatives that are characterized by multiple, usually conflicting, attributes. The main peculiarity of multiattribute decision-making problems is that there are usually a limited number of predetermined alternatives that are associated with a level of achieving the attributes. Based on the attributes, the final decision is to be made.

Finally, it is necessary to discuss in detail the concept of alternatives. How to generate alternatives is an important moment of the process of multiobjective and multiattribute decision-making model construction (Lu et al. 2007). In almost all multiobjective decision-making models, the alternatives are generated automatically by the models. In the case of multiattribute decision-making models, however, it is necessary to generate alternatives manually. Sometime, the essence of the problem defines the number of alternatives. However, in general, how and when to stop generating alternatives becomes a very important issue. Generating alternatives significantly depends on the availability and cost of information, and also requires reliance on expertise in the problem area. Alternatives can be generated with the use of heuristics as well, and they could come from either individuals or groups.

The issues related to the necessity of setting up and solving multicriteria problems as well as the classification of decision-making situations, which need the application of the multicriteria approach, have been discussed in many works (for instance, Larichev 1984; Gomes et al. 2002). However, it is possible to identify two major types of situation that require the application of a multicriteria approach (Ekel 2002; Pedrycz et al. 2011):

- Problems whose solution consequences cannot be estimated with a single criterion: these problems are associated with the analysis of models including economic as well as physical indices (when alternatives cannot be reduced to comparable form) and also by the need to consider indices whose cost estimation is hampered or impossible (for example, many power engineering problems are considered on the basis of technological, economical, ecological, and social nature criteria: Berredo et al. 2011; Ekel et al. 2016).
- Problems that may be solved on the basis of a single criterion (or several criteria). However, if the uncertainty of information does not permit one to derive unique solutions, it is possible to reduce these problems to multicriteria decision-making by applying additional criteria, including those of a qualitative character (for example, "flexibility of development," "complexity of maintenance," "attractiveness of investments," and so on, whose utilization is based on the knowledge, experience, and intuition of involved experts). This can serve as a convincing means to contract the corresponding decision uncertainty region. It could be regarded as an intuitively appealing approach exercised in the practice of decision-making.

In accordance with the major types of situations outlined before, two classes of models, so-called $<X, F>$ models and $<X, R>$ models (Ekel 2001, 2002; Pedrycz et al. 2011) can be constructed. Both of these classes of models are comprehensively discussed in detail in the book. The $<X, F>$ models correspond to multiobjective decision-making problems, which include a vector of objective functions F. At the same time, the $<X, R>$ models correspond to the multiattribute decision-making problems and include a vector of fuzzy preference relations R (Orlovsky 1981; Fodor and Roubens 1994), which play the role of attributes. In the book, the construction and analysis of both types of models are illustrated by considering diverse practical problems.

The $<X, R>$ models are also used in the statement and solution of problems of group decision-making, which are briefly discussed in the next section.

Finally, two fundamental questions such as “What to do?” and “How to do it?” arise in the design, planning, operation, and control of complex systems of diverse nature (Pedrycz et al. 2011). Taking this into account, it is necessary to indicate the following.

To decide on the first fundamental question, if we speak, for instance, about diverse types of planning (strategic, new business, innovation, expansion, maintenance, operational, etc.) activities, usually, it is necessary to evaluate, compare, choose, prioritize, and/or order solutions or alternatives (strategic actions, new business projects, innovation projects, expansion alternatives, maintenance actions, operational strategies, etc.) that permit one to achieve the stated planning objectives. The answers to the first fundamental question are based on processing of information related to different perspectives, such as “investment attractiveness,” “implementation cost,” “political impact,” “development flexibility,” and so on and can be elaborated by constructing and analyzing $<X, R>$ models (Pedrycz et al. 2011; Ekel et al. 2016).

The second fundamental question is associated with the allocation of various types of resources (financial, human, logistics, etc.) or their shortages between solutions or alternatives (strategic actions, business projects, innovation projects, expansion alternatives, maintenance actions, operational strategies, etc.) to achieve the stated objectives in the maximal degree. The rational allocation of resources and the answers to the second fundamental question can be obtained by constructing and analyzing $<X, F>$ models (Pedrycz et al. 2011; Ekel et al. 2016).

1.5 Group Decision-Making: Basic Notions

Although in this book we do not consider in detail the issues of group decision-making, many of its results are very helpful for group decision-making (first of all, the results of Chapter 5). Taking this into account, we discuss the basic notions of group decision-making next.

Group decision-making is defined as a decision situation in which there is more than one DM.

The group members have their own attitudes and motivations, recognize the existence of a common problem, and attempt to reach a collective decision (Lu et al. 2007). The necessity of applying procedures of group decision-making is associated with the following considerations.

There are many real-world situations, for instance, at the high managerial levels of organizations, when the decision problems involve wide knowledge areas that are beyond a single individual (this is particularly true when the decision environment becomes more complex and multifaceted). As a consequence, it is usually necessary to attract more than one professional to the decision process. This is particularly valid in environments with a diverse workforce, where decisions require multiple perspectives and different areas of expertise of the individuals involved in the decision process. It is possible to indicate the following advantages of group decision-making over individual decision-making (Tan et al. 1995):

- Group decision-making allows more intellectual resources to be gathered to support the decision. The resources available to the group include the individual competencies, intuition, and knowledge.
- With the participation of multiple experts, it becomes possible to distribute among them the labor related to acquiring and processing the vast amount of information pertaining to the decision.
- If the group members exhibit divergent interests, the final decision tends to be more representative of the needs of the organization.

In addition, it is necessary to indicate an important factor of the possibility to share the responsibility for the decision between members of the group.

Lu et al. (2007) highlight the following important characteristics of group decision-making:

- the group performs a decision-making task;
- group decision-making may cover the whole process of transfer from generating ideas for solving a problem to implementing solutions;
- group members may be located at the same place or at different places;
- group members may work at the same or different times;
- group members may work for the same or different departments or organizations;
- the group can be at any managerial level;
- there may be conflicting opinions in the group decision process among group members;
- the decision might have to be accomplished in a short time;
- group members might not have complete information for the decision;
- some required data, information, or knowledge for a decision may be located in many sources and some may be external to the organization.

Considering these characteristics, it should be noted that quite often, the group members may be at different locations and may be working at different times. Thus, they need to communicate, collaborate on, and access a diverse set of information sources, which can be met by the development of the internet and its derivates (intranets and extranets). The questions of constructing and utilizing web-based group decision support systems are discussed, for instance, in Lu et al. (2007) and Kokshenev et al. (2014).

With respect to the common goals and interests of the experts in the group, it is possible to distinguish two environments; namely, cooperative and non-cooperative work (Lu et al. 2007). In cooperative decision-making, all the experts are supposed to work together to achieve a decision for which they will share the responsibility. In non-cooperative decision-making, the experts play the role of antagonists or disputants over some common interest for which they are to negotiate. Taking this into account, it should be made clear that this book addresses problems of group decision-making in the cooperative environment.

As in cooperative work, the experts share responsibility for the decision (and, as indicated before, they also may participate in the implementation of the selected solution), it is important to guarantee that each member is satisfied with the selected solution (Pedrycz et al. 2011). Obviously, the commitment of the group to the implementation of the outcomes depends on the level of consensus achieved by the group. Therefore, a group decision constructed by means of domination and enforced concessions should be considered inferior to an individual decision, because it will probably face more difficulties in its implementation. Therefore, achieving a veritable consensus on the solution is an important task for the group. However, it should be noted that achieving unimprovable concordance among the experts is very difficult or impossible. Although, ideally, the condition for terminating decision-making process under group settings should be the achievement of a unanimous solution, in reality, because that unanimous solution hardly ever exists, it is sufficient to meet the alternative that is the most satisfactory for the group as a whole (Pedrycz et al. 2011). Otherwise, the decision will probably take longer than is admissible or affordable.

Among the reasons for the occurrence of discordance among the group members, it is possible to indicate the following (Pedrycz et al. 2011):

- Although members of the group are to share the primary goal, which obviously is to generate the solution that provides the maximum benefit for the organization, their secondary goals may be just partially shared. For instance, when experts are representatives of different departments, it is natural that they would have specific interests associated with the priorities and needs of the corresponding departments.
- Each group member usually has a distinct perception of the problem and own intuition that may be difficult to formalize and communicate to the other group members.
- Generally, no single expert knows the entire domain of the decision problem. Each expert usually has access to different sources and different profiles of information. In particular, certain group members may have privileged access to secure information.

In general, the influence of these factors can be reduced by realizing discussions among the experts, trying to pool all relevant information related to the decision. In fact, by pooling the undistributed information, it is possible to raise chances to achieve decisions better than each group member could obtain without help. However, the availability of rich intellectual resources is not sufficient to provide high-quality decisions, as some group members may fail to wisely consider, evaluate, and integrate the profiles of information and perspectives held by the other group members (Bonner et al. 2002; van Ginkel and van Knippenberg 2009). The existing literature identifies some factors that can adversely affect the decision process, leading to low-quality decisions. In particular, the authors of Pedrycz et al. (2011) distinguish the following factors:

- The pressure for early consensus that is due to the need to obtain a solution rapidly.
- The pressure of concordant majorities on the other experts, which is reflected by the group's tendency to prematurely converge on a single solution, once a majority supports a position (even if such a solution is not high-quality).
- The problem of critical pooling of non-distributed information, which can be described as follows: the information supporting the best alternative is not shared among all experts, whereas all group members have information supporting the inferior alternatives. In this case, the group may prematurely achieve a consensus on a bad solution that is apparently good, as the information shared among most experts has more chance of being recalled. The authors Stasser and Vaughan (2000) indicate that one way to reduce this specific problem is to stimulate the group members to focus on information related to their proper areas of expertise during the discussion.

Taking this into account, it is necessary to stress the importance of the moderator (or facilitator) in the discussion among the group members. As indicated in Wong and Aiken (2003), the participation of a moderator (human or automated) in the decision process always generates better outcomes. The moderator is to act as an arbiter responsible for controlling the information flows across the group. Thus, the moderator does not participate directly in the decision, but is supposed to enhance the ability of the group to make decisions (Griffith et al. 1998).

In real-world applications (Pedrycz et al. 2011), sometimes it is impossible to promote the consensus and thereby the exchange of information among the experts, for instance, due to logistic, timing, monetary, or other constraints. In this case, the invited professionals may give their opinions individually and then the group decision is dictatorially built with the use of aggregation rules, despite the existence of significant discordances among the experts. The authors of (Pedrycz et al. 2011) indicate the following approaches for dealing with this situation:

- the use of a majority rule, according to which the group decision is constructed in concordance with the opinion of the majority in the group (Lu et al. 2007);
- the use of a rule determined by a member of the group with authority to make the ultimate decision for the group (Lu et al. 2007);

- the search for a collective opinion that minimizes the major discordance in the group, in such a way that no expert is extremely dissatisfied with the group outcomes (Parreiras et al. 2010).

1.6 Fuzzy Sets in Problems of Decision-Making

As elaborated in Section 1.1, various types of uncertainty are commonly arising in a wide range of problems of an optimization character, which are inherently encountered in the design, planning, operation, and control of complex systems. Besides, as indicated in Section 1.3, taking these types of uncertainty into account when constructing mathematical models serves as a vehicle for increasing the adequacy of the models and, as a result, the credibility and factual efficiency of decisions based on their analysis.

Taking this into account, it is necessary to indicate two fundamental points, related to applying fuzzy set theory to solve the problems of the design, planning, operation, and control of complex systems:

- The theory of fuzzy sets can serve as a basis for considering diverse types of uncertainty, as well as diverse combinations of diverse types of uncertainty.
- Traditional models used in the design, planning, operation, and control of complex systems (for example, power systems, transportation systems, etc.) exhibit a high level of complexity. However, the professionals who utilize these models not are always satisfied with the results obtained on the basis of their utilization. At the same time, experienced planners and operators can plan or operate complex systems without the use of these models, applying the proper knowledge, experience, and intuition. However, their capacity to perceive and to process large volumes of information is too limited. So, for instance, operators are weak elements in the control loops. Considering this, it is necessary to create systems where the technology of thinking of operators may work, but which have no these limitations. The fuzzy sets (Dubois and Prade 1980; Zimmermann 1996; Pedrycz and Pedrycz and Gomide 1998) and/or their modifications (for instance, Pavlak 1982; Atanassov 1986) may serve as an universal language for transferring the knowledge, experience, and intuition to the computer that does not have these limitations.

The starting point in the formation of mathematical models is the requirement of a strict correspondence of these models to the level of uncertainty of information used for their building. Observing just this correspondence, we can talk about the adequacy of the presentation of the object, system, or process and the possibility of obtaining a real effect as a result of solving the corresponding problems of an optimization character. Any simplification of reality or its idealization, undertaken with the goal of using rigorous mathematical models, distorts the nature of many problems and reduces the practical value of results obtained on the basis of analyzing these models. Considering this, researchers, for instance Belyaev and Krumm (1983) and Rommelfanger (2004), for a number of reasons, have doubts about the validity or, at least, the expediency of taking into account the uncertainty factor within the framework of traditional approaches (first of all, approaches based on probability theory; for instance, Dantzig 1955; Grassman 1981; Wagner 1982). In particular, the authors of Belyaev and Krumm (1983) indicate that, similar to the solution of problems on the basis of deterministic methods, when we assume exact knowledge of the information, which usually does not correspond to reality, the application of probabilistic methods also supposes exact knowledge of the distribution laws and their parameters, which does not always correspond to the real possibilities of obtaining the entire spectrum of the probabilistic description.

In general, the approaches indicated previously do not ensure an adequate or sufficiently rational consideration of the uncertainty factor along with an entire spectrum of its manifestations.

Giving up the traditional approaches to the construction of mathematical models (Ekel and Popov 1985; Popov and Ekel 1987) and the application of the concepts of fuzziness to the studied objects and systems and processes associated with them, the application of fuzzy set theory (Dubois and Prade 1980; Zimmermann 1996; Pedrycz and Gomide 1998), established by Zadeh (1965), may play and plays a significant positive role in overcoming the existing difficulties. The use of this theory opens a convincing avenue of giving up "excessive" precision, which is inherent in the traditional modeling approaches, while preserving reasonable rigor. The principle of incompatibility coined by Zadeh (1973) offers an interesting view of the tradeoffs between precision and relevance of the models: "As the complexity of a system increases, our ability to make precise and yet significant statements about its behavior diminishes until a threshold is reached beyond which precision and significance (or relevance) become almost mutually exclusive characteristics."

Furthermore, operating in a fuzzy parameter space allows one not only to be oriented toward the contextual or intuitive aspect of the qualitative analysis as a fully substantial aspect, but, by means of fuzzy set theory, to use this aspect as a sufficiently reliable source for obtaining quantative information (Popov and Ekel 1987). Finally, fuzzy sets allow one to reflect in an adequate way the essence of the decision-making process. In particular, since the "human factor" has a noticeable effect and occupies a very visible position in making decisions in many real-world problems, we can capitalize on the way in which fuzzy sets help quantify the linguistic facets of available data and preferences (Dubois and Prade 1980; Zimmermann 1996; Pedrycz and Gomide 1998).

Besides, we also have to bear in mind that the aspiration for attaining the maximum effectiveness in decision-making in the presence of uncertainty requires, first of all, that a significant effort be directed toward finding ways to remove or, at least, partially overcome the uncertainty factor (Popov and Ekel 1987). In particular, this can be attained by aggregating information coming from diverse sources of both a formal and informal nature. This aggregation allows one (Ekel and Popov 1985) to supplement the characteristics of the uncertain initial information by justified assumptions about the differentiated confidence (reliability) of its various values that could be reflected by choosing appropriate membership functions (Dubois and Prade 1980; Zimmermann 1996; Pedrycz and Gomide 1998).

However, taking this into account, it is necessary to indicate that the issues related to the relationships between probability theory and fuzzy set theory, as well as an interpretation of membership functions, have been the subject of intensive discussions of methodological and philosophical character over the years. Considering this, it should be emphasized (Pedrycz et al. 2011) that the decision-making approaches based on fuzzy set theory do not compete with probabilistic methods, but these two approaches can complement each other. In particular, we can observe the appearance of some hybrid approaches in which fuzzy sets and probability are used synergistically.

Recent years saw intensive investigations, which took advantage of applying fuzzy set theory directly or in combination with other branches of mathematics of uncertainty to deal with diverse manifestations of the uncertainty factor. Its use in problems of an optimization character offers advantages both of the fundamental nature (the possibility of validly obtaining more effective, less "cautious" solutions, as well as the possibility of considering simultaneously different manifestations of the uncertainty factor) and of a computational character (Ekel 2002; Pedrycz et al. 2011).

Besides, it is possible to indicate two principal ways for solving problems under conditions of uncertainty. In the use of the first way, one obtains (at least, theoretically) an exact solution for fixed values of the uncertain parameters, and then estimates its stability for variations of such parameters (for example, by performing multi-variant computations). The second way presupposes the tracking of the effect of the uncertainty at all stages along the path toward the final decision. This approach can be implemented on the basis of the theory of fuzzy sets. It is more complicated than the first one, but is also more fruitful and highly promising (Ekel and Popov 1985).

As mentioned before, in many real-world problems we have to take into account the criteria, constraints, indices, and so on of a qualitative character. Thus, it should be emphasized that this type of information was taken into account in the past. However, it was used only after obtaining solutions on the basis of the use of formal models, with the disruption of the solutions obtained on their basis (to consider information of a qualitative character) without any sufficient justification (Pedrycz et al. 2011). Since this approach significantly reduces the value of the obtained solutions, it remains necessary to develop ways of introducing this type of information directly into the decision-making processes. This book demonstrates that fuzzy sets can be considered as a sound way of proceeding along this path.

Returning to the considerations of Section 1.1, it is necessary to highlight that one of the most important criteria for classifying decision-making problems (Larichev 1987) is the existence or lack of an objective model for the problem. Considering this, it should be noted that it is not uncommon to encounter situations, as mentioned in Section 1.1, where it is impossible to talk about the existence of objective functions in decision-making problems. Thus, the models corresponding to these problems are to reflect the "world outlook" of a DM. In this case, an obvious question is how to choose actions that correspond, in the best way, to the preferences of the individual (Keeney and Raifa 1976). Considering that the manner of human thinking, including the perception of preferences, is vague and subjective, fuzzy set theory can play an important role in individual and group preference modeling (Fedrizzi and Kacprzyk 1990; Fodor and Roubens 1994).

The application of fuzzy sets to preference modeling and analysis of the corresponding decision-making problems provides a flexible environment that permits one to deal with the inherent fuzziness of perception and, in this manner, to incorporate more human consistency into preference models. Besides, a stimulus for using fuzzy set theory stems, as indicated before, from one of its most important facets that concerns the linguistic aspect commonly applied to different decision-making problems and different preference structures (Herrera and Viedma 2000; Xu 2005). In particular, it is possible to distinguish among several directions in decision-making by applying the linguistic aspect of fuzzy set theory, such as multicriteria decision-making (Buckley 1995; Rasmy et al. 2002), group decision-making (Yager 1993b; Herrera et al. 1995), diverse consensus schemes (Herrera et al. 1995; Bordogna et al. 1997), decision-making on the basis of information granularity (Borisov et al. 1989; Herrera et al. 2000), and others. In principle, all these directions are associated with analyzing the $<X, R>$ models mentioned before. Taking into account the rationality of analyzing the $<X, R>$ models on the basis of fuzzy sets as well, it is possible to assert that their use in the statement and solution of decision-making problems, as indicated previously, provides answers to the fundamental questions "What to do?" and "How to do it?" arising in the project, planning, operation, and control of complex systems of a diverse nature.

Finally, it is important to note that some specific advances are created by the application of fuzzy set theory in solving decision-making problems:

- Any expert involved in the process of decision-making or any considered criterion may demand different forms (or formats) of preference representation (for instance, direct ordering of alternatives, levels of utility functions, multiplicative preference, fuzzy estimates, etc., see Zhang et al. 2004, 2007). Diverse preference formats can be reduced to fuzzy preference relations, using so-called transformation functions (for example, Chiclana et al. 1998; Herrera-Viedma et al. 2002) to prepare homogeneous information for decision-making processes.

- Any expert can give no information on some preferences, due to the lack of knowledge or because of unwillingness to respond (for example, because of political considerations), creating information "holes." Applying the interpolative nature of fuzzy set theory, it is possible to fill these "holes" (for instance, Herrera-Viedma et al. 2007; Chen et al. 2014).
- It is not uncommon that information related to the preferences is inconsistent (for instance, this information does not have the necessary level of transitivity). The theory of fuzzy sets can offer efficient means to identify and to correct these situations (for example, Ma et al. 2006; Xu et al. 2013).
- The use of aggregation operators, including those offered by the theory of fuzzy sets, offered by the theory of fuzzy sets (for instance, Yager 1993a; Yager and Kacprzyk 1997), permits one to reflect all requirements of DMs; for example, related to the level of mutual compensation among criteria in the decision-making processes.
- Diverse indicators created with the use of fuzzy sets can serve for constructing flexible consensuses to help a human or computational moderator (Lu et al. 2006: Parreiras et al. 2012).

1.7 Conclusions

In this chapter, we have discussed the fundamental questions of the appearance and essence of problems of an optimization character related to the design, planning, operation, and control of complex systems of diverse nature. The fundamental differences between optimization and decision-making problems as well as between the notions of optimal solutions and robust solutions have been emphasized. The relevance and omnipresence of the uncertainty factor and its influence on the character of the analyzed decision-making models have been considered. The structured, semi-structured, and unstructured problems of decision-making have been classified with a focus on unstructured problems. The attention has been drawn to the necessity of constructing robust solutions in the multicriteria analysis under conditions of uncertainty. The models of multicriteria decision-making have been characterized and classified with the split into two main categories of so-called $<X, F>$ models (as multiobjective models) and $<X, R>$ models (as multiattribute models), which are the subject of comprehensive considerations in this book. The essence, main concepts, and characteristics of group decision-making have been discussed. Finally, the role of fuzzy set theory in decision-making processes has been discussed, including consideration of its advantages. First of all, we have stressed the fundamental benefit stemming from the use of fuzzy sets that is the possibility of obtaining more effective, less "cautious" solutions to the decision-making problems as well as the abilities of the incorporation, including mutual incorporation, of different manifestations of the uncertainty factor.

References

Atanassov, K. (1986). Intuitionistic fuzzy sets. *Fuzzy Sets and Systems* 20 (1): 87–96.

Belyaev, L.S. (1977). *A Practical Approach to Choosing Alternative Solutions to Complex Optimization Problems under Uncertainty*. Luxemburg: IIASA.

Belyaev, L.S. and Krumm, L.A. (1983). Applicability of probablistic methods in energy calculations. *Power Engineering* 21 (2): 3–10.

Berredo, R.C., Ekel, P.Y., Martini, J.S.C. et al. (2011). Decision making in fuzzy environment and multicriteria power engineering problems. *International Journal of Electrical Power & Energy Systems* 33 (3): 623–632.

Bonner, B.L., Baumannb, M.R., and Dalal, R.S. (2002). The effects of member expertise on group decision-making and performance. *Organizational Behavior and Human Decision Processes* 88 (2): 719–736.

Bordogna, G., Fedrizzi, M., and Passi, G. (1997). A linguistic modeling of consensus in group decision making based on OWA operators. *IEEE Transactions on Systems, Man, and Cybernetics, A: Systems and Humans* 27 (1): 126–132.

Borisov, A.N., Alekseev, A.V., Merkur'eva, G.V. et al. (1989). *Fuzzy Information Processing in Decision Making Systems*. Moscow (in Russian): Radio i Svyaz.

Buckley, J.J. (1995). The multiple judge, multiple criteria ranking problem: a fuzzy set approach. *Fuzzy Sets and Systems* 13 (1): 23–37.

Chen, S.M., Lin, T.E., and Lee, L.W. (2014). Group decision making using incomplete fuzzy preference relations based on the additive consistency and the order consistency. *Information Sciences* 259 (1): 1–15.

Chiclana, F., Herrera, F., and Herrera-Viedma, E. (1998). Integrating three representation models in fuzzy multipurpose decision making based on fuzzy preference relations. *Fuzzy Sets and Systems* 97 (1): 33–48.

Dantzig, G.B. (1955). Linear programming under uncertainty. *Management Science* 1 (2): 197–207.

Dubois, D. and Prade, H. (1980). *Fuzzy Sets and Systems: Theory and Applications*. New York: Academic Press.

Durbach, I.N. and Stewart, T.J. (2012). Modeling uncertainty in multi-criteria decision analysis. *European Journal of Operational Research* 223 (1): 1–14.

Edgeworth, F.Y. (1881). *Mathematical Physics*. London: P. Keegan.

Ekel, P.Y. (1999). Approach to decision making in fuzzy environment. *Computers and Mathematics with Applications* 37 (1): 59–71.

Ekel, P.Y. (2001). Methods of decision making in fuzzy environment and their applications. *Nonlinear Analysis: Theory, Methods and Applications* 47 (5): 979–990.

Ekel, P.Y. (2002). Fuzzy sets and models of decision making. *Computers and Mathematics with Applications* 44 (7): 863–875.

Ekel, P., Kokshenev, I., Parreiras, R. et al. (2016). Multiobjective and multiattribute decision making in a fuzzy environment and their power engineering applications. *Information Sciences* 361: 100–119.

Ekel, P.Y. and Popov, V.A. (1985). Consideration of the uncertainty factor in problems of modelling and optimizing electrical networks. *Power Engineering* 23 (2): 45–52.

Eom, S.B. (1995). Decision support systems research: reference disciplines and a cumulative tradition. *Omega* 23 (5): 511–523.

Fedrizzi, M. and Kacprzyk, J. (eds.) (1990). *Multiperson Decision Making Models Using Fuzzy Sets and Possibility Theory*. Boston, MA: Kluwer.

Fodor, J. and Roubens, M. (1994). *Fuzzy Preference Modelling and Multicriteria Decision Support*. Boston, MA: Kluwer.

Gomes, L.F.A.M., Gomes, C.F.S., and Almeida, A.T. (2002). *Managerial Decision Making: Multicriteria Approach*. Sao Paulo (in Portuguese): Atlas.

Grassman, W.K. (1981). *Stochastic Systems for Management*. New York: North-Holland.

Griffith, T.L., Fuller, M.A., and Northcraft, G.B. (1998). Facilitator influence in group support systems: intended and unintended effects. *Information Systems Research* 9 (1): 20–36.

Herrera, F. and Viedma, E.H. (2000). Linguistic decision analysis: steps for solving decision problems under linguistic information. *Fuzzy Sets and Systems* 115 (1): 67–82.

Herrera, F., Herrera-Viedma, E., and Verdegay, J.L. (1995). A sequential selection process in group decision making with linguistic assessment. *Information Sciences* 85 (2): 223–239.

Herrera, F., Herrera-Viedma, E., and Martinez, L. (2000). A fusion approach for managing multi-granularity linguistic term sets in decision making. *Fuzzy Sets and Systems* 114 (1): 43–58.

Herrera-Viedma, E., Herrera, F., and Chiclana, F. (2002). A consensus model for multiperson decision making with different preferences structures. *IEEE Transactions on Systems, Man, and Cybernetics, A: Systems and Humans* 32 (3): 394–402.

Herrera-Viedma, E., Chiclana, F., Herrera, F., and Alonso, S. (2007). Group decision-making model with incomplete fuzzy preference relations based on additive consistency. *IEEE Transactions on Systems, Man, and Cybernetics. B: Cybernetics* 37 (1): 176–189.

Hwang, C.L. and Masud, A.S. (1979). *Multiple Objective Decision Making: Methods and Applications*. Berlin: Springer-Verlag.

Hwang, C.L. and Yoon, K. (1981). *Multiple Attribute Decision Making: Methods and Applications – A State-of-the-Art Survey*. Berlin: Springer-Verlag.

Keeney, R. and Raifa, H. (1976). *Decisions with Multiple Objectives: Preferences and Value Trade-Offs*. New York: Wiley.

Kokshenev, I., Parreiras, R.O., Ekel, P.Y. et al. (2014). A web-based decision support center for electrical energy companies. *IEEE Transactions on Fuzzy Systems* 23 (1): 16–28.

Kuhn, T.S. (1962). *The Structure of Scientific Revolutions*. Chicago: University of Chicago Press.

Larichev, O.I. (1984). Psychological validation of decision methods. *Journal of Applied Systems Analysis* 11 (1): 37–46.
Larichev, O.I. (1987). *Objective Models and Subjective Decisions*. Moscow (in Russian): Nauka.
Larichev, O.I. and Moshkovich, E.M. (1996). *Qualitative Methods of Decision Making*. Moscow (in Russian): Nauka.
Lu, C., Lan, J., and Wang, Z. (2006). Aggregation of fuzzy opinions under group decision-making based on similarity and distance. *Journal of Systems Science and Complexity* 19 (1): 63–71.
Lu, J., Zhang, G., Ruan, D., and Wu, F. (2007). *Multi-Objective Group Decision Making: Methods, Software and Applications with Fuzzy Set Techniques*. London: Imperial College Press.
Luce, R.D. and Raiffa, H. (1957). *Games and Decisions*. New York: Wiley.
Lyapunov, A.A. (ed.) (1972). *Operational Research: Methodological Aspects*. Moscow (in Russian): Nauka.
Ma, J., Fan, Z.P., Jiang, Y.P. et al. (2006). A method for repairing the inconsistency of fuzzy preference relations. *Fuzzy Sets and Systems* 157 (1): 20–33.
Orlovsky, S.A. (1981). *Problems of Decision Making with Fuzzy Information*. Moscow (in Russian): Nauka.
Pareto, V. (1886). *Cours d'Economie Politique*. Lousanne: Lousanne Rouge.
Parreiras, R.O., Ekel, P.Y., Martini, J.S.C., and Palhares, R.M. (2010). A flexible consensus scheme for multicriteria group decision making under linguistic assessments. *Information Sciences* 180 (7): 1075–1089.
Parreiras, R.O., Ekel, P.Y., and Morais, D.C. (2012). Fuzzy set based consensus schemes for multicriteria group decision making applied to strategic planning. *Group Decision and Negotiation* 21 (2): 153–183.
Pavlak, Z. (1982). Rough sets. *International Journal of Computer and Information Sciences* 11 (5): 341–356.
Pedrycz, W. and Gomide, F. (1998). *An Introduction to Fuzzy Sets: Analysis and Design*. Cambridge, MA: MIT Press.
Pedrycz, W., Ekel, P., and Parreiras, R. (2011). *Fuzzy Multicriteria Decision-Making: Models, Methods and Applications*. Chichester: Wiley.
Popov, V.A. and Ekel, P.Y. (1987). Fuzzy set theory and problems of controlling the design and operation of electric power systems. *Soviet Journal of Computer and System Sciences* 25 (4): 92–99.
Pospelov, D.A. and Pushkin, V.M. (1972). *Thinking and Machines*. Moscow (in Russian): Sovetskoe Radio.
Raiffa, H. (1968). *Decision Analysis*. Reading: Addison-Wesley.
Ramesh, R., Zionts, S., and Karwan, M.H. (1986). A class of practical interactive branch and bound algorithms for multicriteria integer programming. *European Journal of Operational Research* 26 (1): 161–172.
Rasmy, M.H., Lee, S.M., Abd El-Wahed, W.F. et al. (2002). An expert system for multiobjective decision making: application of fuzzy linguistic preferences and goal programming. *Fuzzy Sets and Systems* 127 (2): 209–220.

Rommelfanger, H. (2004). The advantages of fuzzy optimization models in practical use. *Fuzzy Optimization and Decision Making* 3 (4): 295–309.
Simon, H.A. (1977). *The New Science of Management Decision*. Englewood Cliffs, NJ: Prentice Hall.
Slovic, P., Fischhoff, B., and Lichtenstein, S. (1977). Behavioral decision theory. *Annual Review of Psychology* 28 (1): 1–39.
Stasser, G. and Vaughan, S.I. (2000). Pooling unshared information: the benefits of knowing how access to information is distributed among group members. *Organizational Behavior and Human Decision Processes* 82 (1): 102–116.
Stewart, T. (2005). Dealing with uncertainties in MCDA. In: *Multiple Criteria Decision Analysis – State of the Art Annotated Surveys, International Series in Operations Research and Management Science* (eds. J. Figueira, S. Greco and M. Ehrgott), 445–470. New York: Springer.
Tan, B.C.Y., Teo, H.H., and Wei, K.K. (1995). Promoting consensus in small decision making groups. *Information & Management* 28 (4): 251–259.
Trachtengerts, E.A. (1998). *Computer Support of Decision Making*. Moscow (in Russian): SINTEG.
van Ginkel, W.P. and van Knippenberg, D. (2009). Knowledge about the distribution of information and group decision-making: when and why does it work? *Organizational Behavior and Human Decision Processes* 108 (2): 218–229.
Villareal, B. and Karwan, B.H. (1982). Multicriteria dynamic programming with an application to the integer case. *Journal of Optimization Theory and Applications* 38 (1): 43–69.
Wagner, H.M. (1982). *Operations Research: An Introduction*. New York: Macmillan.
Webster, T.J. (2003). *Managerial Economics: Theory and Practice*. San Diego: Academic Press.
Wong, Z. and Aiken, M. (2003). Automated facilitation of electronic meetings. *Information & Management* 41 (2): 125–134.
Xu, Z. (2005). On method for uncertain multiple attribute decision making problems with uncertain multiplicative preference information on alternatives. *Fuzzy Optimization and Decision Making* 4 (2): 131–139.
Xu, Y., Patnayakuni, R., and Wang, H. (2013). The ordinal consistency of a fuzzy preference relation. *Information Sciences* 224 (2): 152–164.
Yager, R.R. (1993a). Non-numeric multi-criteria multi-person decision making. *Group Decision and Negotiation* 2 (11): 81–93.
Yager, R.R. (1993b). A general approach to criteria aggregation using fuzzy measures. *International Journal of Man-Machine Studies* 38 (2): 187–213.
Yager, R.R. and Kacprzyk, J. (1997). *The Ordered Weighted Averaging Operators: Theory and Applications*. Norwell: Kluwer Academic Publishers.
Zadeh, L.A. (1965). Fuzzy sets. *Information and Control* 8 (3): 338–353.
Zadeh, L.A. (1973). Outline of a new approach to the analysis of complex systems and decision processes. *IEEE Transactions on Systems, Man, and Cybernetics* 3 (1): 28–44.

Zhang, Q., Chena, J.H.C., and Chong, P. (2004). Decision consolidation: criteria weight determination using multiple preference formats. *Decision Support Systems* 38 (2): 247–258.
Zhang, Q., Wang, Y., and Yang, Y. (2007). Fuzzy multiple attribute decision making with eight types of preference information. In: *Proceedings of the 2007 IEEE Symposium on Computational Intelligence in Multicriteria Decision Making* (ed. P.P. Bonissoni), 288–293. IEEE, Piscataway.
Zimmermann, H.J. (1996). *Fuzzy Set Theory and Its Application*. Boston: Kluwer.

2

Notions and Concepts of Fuzzy Sets

An Introduction

In this chapter, we introduce the fundamental concepts of fuzzy sets. We focus on the fundamental ideas of partial membership, which are conveniently quantified through membership functions and membership degrees. Fuzzy sets come as a manifestation of a general idea of information granule. Processing of information granules is realized in the framework of Granular Computing. We present the underlying rationale and next move on to the detailed description of fuzzy sets by discussing the most commonly encountered classes of membership function, and relating these classes to the semantics of fuzzy sets. We elaborate on the basic operations on fuzzy sets (intersection, union, complement, negations) and discuss concepts of fuzzy relations and their main properties, which are of direct relevance in the context of decision-making.

2.1 Sets and Fuzzy Sets: A Fundamental Departure from the Principle of Dichotomy

Conceptually and algorithmically, fuzzy sets constitute one of the most fundamental and widely influential concepts in science and engineering. The notion of a fuzzy set is highly intuitive and transparent since it captures what really becomes an essence of a way in which a real world is being perceived and described in our everyday activities. We are faced with categories of objects whose belongingness to a given category (concept) is always a matter of degree. There are numerous examples in which we encounter elements whose allocation to the concept we want to define can be satisfied to some degree. One may eventually claim that continuity of transition from full belongingness and full exclusion is the major and ultimate feature of the physical world and natural systems. For instance, we may qualify an in-door environment as *comfortable* when its temperature is kept *around* 20 °C. If we observe a value of 19.5 °C it is very likely we still feel quite *comfortable*. The same holds if we encounter

Multicriteria Decision-Making under Conditions of Uncertainty: A Fuzzy Set Perspective,
First Edition. Petr Ekel, Witold Pedrycz, and Joel Pereira, Jr.

20.5 °C – humans usually do not discriminate (distinguish) between changes in temperature within the range of 1 °C. A value of 20 °C would be fully compatible with the concept of *comfortable* temperature yet 0 or 30 °C would not. In these two cases as well for temperatures close to these two values, we would describe them as being *cold* and *warm*, respectively. We could question whether the temperature of 25 °C is viewed as *warm* or *comfortable* or, similarly, if 15 °C is *comfortable* or *cold*. Intuitively, we know that 25 °C is somehow between *comfortable* and *warm* while 15 °C is between *comfortable* and *cold*. The value 25 °C is partially compatible with the term *comfortable* and *warm*, and somewhat compatible or, depending on observer's perception, incompatible with the term of *cold* temperature. Similarly, we may say that 15 °C is partially compatible with the comfortable and cold temperature, and slightly compatible or incompatible with the warm temperature. In spite of this highly intuitive and apparent categorization of environment temperatures into the three classes, namely *cold*, *comfortable*, and *warm*, we note that the transition between the classes is not instantaneous and sharp. Simply, when moving across the range of temperatures, these values become gradually perceived as *cold*, *comfortable*, or *warm*. Similar phenomenon happens when we are dealing with the concept of height of people. An individual of height of 1 m is *short* whereas a person of 1.90 m is perceived to be *tall*. Again, the question is, what is the range of height values that could qualify a person to be *tall*? Does a height of 1.85 discriminate between *tall* and *short* individuals? Or maybe 1.86 m would be the right choice? Asking these questions, we sense that they do not make too much sense. We realize that the nature of these concepts is such that we cannot use a single number – a transition between the notion of tall and short is no abrupt in any way. Hence, we cannot assign a single number that does a good job. This sends a clear message: the concept of dichotomy does not apply when defining even simple concepts. The illustration of the concept of dichotomy is included in Figure 2.1a. In contrast, defining a concept using that we do not confine ourselves to the dichotomy is illustrated in Figure 2.1b.

Fuzzy sets and the corresponding membership functions form a viable and mathematically sound framework to formalize these concepts. When talking about heights of Europeans we may refer to real numbers within the interval [0, 3] to represent a universe of heights that range in between 0 and 3 m. This universe of discourse is suitable for describing the concept of *tall* people.

Let us denote by $\mathbf{X}$ a universe of discourse (space) of all elements. The universe can be either continuous or discrete. For instance, the closed interval [0, 3] constitutes a continuous and bounded universe whereas the set $\mathbf{N} = \{0,1,2, \ldots.\}$ of natural numbers is discrete and countable.

Consider the universe of discourse $\mathbf{X} = [0, 3]$ and the collection S of values in $\mathbf{X}$ that are less than a threshold value τ in $\mathbf{X}$, for example $\tau = 1.8$. Consider the sets $S = \{x \in \mathbf{X} \mid 0 < x < 1.8\}$ and $T = \{x \in \mathbf{X} \mid 1.8 \leq x \leq 3.0\}$, Figure 2.2. Each set is a

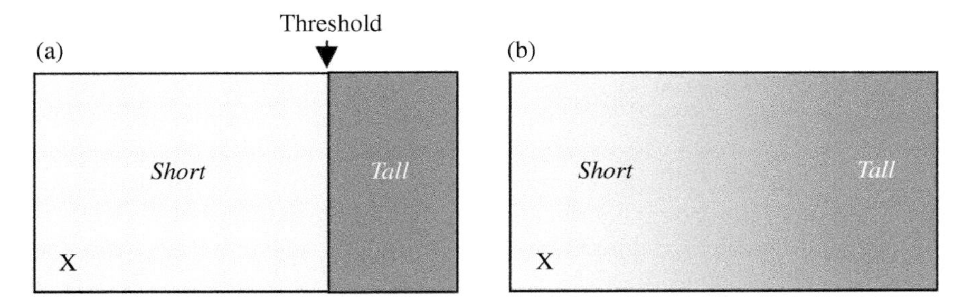

Figure 2.1 Contrasting a concept of a set and the principle of dichotomy itself versus a relaxation of the concept of complete inclusion and exclusion.

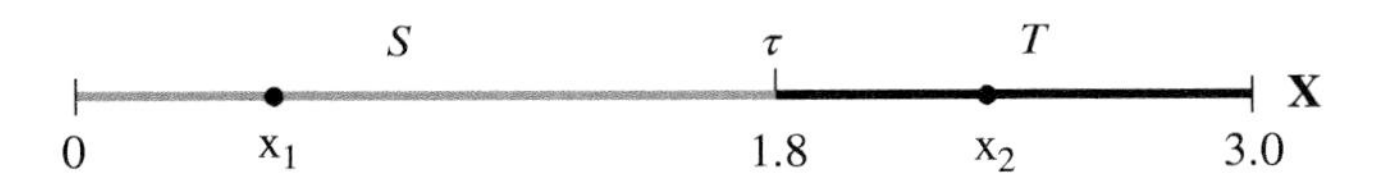

Figure 2.2 Set as a collection of numeric values located in the corresponding intervals.

class whose members are elements of the universe that satisfy the same property. This set is equivalent to a list of elements of the universe that are members of the set.

Given a certain element in **X**, the process of dichotomization (binarization) imposes a binary, all-or-none classification decision: either accept or reject the element as belonging to a given collection. For instance, consider the set S shown in Figure 2.2. Clearly, the point x_1 belongs to S whereas x_2 does not, that is, $x_1 \in S$ and $x_2 \notin S$. Similarly, for the set T we have $x_1 \notin$ T, and $x_2 \in$ T. If we denote the accept decision by 1 and the reject decision by 0, for short, we may express the classification (assignment) decision of $x \in \mathbf{X}$ through a characteristic function as follows:

$$S(x) = \begin{cases} 1, if\ x \in S \\ 0, if\ x \notin S \end{cases} \qquad T(x) = \begin{cases} 1, if\ \ x \in T \\ 0, if\ x \notin T \end{cases} \tag{2.1}$$

In general, a characteristic function of set A defined in **X** assumes the following form

$$A(x) = \begin{cases} 1, \text{if } x \in A \\ 0, if\ x \notin A \end{cases} \tag{2.2}$$

The empty set ∅ has a characteristic function that is identically equal to zero, $\varnothing(x) = 0$ for all x in **X**. The universe **X** itself comes with the characteristic function that is identically equal to one, that is $\mathbf{X}(x) = 1$ for all x in **X**. Also, a singleton $A = \{a\}$, a set with only a single element, has a characteristic function such that $A(x) = 1$ if x = a and $A(x) = 0$ otherwise.

Characteristic functions $A: \mathbf{X} \rightarrow \{0, 1\}$ induce a constraint with well-defined boundaries on the elements of the universe **X** that can be assigned to a set *A*. The fundamental idea of fuzzy set is to relax this requirement by admitting intermediate values of class membership. Therefore, we may assign intermediate values between 0 and 1 to quantify our perception on how compatible these values are with the class with 0 meaning incompatibility (complete exclusion) and 1 compatibility (complete membership). Membership values thus express the degrees to which each element of the universe is compatible with the properties distinctive to the class. Intermediate membership values mean that no "natural" threshold exists and that elements of a universe can be members of a class and at the same time belong to other classes with different degrees. Allowing for gradual, hence less strict membership degrees is the crux of fuzzy sets.

Formally, a fuzzy set *A* is described by a membership function mapping the elements of a universe **X** to the unit interval [0,1] (Zadeh 1965, 1975):

$$A: \mathbf{X} \rightarrow [0,1] \tag{2.3}$$

The membership functions are therefore synonymous of fuzzy sets. In a nutshell, membership functions generalize characteristic functions in the same way as fuzzy sets generalize sets.

Fuzzy sets can be also be viewed as a set of ordered pairs of the form $\{x, A(x)\}$ where x is an element of **X** and $A(x)$ denotes its corresponding degree of membership. For a finite universe of discourse $\mathbf{X} = \{x_1, x_2, \ldots, x_n\}$, *A* can be represented by a *n*-dimensional vector $A = (a_1, a_2, \ldots, a_n)$ with the element $a_i = A(x_i)$. Figure 2.3 illustrates a fuzzy set whose membership function captures the concept of integer *around* 5. Here, $n = 10$ and linguistically expressing the integer quantity around 5 in a finite universe formed of 10 integers $A = (0, 0, 0, 0.2, 0.5, 1.0, 0.5, 0.2, 0, 0, 0)$. An equivalent notation of *A* can be read as $A = \{0/1, 0/2, 0/3, 0.2/4, 0.5/4, \ldots, 0/10\}$.

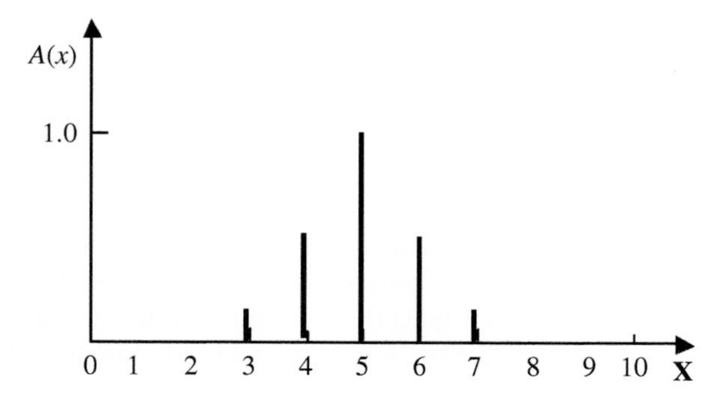

Figure 2.3 Fuzzy set *A* defined in a discrete universe **X**.

The choice of the unit interval for the values of membership degrees is usually a matter of convenience. The choice of the very detailed membership values (up to several decimal digits), say $A(4) = 0.9865$, is not crucial; in describing membership grades we are predominantly after the reflecting an order of the elements in A in terms of their belongingness to the fuzzy set (Dubois and Prade 1979, 1997).

Being more descriptive, we may view fuzzy sets as elastic constraints imposed on the elements of a universe. As emphasized before, fuzzy sets deal primarily with the concept of elasticity, graduality, or absence of sharply defined boundaries. In contrast, when dealing with sets we are concerned with rigid boundaries, lack of graded belongingness and sharp, binary boundaries. Gradual membership means that no natural Boolean boundary exists and that some elements of the universe of discourse can, contrary to sets, coexist (belong) to different fuzzy sets with different degrees of membership.

2.2 Interpretation of Fuzzy Sets

In fuzzy sets, the concept of fuzziness comes with a precise meaning. Fuzziness primarily means lack of precise boundaries of a collection of objects and, as such, it is an evident manifestation of imprecision and a particular type of uncertainty. Let us make some observations with this regard.

First, it is worth indicating that fuzziness is both conceptually and formally different from the fundamental concept of probability. In general, it is difficult to foresee the result of tossing a fair coin once it is impossible to know if either head or tail will occur for certain. We may, at most, say that there is 50% chance to have head or tail occur, but as soon as the coin falls, uncertainty vanishes. But, in the case of the person height, imprecision remains. Formally, fuzzy sets are membership functions that are treated as mappings from a given universe of discourse to the unit interval as presented in Eq. (2.3). In contrast, probability is a set function, a mapping whose universe is a set of subsets of a domain.

Second, there are differences between fuzziness, generality, and ambiguity. A notion is general when it applies to a multiplicity of objects and keeps only a common essential property. An ambiguous notion stands for several unrelated objects. Therefore, from this point of view fuzziness does not mean neither generality nor ambiguity and applications of fuzzy sets exclude these categories. Fuzzy set theory assumes that the universe is well-defined and has its elements assigned to classes by means of a numerical scale.

Applications of fuzzy set to areas such as data analysis, reasoning under uncertainty, and decision-making suggest different interpretations of membership grades in terms of similarity, uncertainty, and preference (Dubois and Prade 1997, 1998). From the similarity point of view, $A(x)$ means the degree of compatibility of an element $x \in \mathbf{X}$ with representative elements of A. This is the primary and most intuitive interpretation of a fuzzy set, one that is particularly suitable for data analysis. An example is the case when we question how to qualify an environment as *comfortable* when we know that current temperature is 25 °C. As discussed at the beginning of this chapter, such quantification is a matter of degree. For instance, assuming a universe of discourse to $\mathbf{X} = [0, 40]$ and choosing 20 °C as representative of *comfortable* temperature, we note, Figure 2.4, that the degree at which 25 °C is comfortable to the degree of 0.2. In the example, we have adopted piecewise linearly decreasing functions of the distance between temperature values and the representative value 20 °C to determine the corresponding membership degree.

Now let us assume that values of a variable x is located within the support of a fuzzy set A. Then given a value "v" of $\mathbf{X}$, $A(\mathrm{v})$ expresses a possibility that $x = v$ given that "x" is in A is all that is known about. In this situation, the membership degree of a given tentative value "v" to the class A reflects the degree of plausibility that this value is the same as "x." This idea reflects a type of uncertainty because if the membership degree is high, our confidence about the value of "x" may still be low, but if the degree is low, than the tentative value may be rejected as an implausible candidate. The variable labeled by the class A is uncontrollable. This allows assignment of fuzzy sets to possibility distributions as presented in possibility theory (Zadeh 1978, 1999). For instance, suppose someone said he felt comfortable when watching a soccer game. In this situation the membership degree of a given tentative temperature value, say 25 °C, reflects the degree of plausibility that this value of temperature is the same as the one when he felt comfortable. Note that the temperature value felt is unknown, but there is no question if it did occur or not. Possibility is whether an event may occur and to what degree. On the contrary, probability is about whether an event will occur.

Finally, assume that A reflects a preference on the values of a variable "x" in $\mathbf{X}$. For instance, "x" can be a decision variables and fuzzy set A is an elastic constraint characterizing feasible values and decision-maker preferences. In this

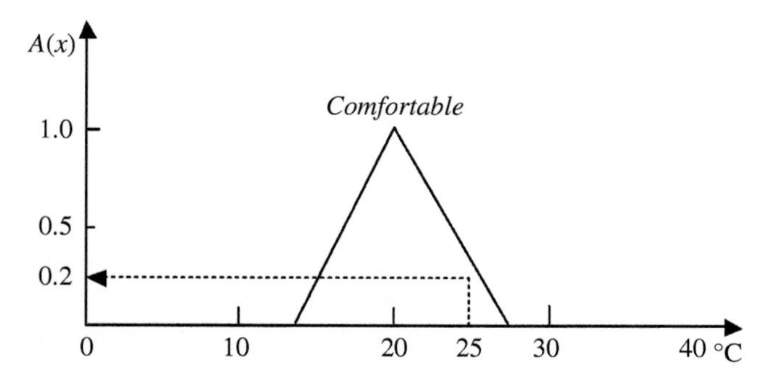

Figure 2.4 Membership function of a fuzzy set of *comfortable* temperature.

case, $A(v)$ denotes the grade of preference in favor of "v" as the value of "x." This interpretation prevails in fuzzy optimization and decision analysis. For instance, we may be interested in finding a *comfortable* value of temperature. The membership degree of a candidate temperature value "v" reflects our degree of satisfaction with the particular temperature value chosen. In this situation, the choice of the value is controllable in the sense that the value being adopted depends on our choice.

2.3 Membership Functions and Classes of Fuzzy Sets

Formally speaking, any function $A: \mathbf{X} \rightarrow [0, 1]$ could be qualified to serve as a membership function describing the corresponding fuzzy set. In practice, the form of the membership functions should be reflective of the problem at hand for which we construct fuzzy sets. They should reflect our perception (semantics) of the concept to be represented and further used in problem solving, the level of detail we intend to capture, and a context, in which the fuzzy set are going to be used. It is also essential to assess the type of a fuzzy set from the standpoint of its suitability when handling the ensuing optimization procedures. Given these criteria in mind, we elaborate on the most commonly used categories of membership functions. All of them are defined in the universe of real numbers, that is $\mathbf{X} = \mathbf{R}$.

Triangular membership functions. These are described by their piecewise linear segments in the form

$$A(x,a,m,b) = \begin{cases} 0 & \text{if } x \le a \\ \dfrac{x-a}{m-a} & \text{if } x \in [a,m] \\ \dfrac{b-x}{b-m} & \text{if } x \in [m,b] \\ 0 & \text{if } x \ge b \end{cases} \tag{2.4}$$

Using more concise notation, expression (2.4) can be written down in the following form $A(x, a, m, b) = \max\{\min[(x-a\}/(m-a), (b-x)/(b-m)], 0\}$. The meaning of the parameters is straightforward: "m" denotes a modal (typical) value of the fuzzy set while "a" and "b" are the lower and upper bounds, respectively. They could be sought as the extreme elements of the universe of discourse that delineate the elements belonging to A with nonzero membership degrees.

Triangular fuzzy sets (membership functions) are the simplest possible models of grades of membership as they are fully defined by only three parameters. As mentioned, the semantics is evident as the fuzzy sets are expressed on a basis of knowledge of the spreads of the concepts and their typical values. The linear change in the membership grades is the simplest possible model of membership one could think of. Taking the absolute value of the derivative of the triangular membership function, which could be sought as a measure of sensitivity of A, $|dA/dx|$, we conclude that its sensitivity is constant for each of the linear segments of the fuzzy set.

Trapezoidal membership functions. These are piecewise linear function characterized by four parameters, "a," "m," "n," and "b," each of which defines one of the four linear parts of the membership function. They assume the following form

$$A(x) = \begin{cases} 0 & \text{if } x < a \\ \dfrac{x-a}{m-a} & \text{if } x \in [a, m] \\ 1 & \text{if } x \in [m, n] \\ \dfrac{b-x}{b-n} & \text{if } x \in [n, b] \\ 0 & \text{if } x > b \end{cases} \tag{2.5}$$

Using an equivalent notation, we can rewrite A as follows: $A(x, a, m, n, b) = \max\{\min[(x-a\}/(m-a), 1, (b-x)/(b-n)], 0\}$

Γ-membership functions. These are expressed in the following form:

$$A(x) = \begin{cases} 0 & \text{if } x \le a \\ 1 - e^{-k(x-a)^2} & \text{if } x > a \end{cases} \quad or \quad A(x) = \begin{cases} 0 & \text{if } x \le a \\ \dfrac{k(x-a)^2}{1 + k(x-a)^2} & \text{if } x > a \end{cases} \tag{2.6}$$

where $k > 0$.

S-membership functions. These are functions of the form

$$A(x) = \begin{cases} 0 & \text{if } x \le a \\ 2\left(\dfrac{x-a}{b-a}\right)^2 & \text{if } x \in [a, m] \\ 1 - 2\left(\dfrac{x-b}{b-a}\right)^2 & \text{if } x \in [m, b] \\ 1 & \text{if } x > b \end{cases} \tag{2.7}$$

The point $m = (a + b)/2$ is the crossover point of the S-function.

Gaussian membership functions. These membership functions are described by the following relationship:

$$A(x, m, \sigma) = \exp\left(-\frac{(x-m)^2}{\sigma^2}\right) \tag{2.8}$$

Gaussian membership functions are described by two important parameters. The modal value (m) represents the typical element of A while σ denotes a spread of A. Higher values of s corresponds to larger spreads of the fuzzy sets.

Exponential-like membership functions. These membership functions are described in the form

$$A(x) = \frac{1}{1 + k(x-m)^2} \quad k > 0 \tag{2.9}$$

The spread of the exponential-like membership function increases as the values of "k" get lower.

2.4 Information Granules and Granular Computing

Information granules are intuitively appealing constructs, which play a pivotal role in human cognitive and decision-making activities (Bargiela and Pedrycz 2003, 2005, 2008; Zadeh 1997, 1999, 2005). We perceive complex phenomena by organizing existing knowledge along with available experimental evidence and structuring them in a form of some meaningful, semantically sound entities, which are central to all ensuing processes of describing the world, reasoning about the environment, and support decision-making activities.

The term information granularity itself has emerged in different contexts and numerous areas of application, it carries various meanings. One can refer to Artificial Intelligence in which case information granularity is central to a way of problem solving through problem decomposition, where various subtasks could be formed and solved individually. Information granules and the area of intelligent computing revolving around them being termed Granular Computing are quite often presented with a direct association with the pioneering studies by Zadeh (1997). He coined an informal yet highly descriptive and compelling concept of information granules. In a general way, by information granules one regards a collection of elements drawn together by their closeness (resemblance, proximity, functionality, etc.) articulated in terms of some useful spatial, temporal, or functional relationships. Subsequently, Granular Computing is about representing, constructing, processing, and communicating information granules.

It is again worth emphasizing that information granules permeate almost all human endeavors. No matter which problem is taken into consideration, we usually set it up in a certain conceptual framework composed of some generic and conceptually meaningful entities – information granules, which we regard to be of relevance to the problem formulation, further problem solving, and a way in which the findings are communicated to the community. Information granules realize a framework in which we formulate generic concepts by adopting a certain level of abstraction. Let us refer here to some areas, which offer compelling evidence as to the nature of underlying processing and interpretation in which information granules play a pivotal role.

2.4.1 Image Processing

In spite of the continuous progress in the area, a human being assumes a dominant and very much uncontested position when it comes to understanding and interpreting images. Surely, we do not focus our attention on individual pixels and process them as such but group them together into a hierarchy of semantically meaningful constructs – familiar objects we deal with in everyday life. Such objects involve regions that consist of pixels or categories of pixels drawn together because of their proximity in the image, similar texture, color, and so on. This remarkable and unchallenged ability of humans dwells on our effortless ability to construct information granules, manipulate them, and arrive at sound conclusions.

2.4.2 Processing and Interpretation of Time Series

From our perspective we can describe them in a semi-qualitative manner by pointing at specific regions of such signals. Medical specialists can effortlessly interpret various diagnostic signals including ECG or EEG recordings as representative examples. They distinguish some segments of such signals and interpret their combinations. On the stock market, one analyzes numerous time series by looking at amplitudes, trends, and patterns. Experts can interpret temporal readings of sensors and assess a status of the monitored system. Again, in all these situations, the individual samples of the signals are not the focal point of the analysis, synthesis, and the signal interpretation. We always granulate all phenomena (no matter if they are originally discrete or analog in their nature). When working with time series, information granulation occurs in time and in the feature space where the data are described.

2.4.3 Granulation of Time

Time is another important and omnipresent variable that is subjected to granulation. We use seconds, minutes, days, months, and years. Depending on a specific problem we have in mind who the user is, the size of information granules (time intervals) could vary quite significantly. To the high-level management time intervals of quarters of year or a few years could be meaningful temporal information granules on basis of which one develops any predictive model. For those in charge of everyday operation of a dispatching center, minutes, and hours could form a viable scale of time granulation. Long-term planning is very much different from day-to-day operation. For the designer of high-speed integrated circuits and digital systems, the temporal information granules concern nanoseconds, microseconds, and perhaps milliseconds. Granularity of information (in this case time) helps us focus on the most suitable level of detail.

2.4.4 Data Summarization

Information granules naturally emerge when dealing with data, including those coming in the form of data streams. The ultimate objective is to describe the underlying phenomenon in an easily understood way and at a certain level of abstraction. This requires that we use a vocabulary of commonly encountered terms (concepts) and discover relationships between them and possible linkages among the underlying concepts. Consider some meteorological data. Having a collection of detailed numeric weather data concerning temperature, humidity, wind speed, they are transformed into a linguistic description at the higher level of abstraction. It is noticeable that information granularity emerges with regard to several variables present in the data.

2.4.5 Design of Software Systems

We develop software artifacts by admitting a modular structure of an overall architecture of the designed system where each module is a result of identifying essential functional closeness of some components of the overall system. Modularity (granularity) is a holy grail of the systematic software design supporting a production and maintenance of high-quality software products.

Even such commonly encountered and simple examples presented before are convincing enough to lead us to ascertain that (i) information granules are the key components of knowledge representation and processing, (ii) the level of granularity of information granules (their size, to be more descriptive) becomes crucial to the problem description and an overall strategy of problem solving, (iii) hierarchy of information granules supports an important aspect of perception of phenomena and delivers a tangible way of dealing with complexity by focusing on the most essential facets of the problem, and (iv) there is no universal level of granularity of information; commonly the size of granules is problem-oriented and user dependent.

2.5 Formal Platforms of Information Granularity

Along with fuzzy sets, there is a spectrum of formal platforms in which information granules are conceptualized, defined, and processed.

Sets (intervals) realize a concept of abstraction by introducing a notion of dichotomy: we admit element to belong to a given information granule or to be excluded from it. Along with the set theory comes a well-developed discipline of interval analysis (Moore 1966; Moore et al. 2009; Alefeld and Herzberger

1983). Alternative to an enumeration of elements belonging to a given set, sets are described by characteristic functions taking on values in {0,1}. Formally, a characteristic function describing a set A is defined as follows

$$A(x) = \begin{cases} 1, \text{if } x \in A \\ 0, \text{if } x \notin A \end{cases} \tag{2.10}$$

where $A(x)$ stands for a value of the characteristic function of set A at point x. With the emergence of digital technologies, interval mathematics has appeared as an important discipline encompassing a great deal of applications. A family of sets defined in a universe of discourse **X** is denoted by **P**(**X**). Well-known set operations – union, intersection, and complement are the three fundamental constructs supporting a manipulation on sets. In terms of the characteristic functions they result in the following expressions

$$(A \cap B)(x) = \min(A(x), B(x)) \quad (A \cup B)(x) = \max(A(x), B(x)) \; \bar{A}(x) = 1 - A(x) \tag{2.11}$$

where $A(x)$ and $B(x)$ are the values of the characteristic functions of A and B at x and $\bar{A}$ denotes the complement of A.

Shadowed sets (Pedrycz 1998, 2005) offer an interesting description of information granules by distinguishing among three categories of elements. Those are the elements, which (i) fully belong to the concept, (ii) are excluded from it, and (iii) their belongingness is completely *unknown*. Formally, these information granules are described as a mapping $X: \mathbf{X} \rightarrow \{1, 0, [0,1]\}$ where the elements with the membership quantified as the entire [0,1] interval are used to describe a shadow of the construct. Given the nature of the mapping here, shadowed sets can be sought as a granular description of fuzzy sets where the shadow is used to localize unknown membership values, which in fuzzy sets are distributed over the entire universe of discourse. Note that the shadow produces non-numeric descriptors of membership grades. A family of fuzzy sets defined in **X** is denoted by **S**(**X**)

Probability-oriented information granules are expressed in the form of some probability density functions or probability functions (in continuous cases). They capture a collection of elements resulting from some experiment. In virtue of the fundamental concept of probability, the granularity of information associates with a manifestation of occurrence of some elements. Probability function or probability density function are commonly encountered descriptors of experimental data. The abstraction offered by probabilities is apparent: instead of coping with huge masses of data, one produces their abstract manifestation in the form of a single or a few probability functions. Histograms are examples of probabilistic information granules arising as a concise characterization of

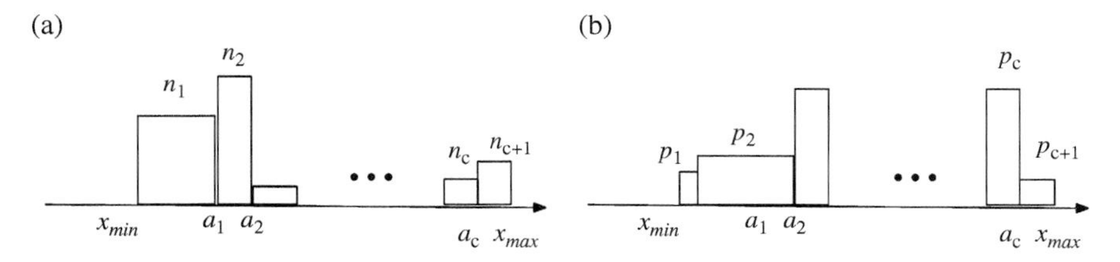

Figure 2.5 Histogram as an example of information granule with data coming from a (a) single class and (b) two-class problem.

one-dimensional data; see Figure 2.5. In case of data belonging to a single class, an information granule – histogram composed of $c + 1$ bins is described by vectors of cutoff points **a**, $\mathbf{a} = [a_1, a_2 \ldots a_c]$ and the corresponding vector of counts **n**, $\mathbf{n} = [n_1, n_2, \ldots, n_c, n_{c+1}]$; that is, $H = (\mathbf{a}, \mathbf{n})$. There description could be provided in different ways. For instance, if the data come from a two-class problem, a histogram is an information granule containing probabilities (frequencies) of data belonging to a certain class and falling into a given interval (bin). In this case, a histogram H is an information granule in the form $H = (\mathbf{a}, \mathbf{p})$ with **p** standing for a vector of the corresponding probabilities, see Figure 2.5.

When dealing with data coming from several classes, a suitable description of a histogram is composed of the vector of cutoff points and the associated entropy values, $H = (\mathbf{a}, \mathbf{h})$ where **h** denotes a vector of entropies computed for data falling within the corresponding bins.

Rough sets emphasize a roughness of description of a given concept X when being realized in terms of the indiscernibility relation provided in advance. The roughness of the description of X is manifested in terms of its lower and upper approximations of a certain rough set. A family of fuzzy sets defined in **X** is denoted by **R**(**X**).

The choice of a certain formal setting of information granulation is mainly dictated by the formulation of the problem and the associated specifications coming with the problem. There is an interesting and a quite broad spectrum of views at information granules and their processing. The two extremes are quite visible here:

2.5.1 Symbolic Perspective

A concept–information granule is viewed as a single symbol (entity). This view is very much present in the AI community, where computing revolves around symbolic processing. Symbols are subject to processing rules giving rise to results, which are again symbols coming from the same vocabulary one has started with.

2.5.2 Numeric Perspective

Here information granules are associated with a detailed numeric characterization. Fuzzy sets are profound examples with this regard. We start with numeric membership functions. All ensuing processing involves *numeric* membership grades, so in essence it focuses on number crunching. The results are inherently numeric. The progress present here has resulted in a diversity of numeric constructs. Because of the commonly encountered numeric treatment of fuzzy sets, the same applies to logic operators (connectives) encountered in fuzzy sets.

There are a number of alternatives of describing information that are positioned in-between these two extremes or the descriptions could be made more in a multilevel fashion. For instance, one could have an information granule described by a fuzzy set whose membership grades are symbolic (ordered terms, say *small, medium, high*; all defined in the unit interval).

With regard to the formal settings of information granules as briefly highlighted before, it is instructive to mention that all of them offer some operational realization (in different ways, though) of implicit concepts by endowing them by a well-defined semantics. For instance, treated implicitly an information granule *small error* is regarded just as a symbol (and could be subject to symbolic processing as usually realized in Artificial Intelligence); however, once explicitly articulated as an information granule, it comes associated with some semantics (becomes calibrated), thus coming with a sound operational description, say characteristic or membership function. In the formalism of fuzzy sets, the symbol *small* comes with the membership description, which could be further processed.

2.6 Intervals and Calculus of Intervals

Sets and set theory are the fundamental notions of mathematics and science. They are in common usage when describing a wealth of concepts, describing relationships, and formalizing solutions. The underlying fundamental notion of set theory is that of *dichotomy*: a certain element belongs to a set or is excluded from it. A universe of discourse **X** over which a set or sets are formed could be very diversified depending on the nature of the problem.

Given a certain element in a universe of discourse **X**, a process of dichotomization (binarization) imposes a binary, *all-or-none* classification decision: we either accept or reject the element as belonging to a given set. If we denote the acceptance decision about the belongingness of the element by 1 and the reject decision (non-belongingness) by 0, we express the classification (assignment) decision of $x \in \mathbf{X}$ to some given set (S or T) through a characteristic function:

$$S(x) = \begin{cases} 1, \text{if } x \in S \\ 0, \text{if } x \notin S \end{cases} \qquad T(x) = \begin{cases} 1, \text{if } x \in T \\ 0, \text{if } x \notin T \end{cases} \tag{2.12}$$

The empty set Ø has a characteristic function that is identically equal to zero, $\varnothing(x) = 0$ for all x in $\mathbf{X}$. The universe $\mathbf{X}$ itself comes with the characteristic function that is identically equal to one, that is $\mathbf{X}(x) = 1$ for all x in $\mathbf{X}$. Also, a singleton $A = \{a\}$, a set comprising only a single element, has a characteristic function such that $A(x) = 1$ if $x = a$ and $A(x) = 0$ otherwise.

Characteristic functions $A : \mathbf{X} \rightarrow \{0, 1\}$ induce a constraint with well-defined binary boundaries imposed on the elements of the universe X that can be assigned to a set A. By looking at the characteristic function, we see that all elements belonging to the set are non-distinguishable – they come with the same value of the characteristic function so by knowing that $A(x_1) = 1$ and $A(x_2) = 1$ we cannot tell these elements apart. The operations of union, intersection, and complement are easily expressed in terms of the characteristic functions. The characteristic function of the union comes as the maximum of the characteristic functions of the sets involved in the operation. The complement of A denoted by $\bar{A}$, comes with a characteristic function equal to $1 - A(x)$.

Interval analysis has emerged with the inception of digital computers and was mostly motivated by the models of computations therein, which are carried out for intervals implied by the finite number of bits used to represent any number on a (digital) computer. This interval nature of the arguments (variables) implies that the results are also intervals. This raises awareness about the interval character of the results. Interval analysis is instrumental in the analysis of propagation of granularity residing within the original arguments (intervals).

Here, we elaborate on the fundamentals of interval calculus. It will become apparent that they will be helpful in the development of the algorithmic fabric of other formalisms of information granules.

We briefly recall the fundamental notions of numeric intervals. Two intervals $A = [a, b]$ and $B = [c, d]$ are equal if their bounds are equal, $a = c$ and $b = d$. A degenerate interval $[a, a]$ is a single number. There are two categories of operations on intervals, namely set-theoretic and algebraic operations.

2.6.1 Set-Theoretic Operations

Assuming that the intervals are not disjoint (have some common points), they are defined as follows

$$\begin{aligned} &\text{Intersection } \{z \mid z \in A \text{ and } z \in B\} = [\max(a, c) \ \min(b, d)] \\ &\text{Union } \{z \mid z \in A \text{ or } z \in B\} = [\min(a, c), \ \max(b, d)] \end{aligned} \tag{2.13}$$

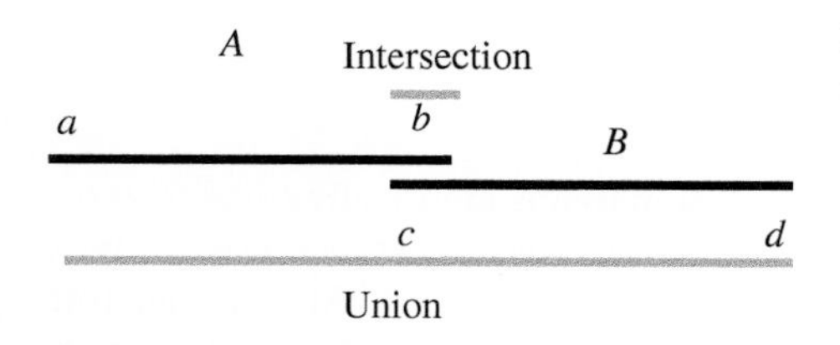

Figure 2.6 Examples of set-theoretic operations on numeric intervals.

For the illustration of the operations, refer to Figure 2.6.

2.6.2 Algebraic Operations on Intervals

The generic algebraic operations on intervals are quite intuitive. As before, let us consider the two intervals $A = [a, b]$ and $B = [c, d]$. The results of addition, subtraction, multiplication, and division are expressed as follows (Moore 1966)

$$\begin{aligned} A + B &= [a + c, b + d] \\ A - B &= [a - d, c - b] = A + [-1 - 1]*B \\ A*B &= [\min(ac, ad, bc, bd), \max(ac, ad, bc, bd)] \\ A/B &= [a, b]\,[1/d, 1/c] \text{ (it is assumed that 0 is not included in the interval } [c, d]) \end{aligned} \tag{2.14}$$

All these formulas result from the fact that these functions are continuous on a compact set; as result they take on the largest and the smallest value as well as the values in-between. The intervals of the obtained values are closed – in all these formulas one computes the largest and the smallest values.

In addition to the algebraic operations, for a continuous unary operation $f(x)$ on the space of real numbers $\boldsymbol{R}$ the mapping of the interval $A = [a, b]$ produces an interval $f(A)$

$$f(A) = [\min f(x), \max f(x)] \tag{2.15}$$

where the minimum (maximum) are taken for all xs belonging to A. Examples of such mappings are x^k, $\exp(x)$, $\sin(x)$, and so on. For the monotonically increasing or decreasing functions, formula (2.15) significantly simplifies:

– monotonically increasing functions

$$f(A) = [f(a), f(b)] \tag{2.16}$$

– monotonically decreasing functions

$$f(A) = [f(b), f(a)] \tag{2.17}$$

Let us consider two intervals $A = [-1, 4]$ and $B = [1, 6]$. The algebraic operations applied to A and B produce the following results

– addition $A+B = [-1 + 1, 4 + 6] = [0, 10]$

– subtraction $A–B = [-1-6, 4-6] = [-7, -2]$

– multiplication $A*B = [\min(-1, -6, 4, 24), \max(-1, -6, 4, 24)] = [-6, 24]$

– division $A/B = [-1, 4]*[1/6, 1/1] = [\min(-1/6, -1, 4/6, 4), \max(-1/6, -1, 4/6, 4)] = [-1/6, 4]$

2.6.3 Distance Between Intervals

The distance between two intervals A and B is expressed as

$$d(A,B) = max\,(|a-c|,\ |\,b-d\,|) \tag{2.18}$$

One can easily show that the properties of distances are satisfied: $d(A, B) = d(B, A)$ (symmetry), $d(A,B)$ is nonnegative with $d(A, B) = 0$ (non-negativity) if and only if $A = B$, $d(A, B) \le d(A,C) + d(B,C)$ (triangle inequality). For real numbers the distance is reduced to the Hamming one.

2.7 Fuzzy Numbers and Intervals

In practice, exact values of parameters of models are not so common. Normally uncertainty and imprecision arise due to lack of knowledge and incomplete information reflected in system structure, parameters, inputs, and possible bounds.

Fuzzy numbers and intervals model imprecise quantities and capture our innate conception of approximate numbers such as about five, around 10, and intervals such as below 100, around two and three, above 10. Fuzzy quantities are intended to model our intuitive notions of approximate numbers and intervals as a generalization of numbers and intervals, as Figure 2.7 suggests. In general, fuzzy quantities summarize numerical data by means of linguistically labeled fuzzy sets whose universe is **R**, the set of real numbers. For instance, if a value of a real variable is certain, say $x = 2.5$, then we can represent it as a certain quantity, a singleton whose characteristic function is $A_{2.5}(x) = 1$ if $x = 2.5$ and $A_{2.5}\ (0) = 0$ otherwise, as shown in Figure 2.7. In this situation,

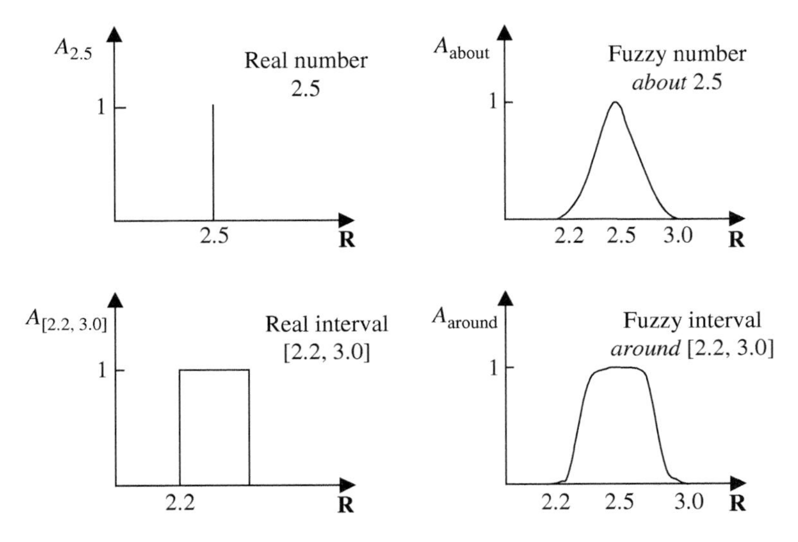

Figure 2.7 Examples of quantities and fuzzy quantities.

the quantity has both, precise value and precise meaning. If we are uncertain on the value of the variable, but certain about its bounds, then the quantity is uncertain and can be represented, for instance, by the closed interval [2.2, 3.0], a set whose characteristic function is $A_{[2.2,\,3.0]}(x) = 1$ if $x \in [2.2, 3.0]$, and $A_{[2.2,\,3.0]}(x) = 0$ otherwise. Here the variable is characterized by an imprecise value, but its meaning is precise. When bounds also are not sharply defined, the quantities become fuzzy numbers or intervals, respectively, as Figure 2.7 illustrates. In both these cases, fuzzy numbers and intervals also are quantities with precise meaning, but with imprecise values.

2.8 Linguistic Variables

One can often deal with variables describing phenomena of physical or human systems assuming a finite, quite small number of descriptors.

We often describe observations about a phenomenon by characterizing its states, which we naturally translate in terms of the idea of a variable. For instance, we may refer to an environment through words such as comfortable, sunny, and neat. In particular, we can qualify the environment condition through the variable temperature with values chosen in a range such as the interval $X = [0, 40]$. Alternatively, temperature could be qualified using labels such as cold, comfortable, and warm. A precise numerical value such as 20 °C seems simpler to characterize the environment than the ill-defined term comfortable. But the linguistic label comfortable is a choice of one out of three values, whereas 20 °C is a choice of out many. The statement could be strengthened if the underlying meaning of comfortable is conceived as about 20 °C. While the numerical quantity 20 °C can be visualized as a point in a set, the linguistic temperature value comfortable can be viewed as a collection of temperature values in a bounded region centered in 20 °C. The label comfortable can, therefore, be regarded as a form of information summarization, called granulation, because it serves to approximate a characterization of ill-defined or complex phenomena (Zadeh 1975). In these circumstances, fuzzy sets provide a way to map a finite term set to a linguistic scale whose values are fuzzy sets. In general, it is difficult to find incontestable thresholds, such as 15 and 30 °C, for instance, which allows us to assign cold = [0,15], comfortable = [15,30], and warm = [30,40]. Cold, comfortable, and warm are fuzzy sets instead of single numbers or sets (intervals). Since fuzzy sets concern the representation of collections with unclear boundaries by means of membership functions taking values in an ordered set of membership values, they provide a means to interface numerical and linguistic quantities, a way to link computing with words and Granular Computing.

In contrast to the idea of numeric variables as being commonly used, the notion of linguistic variable can be regarded as a variable whose values are fuzzy sets. In general, linguistic variables may assume values consisting of words or sentences expressed in a certain language (Zadeh 1999). Formally, a linguistic variable is characterized by a quintuple $< X, T(X), \mathbf{X}, G, M >$ where its components are as follows:

X – the name of the variable;
$T(\mathbf{X})$ – a term set of $\mathbf{X}$ whose elements are labels L of linguistic values of X;
G – a grammar that generates the names of X;
M – a semantic rule that assigns to each label $L \in T(\mathbf{X})$ a meaning whose realization is a fuzzy set on the universe X whose base variable is x.

Example 2.1 Let us consider the linguistic variable of temperature. Here the linguistic variable is formalized by explicitly identifying all the components of the formal definition:

X = temperature, $\mathbf{X}$ = [0, 40].
T(temperature) = {*cold, comfortable, warm*}.
M(cold) → C, M (comfortable) → F and M (warm) → W where C, F and W are fuzzy sets whose membership functions are illustrated in Figure 2.8.

The notion of the linguistic variable plays a major role in applications of fuzzy sets. In fuzzy logic and approximate reasoning truth values can be viewed as

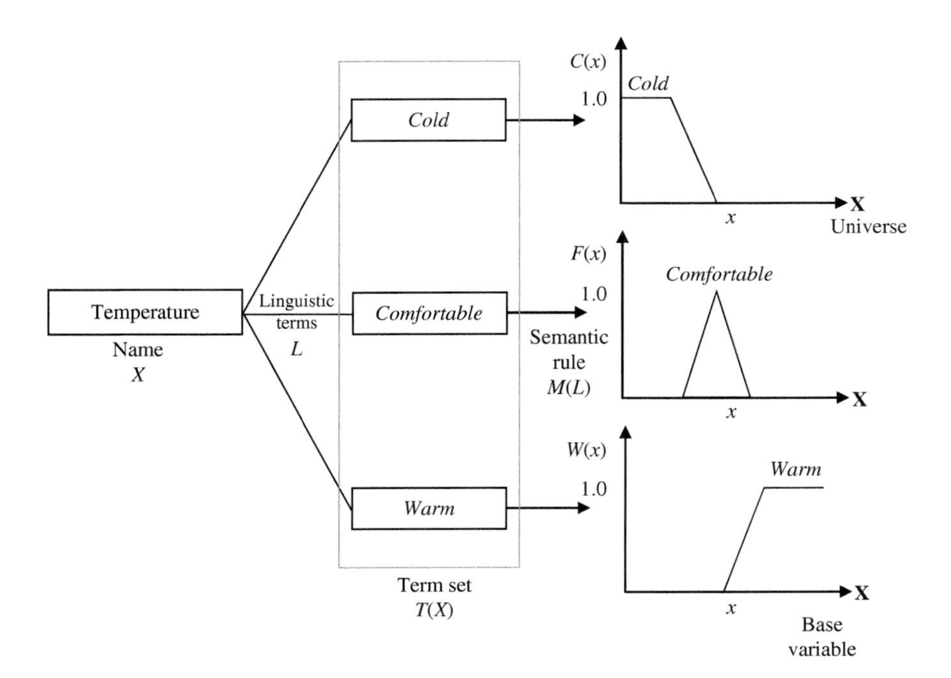

Figure 2.8 An example of the linguistic variable of temperature.

linguistic variables whose truth values form the term set as, for example, *true*, *very true*, *false*, *more or less true*, and the like.

2.9 A Generic Characterization of Fuzzy Sets: Some Fundamental Descriptors

In principle, any function $A : \mathbf{X} \rightarrow [0,1]$ becomes potentially eligible to represent the membership function of fuzzy set A. In practice, however, the type and shape of membership functions should fully reflect the nature of the underlying phenomenon we are interested in describing. Thus, we require that fuzzy sets should be semantically sound, which implies that the selection of membership functions needs to be guided by the character of the application and the nature of the problem we intend to solve.

Given the enormous diversity of potentially useful (namely, semantically sound) membership functions, there are certain common characteristics (descriptors) that are conceptually and operationally qualified to capture the essence of the granular constructs represented in terms of fuzzy sets. In what follows, we provide a list of the descriptors commonly encountered in practice.

Normality: We say that the fuzzy set A is *normal* if its membership function attains 1, that is

$$\sup_{x \in \mathbf{X}} A(x) = 1 \tag{2.19}$$

If this property does not hold, we call the fuzzy set *subnormal*. An illustration of the corresponding fuzzy set is shown in Figure 2.9. The supremum (sup) in expression (2.19) is also referred to as a height of the fuzzy set A, $\text{hgt}(A) = \sup_{x \in \mathbf{X}} A(x) = 1$.

The normality of A has a simple interpretation: by determining the height of the fuzzy set, we identify an element with the highest membership degree. The value of the height being equal to one states that there is at least one element in $\mathbf{X}$ whose typicality with respect to A is the highest one and which could be sought as fully compatible with the semantic category presented by A.

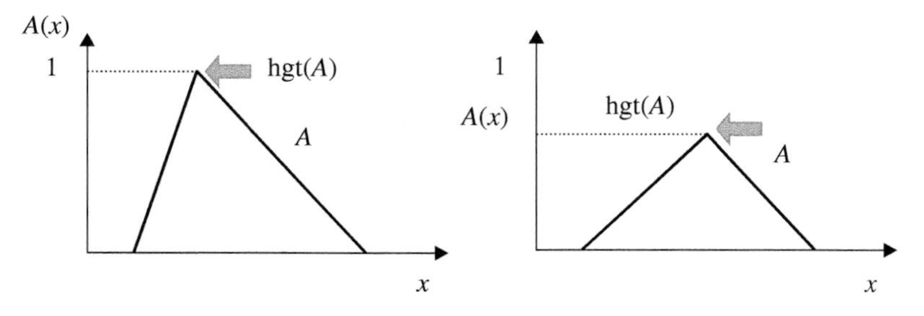

Figure 2.9 Examples of normal and subnormal fuzzy sets.

A subnormal fuzzy set whose height is lower than 1, namely, $\text{hgt}(A) < 1$, means that the degree of typicality of elements in this fuzzy set is somewhat lower (weaker) and we cannot identify any element in **X**, which is fully compatible with the underlying concept. Generally, while forming a fuzzy set we expect its normality (otherwise why would such a fuzzy set for which there are no typical elements come into existence in the first place?).

Normalization: The normalization operation, Norm(A), is a transformation mechanism that is used to convert a subnormal nonempty fuzzy set A into its normal counterpart. This is done by dividing the original membership function by the height of this fuzzy set, that is

$$\text{Norm}(A) = \frac{A(x)}{\text{hgt}(A)} \tag{2.20}$$

While the height describes the global property of the membership grades, the following notions offer an interesting characterization of the elements of **X** vis-à-vis their membership degrees.

Support: Support of a fuzzy set A, denoted by Supp (A), is a set of all elements of **X** with nonzero membership degrees in A.

$$\text{Supp}(A) = \{x \in X \mid A(x) > 0\} \tag{2.21}$$

In other words, support identifies all elements of **X** that exhibit some association with the fuzzy set under consideration (by being allocated to A with nonzero membership degrees).

Core: The core of a fuzzy set A, Core(A), is a set of all elements of the universe that are typical for A, namely, they come with membership grades equal to 1,

$$\text{Core}(A) = \{x \in X \mid A(x) = 1\} \tag{2.22}$$

The support and core are related in the sense that they identify and collect elements belonging to the fuzzy set yet at two different levels of membership. Given the character of the core and support, we note that all elements of the core of A are subsumed by the elements of the support of this fuzzy set. Note that both support and core are sets, not fuzzy sets; see Figure 2.10. We refer to them as the set-based characterizations of fuzzy sets.

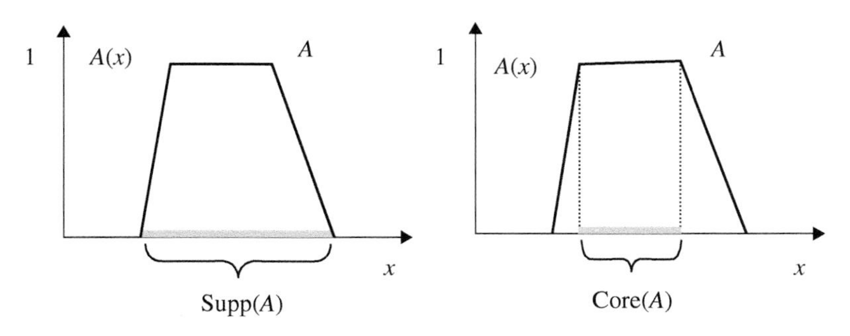

Figure 2.10 Support and core of A.

While core and support are somewhat extreme (in the sense that they identify the elements of A that exhibit the strongest and the weakest linkages with A), we may be also interested in characterizing sets of elements that come with some intermediate membership degrees. A notion of a so-called α-cut offers here an interesting insight into the nature of fuzzy sets.

α-cut: The α-cut of a fuzzy set A, denoted by A_α, is a set consisting of the elements of the universe whose membership values are equal to or exceed a certain threshold level α where $\alpha \in [0,1]$ (Zadeh 1975; Nguyen and Walker 1999). Formally speaking, we have $A_\alpha = \{x \in \mathbf{X} \mid A(x) \geq \alpha\}$. A strong α-cut differs from the α-cut in the sense that it identifies all elements in $\mathbf{X}$ for which we have the following equality $A_\alpha = \{x \in \mathbf{X} \mid A(x) > \alpha\}$. An illustration of the concept of the α-cut and strong α-cut is presented in Figure 2.11. Both support and core are limit cases of α-cuts and strong α-cuts. For $\alpha = 0$ and the strong α-cut, we arrive at the concept of the support of A. The threshold $\alpha = 1$ means that the corresponding α-cut is the core of A.

Convexity: We say that a fuzzy set is convex if its membership function satisfies the following condition:

for all $x_1, x_2 \in \mathbf{X}$ and all $\lambda \in [0,1]$:

$$A[\lambda x_1 + (1-\lambda)x_2] \geq \min[A(x_1), A(x_2)] \tag{2.23}$$

This relationship states that, whenever we choose a point x on a line segment between x_1 and x_2, the point $(x, A(x))$ is always located above or on the line passing through the two points $(x_1, A(x_1))$ and $(x_2, A(x_2))$, refer to Figure 2.12. Note that the membership function is not a convex function in the traditional sense (Klir and Yuan 1995).

Let us recall that a set S is convex if, for all $x_1, x_2 \in S$, then $x = \lambda x_1 + (1-\lambda)x_2 \in S$ for all $\lambda \in [0,1]$. In other words, convexity means that any line segment identified by any two points in S is also contained in S. For instance, intervals of real numbers are convex sets. Therefore, if a fuzzy set is convex, then all of its α-cuts are convex, and conversely, if a fuzzy set has all its α-cuts convex, then it is a convex fuzzy set, refer to Figure 2.13. Thus, we may say that a fuzzy set is convex if all its α-cuts are convex (intervals).

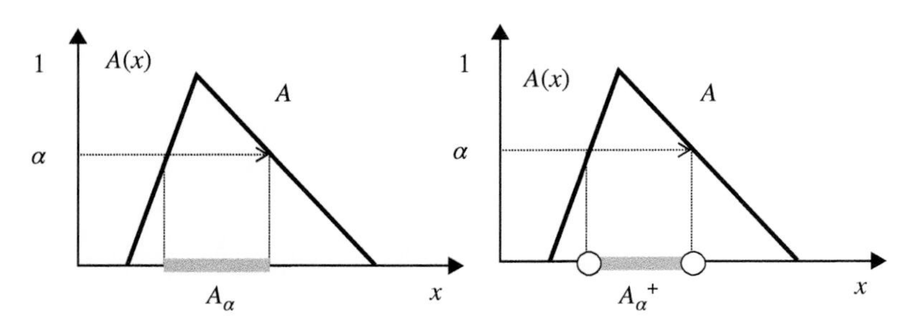

Figure 2.11 Examples of α-cut and strong α-cut.

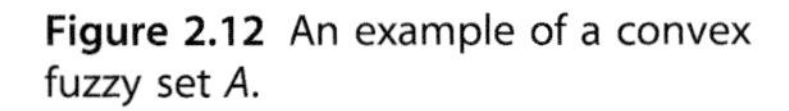
Figure 2.12 An example of a convex fuzzy set *A*.

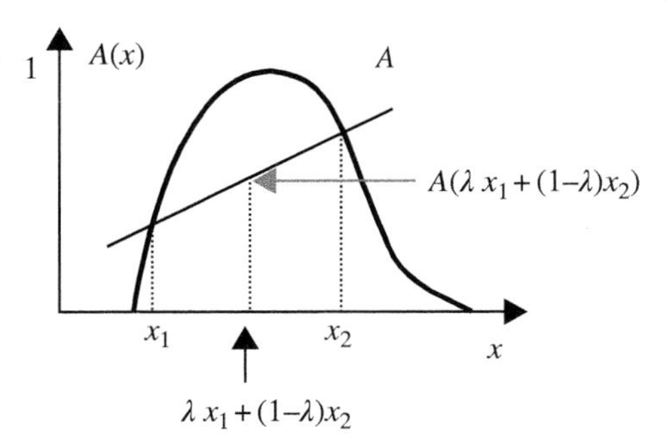

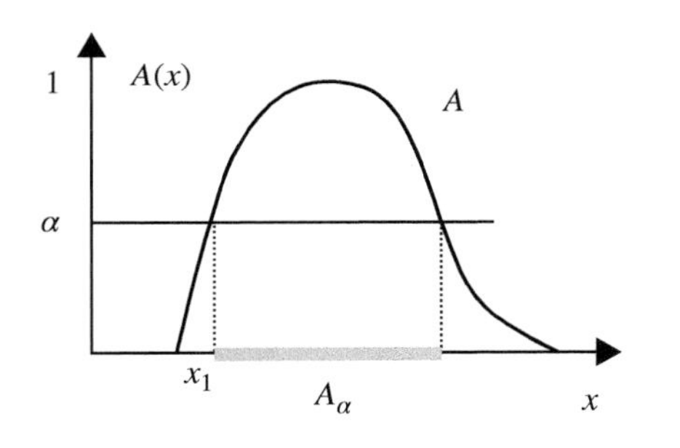

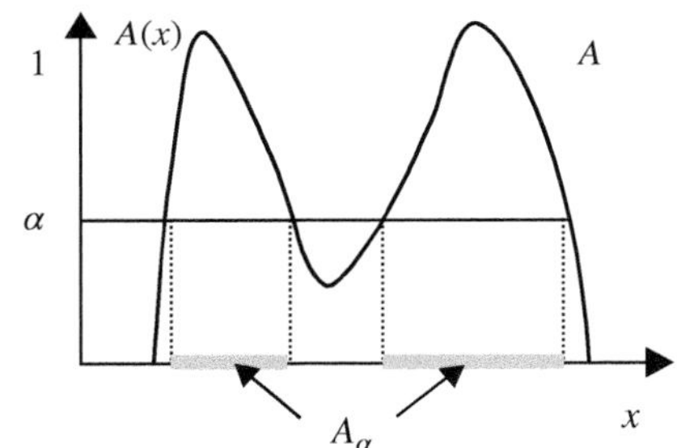

Figure 2.13 Examples of convex and non-convex fuzzy sets.

Fuzzy sets can be characterized by counting their elements and brining a single numeric quantity as a meaningful descriptor of this count. While in the case of sets this sounds convincing, here we have to take into account different membership grades. In the simplest form this counting comes under the name of cardinality.

Cardinality: Given a fuzzy set A defined in a finite or countable universe $\mathbf{X}$, its cardinality, denoted by card(A), is expressed as the following sum

$$\text{card}(A) = \sum_{x \in \mathbf{X}} A(x) \tag{2.24}$$

or alternatively as the following integral

$$\text{card}(A) = \int_X A(x)dx \tag{2.25}$$

(we assume that the integral shown in (2.25) does make sense). The cardinality produces a count of the number of elements in the given fuzzy set. As there are different degrees of membership, the use of the sum here makes sense as we keep adding contributions coming from the individual elements of this fuzzy set. Note that in the case of sets, we count the number of elements belonging to the corresponding sets. We also use the alternative notation of Card(A) = $|A|$, and refer to it as a sigma count (σ-count).

The cardinality of fuzzy sets is explicitly associated with the concept of granularity of information granules realized in this manner. More descriptively, the more the elements of A we encounter, the higher the level of abstraction supported by A and the lower the granularity of the construct. Higher values of cardinality come with the higher level of abstraction (generalization) and the lower values of granularity (specificity).

Example 2.2 Consider fuzzy sets $A = (1.0, 0.6, 0.8, 0.1)$, $B = (0.1, 0.8, 1.0, 0.1)$ and $C = (0.6, 0.9, 1.0, 1.0)$ defined in the same space. We can order them in a linear fashion by computing their cardinalities. Here we obtain: card (A) = 2.5, card (B) = 2.0, and card (C) = 3.5. In terms of the levels of abstraction, C is the most general, A lies in-between, and B is the least general.

So far, we discussed properties of a single fuzzy set. The operations to be studied look into the characterizations of relationships between two fuzzy sets.

Equality: We say that two fuzzy sets A and B defined in the same universe $\mathbf{X}$ are equal if and only if their membership functions are identical, meaning that

$$A(x) = B(x) \;\; \forall x \in \mathbf{X} \tag{2.26}$$

Inclusion: Fuzzy set A is a subset of B (A is included in B), denoted by $A \subseteq B$, if and only if every element of A also is an element of B. This property expressed in terms of membership degrees means that the following inequality is satisfied.

$$A(x) \le B(x) \;\; \forall x \in \mathbf{X} \tag{2.27}$$

An illustration of these two relationships in the case of sets is shown in Figure 2.14. In order to satisfy the relationship of inclusion, we require that the characteristic functions adhere to Eq. (2.18) for all elements of $\mathbf{X}$. If the inclusion is not satisfied even for a single point of $\mathbf{X}$, the inclusion property does not hold.

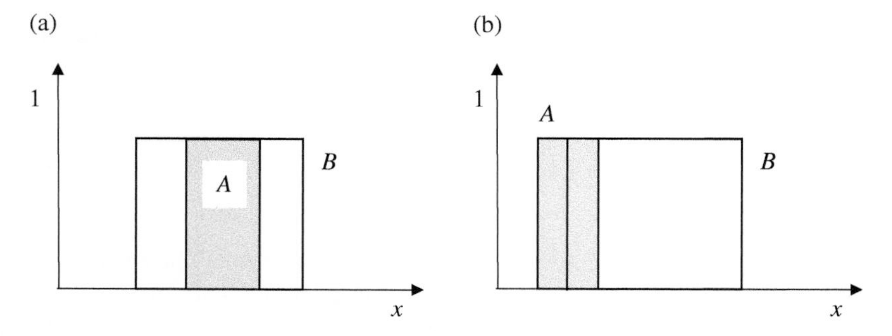

Figure 2.14 Set inclusion: (a) $A \subset B$ and (b) inclusion not satisfied as $A \not\subset B$.

If A and B are fuzzy sets in **X**, we have adopted the same definition of inclusion as being available in set theory.

Interestingly, the definitions of equality and inclusion exhibit an obvious dichotomy as the property of equality (or inclusion) is satisfied or is not satisfied. While this quantification could be acceptable in the case of sets, fuzzy sets require more attention with this regard given that the membership degrees involved in the expressing the corresponding definitions.

Energy measure of fuzziness of a fuzzy set A in **X**, denoted by $E(A)$, is a functional of the membership degrees

$$E(A) = \sum_{i=1}^{n} e[A(x_i)] \tag{2.28}$$

if Card (**X**) = n. In the case of the infinite space, the energy measure of fuzziness is expressed as the following integral.

$$E(A) = \int_X e(A(x))dx \tag{2.29}$$

The mapping e: $[0,1] \rightarrow [0,1]$ is a functional monotonically increasing over $[0,1]$ with the boundary conditions $e(0) = 0$ and $e(1) = 1$.

As the name of this measure stipulates, its role is to quantify a sort of energy associated with the given fuzzy set. The higher the membership degrees are, the more essential are their contributions to the overall energy measure. In other words, by computing the energy measure of fuzziness we can compare fuzzy sets in terms of their overall count of membership degrees.

A particular form of this functional comes with the identity mapping that is $e(u) = u$ for all u in $[0,1]$. We can see that in this case, the expressions Eqs. (2.20) and (2.21) reduce to the cardinality of A,

$$E(A) = \sum_{i=1}^{n} A(x_i) = \text{Card}(A) \tag{2.30}$$

The energy measure of fuzziness forms a convenient way of expressing a total mass of the fuzzy set. Since card(Ø) = 0 and card (**X**) = n, the more a fuzzy set differ from the empty set, the larger its mass is. Indeed, rewriting Eq. (2.30) we obtain

$$E(A) = \sum_{i=1}^{n} A(x_i) = \sum_{i=1}^{n} |A(x_i) - \varnothing(x_i)| = d(A, \varnothing) = \text{Card}(A) \tag{2.31}$$

where $d(A, \varnothing)$ is the Hamming distance between fuzzy set A and the empty set.

While the identity mapping (e) is the simplest alternative one could think of, in general, we can envision an infinite number of possible options. For instance, one could consider the functionals such as $e(u) = u^p, p > 0$ and $e(u) = \sin\left(\frac{\pi}{2}u\right)$.

Note that by choosing a certain form of the functional, we accentuate a varying contribution of different membership grades. For instance, depending on the form of "e," the contribution of the membership grades close to 1 could be emphasized while those located close to 0 could be very much reduced.

Entropy measure of fuzziness of A, denoted by H(A), is built on the entropy functional (h) and comes in the form (De Luca and Termini 1972).

$$H(A) = \sum_{i=1}^{n} h[A(x_i)] \tag{2.32}$$

or in the continuous case of **X**

$$H(A) = \int_X h(A(x))dx \tag{2.33}$$

where h: [0,1] → [0,1] is a functional such that (i) it is monotonically increasing in [0, ½] and monotonically decreasing in [½, 1] and (ii) comes with the boundary conditions $h(0) = h(1) = 0$ and $h(½) = 1$. This functional emphasizes membership degrees around ½; in particular the value of ½ is stressed to be the most "unclear" (causing the highest level of hesitation with its quantification by means of the proposed functional)

Specificity of fuzzy sets: Quite often, we face the issue to quantify how much a single element of a universe could be regarded as a representative of a fuzzy set. If this fuzzy set is a singleton,

$$A(x) = \begin{cases} 1 \text{ if } x = x_0 \\ 0 \text{ if } x \neq x_0. \end{cases} \tag{2.34}$$

then there is no hesitation in selecting x_o as the sole representative of A. We say that A is very *specific* and its choice comes with no hesitation. On the other extreme, if A covers the entire universe **X** and embraces all elements with the membership grade being equal to 1, the choice of the only one representative of A comes with a great deal of hesitation, which is triggered by a lack of specificity being faced in this problem. These two extreme situations are portrayed in Figure 2.15. Intuitively, we sense that the specificity is a concept that relates quite visibly with the cardinality of a set (Yager 1983). The higher the cardinality of the set (namely, the more evident its abstraction) is, the lower its specificity. Having said that, we are interested in developing a measure that could be able to capture this effect of hesitation.

The specificity is about the level of detail being captured by the information granule. As the name stipulates, the specificity indicates how detailed information granule is. In other words, the specificity can be determined by assessing the "size" of this fuzzy set. To explain the notion, we start with a case when information granule is an interval $[a, b]$, see Figure 2.16a. The specificity can be defined in the following way $1 - (b - a)/range$ where *range* is expressed as

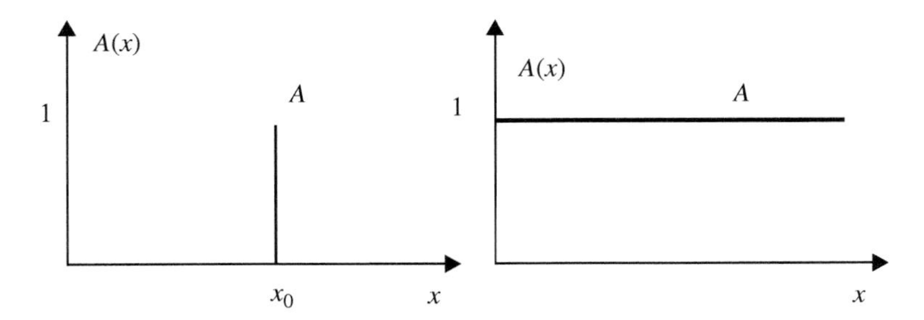

Figure 2.15 Examples of two extreme cases of sets exhibiting distinct levels of specificity.

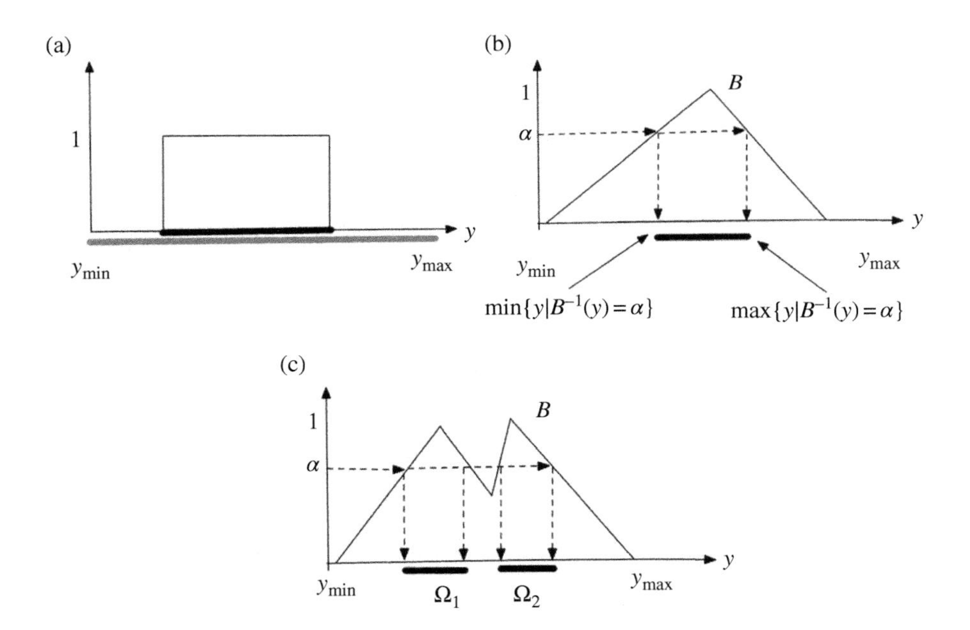

Figure 2.16 Determination of specificity of (a) interval (set), (b) unimodal fuzzy set, and (c) multimodal fuzzy set.

$y_{max}-y_{min}$ with y_{max} and y_{min} being the extreme values of assumed by the variable. This definition adheres to our perception of specificity: the broader the interval, the lower its specificity. In the boundary cases, when $a = b$, the specificity attains its maximal value equal to 1 whereas when the interval spreads across the entire space (range), the specificity value is zero. This definition is a special case. In a general way, specificity is defined as a measure adhering to the conditions we have already outlined:

$$\mathrm{sp} : A \rightarrow [0, 1] \tag{2.35}$$

i) boundary conditions sp({x}) = 1, sp(X) = 0. A single-element information granule is the most specific. The specificity of the entire space is the minimal one (here we can request that it is equal to zero but this is not required in all situations)
ii) monotonicity if $A \subset B$ then sp(A) ≥ sp(B) This reflects our intuition that the more detailed information granule comes with the higher specificity. One has to note here that the requirement is general in the sense that the definition of inclusion and the details of computing the specificity are dependent on the formalism of information granules. In the example, we considered intervals, see Figure 2.16a. Here, the inclusion of intervals is straightforward; apparently $A = [a, b]$ is included in $B = [c, d]$ if the $a \geq c$ and $b \leq d$. The specificity defined here is just an example; any decreasing function of the length of the interval could serve as a viable alternative. For instance, one can consider exp (-|b-a|) as the specificity of A.

The technical details of specificity have to be re-defined when considering some other formalisms of information granules. Consider a given fuzzy set B. The specificity of B can be defined by starting with the already formulated specificity of an interval. In virtue of the representation theorem stating that any fuzzy set can be described through a collection of its α-cuts (which are intervals), we determine the specificity of the α-cut B_α namely, $B_\alpha = \{y | B(y) \geq \alpha\}$ and then integrate the results over all values of the threshold α. Thus, we have

$$\text{sp}(B) = \int_0^{\alpha_{\max}} sp(B_\alpha)d\alpha = \int_0^{\alpha_{\max}} \left(1 - \frac{h(\alpha)}{\text{range}}\right) d\alpha \tag{2.36}$$

where $h(\alpha)$ stands for the length of the interval

$$h(\alpha) = |\max\{y | B^{-1}(y) = \alpha\} - \min\{y | B^{-1}(y) = \alpha\}| \tag{2.37}$$

$\alpha_{\max}$ is the maximal value of membership of B, $\alpha_{\max} = \text{hgt}(B) = \sup_y B(y)$. For normal fuzzy set B, $\alpha_{\max} = 1$. In other words, the specificity comes as an average of specificity values of the corresponding α-cuts. Refer to Figure 2.26b. In practical computing, the integral standing in Eq. (2.36) is replaced by its discrete version involving summation carried out over some finite number of values of α.

For the multimodal membership functions, the calculations need to be refined. We consider the sum of the lengths of the α-cuts as illustrated in Figure 2.16c. In this way, one has

$$h(\alpha) = \text{length}(\Omega_1) + \text{length}(\Omega_2) + \ldots + \text{length}(\Omega_n) \tag{2.38}$$

To visualize coverage and specificity as well as emphasize the relationships between these two characteristics, we assume that the data are governed by the Gaussian probability density function $p(x)$ with a zero mean and some

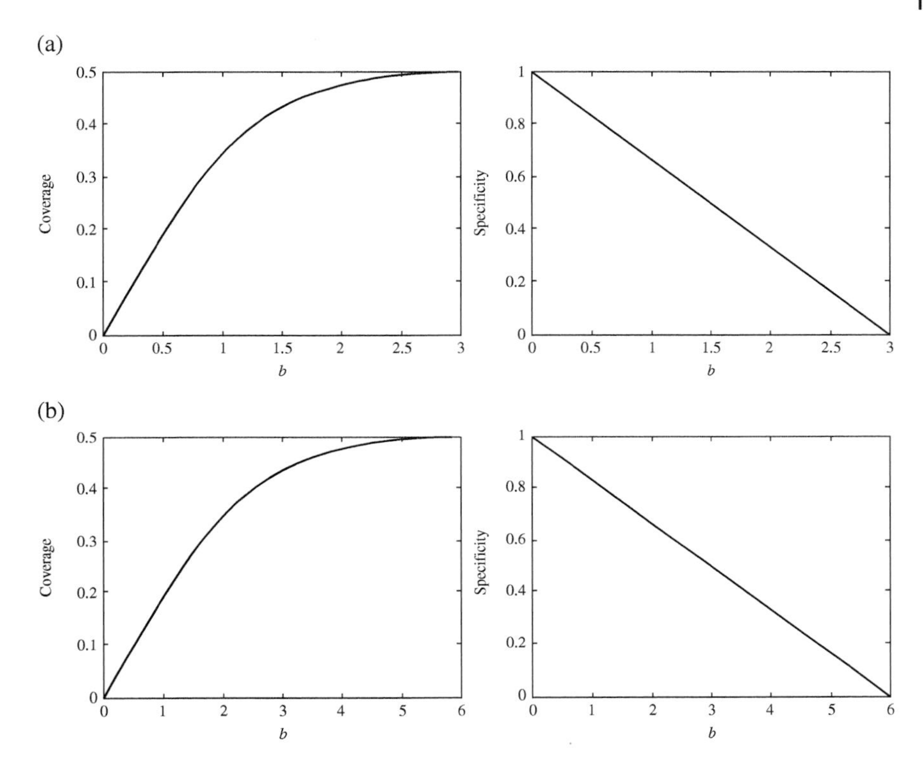

Figure 2.17 Plots of coverage and specificity as a function of b: (a) $\sigma = 1.0$ and (b) $\sigma = 2.0$.

standard deviation. As the interval is symmetric, we are interested in determining an optimal value of its upper bound b. The coverage provided by the interval $[0, b]$ is expressed as the integral of the probability function

$$\text{cov}([0,b]) = \int_0^b p(x)dx \tag{2.39}$$

while the specificity (assuming that the x_{max} is set as 3σ) is defined in the following form

$$\text{sp}([0,b]) = 1 - \frac{b}{x_{\text{max}}} \tag{2.40}$$

The plots of the coverage and the specificity measures being regarded as a function of b are shown in Figure 2.17. One can note that the coverage is a nonlinear and monotonically increasing function of b while the specificity (in the form specified previously) is a linearly decreasing function of b. As becomes intuitively apparent, the increase in the coverage results in lower values of the specificity measure and vice versa. We will consider them as two essential criteria guiding the development of information granules.

2.10 Coverage of Fuzzy Sets

In a descriptive way, a level of abstraction captured by some information granules A is associated with the number of elements (data) embraced by the granule. For instance, these elements are the elements of the space or a collection of some experimental data. A certain measure of cardinality, which counts the number of elements involved in the information granule, forms a sound descriptor of information granularity. The higher the number of elements embraced by the information granule, the higher the abstraction of this granule and the lower its specificity becomes. Starting with a set theory formalism where A is a set, its cardinality is computed in the form of the following sum in the case of a finite universe of discourse $\mathbf{X} = \{x_1, x_2, \ldots, x_n\}$

$$\text{card}(A) = \sum_{i=1}^{n} A(x_i) \tag{2.41}$$

or an integral (when the universe is infinite and the integral of the membership function itself does exists)

$$\text{card}(A) = \int_{x} A(x)dx \tag{2.42}$$

where $\mathrm{A}(x)$ is a formal description of the information granule (say, in the form of the characteristic function or membership function). For a fuzzy set, we count the number of its elements but one has to bear in mind that each element may belong with a certain degree of membership so the calculations carried out previously involve the degrees of membership. In this case, one commonly refers to (2.24) and (2.25) as a σ-count of A. For rough sets, one can proceed in a similar manner as before by expressing the granularity of the lower and upper bounds of the rough set (roughness). In the case of probabilistic information granules, one can consider its standard deviation as a sound descriptor of information granularity. The higher the coverage (cardinality), the higher the level of abstraction being associated with the information granule.

2.11 Matching Fuzzy Sets

The problem of quantifying how close (similar) two information granules A and B defined in the same space are, becomes of paramount relevance. To illustrate the underlying concepts, we start with A and B coming in the form of intervals $A = [a^-, a^+]$ and $B = [b^-, b^+]$. We introduce two operations, namely join and meet, described as follows

$$\text{join}\, A \oplus B = [\min(a^-, b^-), \max(a^+, b^+)]$$

$$\text{meet}\, A \otimes B = [\max(a^-, b^-), \min(a^+, b^+)] \tag{2.43}$$

For the meet operation, we consider that these two intervals are not disjoint; otherwise it is said that the meet is empty. The essence of these two operations is clarified in Figure 2.18.

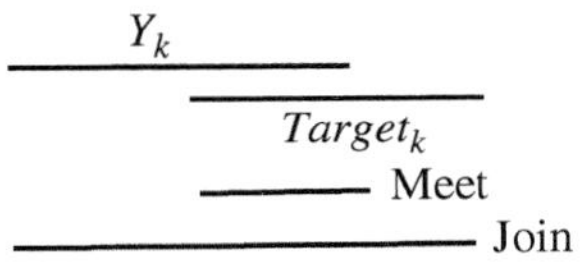

Y_k
$Target_k$
Meet
Join

Figure 2.18 Join and meet of interval information granules.

To quantify a degree to which these intervals A and B are similar (match), we propose the following definition of a degree of matching, $\xi(A, B)$, coming in the form of the following ratio.

$$\xi(A, B) = \frac{|A \otimes B|}{|A \oplus B|} \tag{2.44}$$

The crux of this definition is to regard a length of the meet, $|A \otimes B|$ as a measure of overlap of the intervals and calibrate this measure by taking the length of the join, $|A \oplus B|$. Interestingly, the essence of expression (2.44) associates with the Jaccard's coefficient being used to quantify similarity.

The definition, originally, developed for intervals, can be generalized for other formalism of information granules. For instance, in case of fuzzy sets we engage the concept of α-cuts. We start with a family of α-cuts A_α and B_α, determine the corresponding sequence $\xi(A_\alpha$ and $B_\alpha)$. The meaning the ξ will become clear from the discussion that follows. There are two possible ways of describing the matching as a single numeric descriptor.

Numeric quantification. We aggregate the elements of the sequence by forming a single numeric descriptor. The use of the integral is a sound aggregation alternative considered here

$$\xi(A,B) = \int_0^1 \xi(A_\alpha, B_\alpha) d\alpha \tag{2.45}$$

Granular quantification. In contrast to the previous matching approach, the result of matching is a fuzzy set defined over the unit interval. This characterization is more comprehensive as we obtain a non-numeric result.

There is a certain disadvantage, though. As the results are information granules, one has to invoke some techniques of ranking information granules.

In case two information granules are described in a highly dimensional space, these constructs can be involved here by determining the matching for each variable and then computing the average of the partial results obtained in this way.

2.12 Geometric Interpretation of Sets and Fuzzy Sets

In the case of finite universes of discourse **X**, we can arrive at an interesting and geometrically appealing interpretation of sets and fuzzy sets (Kosko 1992). Such an interpretation is also helpful in contrasting between sets and fuzzy sets. It also visualizes interrelationships between them. The geometric interpretation is also helpful in casting the decision-making problems and their solutions in some illustrative geometric setting. For the n-elements space **X**, any set there can be represented as an n-dimensional vector **x** with the 0–1 values. The cardinality of the family of all sets defined in **X** is 2^n. The i-th component of vector **x** is the value of the corresponding characteristic function of the i-th element in the respective set. In the simplest case when $\mathbf{X} = \{x_1, x_2\}$; $n = 2$, the family of sets comprises the following elements, namely Ø, $\{x_1\}$, $\{x_2\}$, and $\{x_1, x_2\}$ The cardinality of **X** is $2^2 = 4$. Thus each of the four elements of this family can be represented by a two-dimensional vector, say Ø = (0,0), $\{x_1\} = (1,0)$, $\{x_2\} = (0,1)$, and $\{x_1, x_2\} = (1,1)$. Those sets are located at the corners of the unit square, as illustrated in Figure 2.19.

Owing to the values of the membership grades assuming any values in [0,1], fuzzy sets being two-dimensional vectors are distributed throughout the entire unit square. For instance, refer to Figure 2.19, fuzzy set A is represented as vector $\mathbf{x} = (0.25, 0.75)$. A family of fuzzy sets over $\mathbf{X} = \{x_1, x_2\}$, occupies the whole

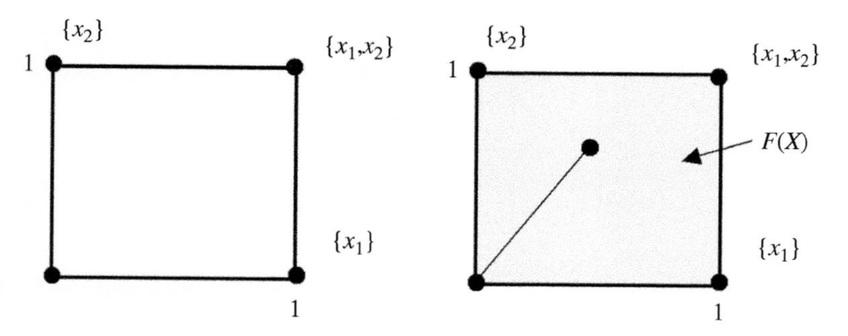

Figure 2.19 Sets and fuzzy sets represented as points in the unit square.

shaded area, including the borders and corners of the unit square. In general, proceeding with higher dimensionality of the space, we end up with a unit cube ($n = 3$) and unit hypercubes (for the dimensionality of the space of dimensionality higher than 3).

2.13 Fuzzy Set and Its Family of α-Cuts

Fuzzy sets offer an important conceptual and operational feature of information granules by endowing their formal models by gradual degrees of membership. We are interested in exploring relationships between fuzzy sets and sets. While sets come with the binary (yes-no) model of membership, it could be worth investigating whether they are indeed some special cases of fuzzy sets and if so, in which sense a set could be treated as a suitable approximation of some given fuzzy set. This could shed light on some related processing aspects. To gain a detailed insight into this matter, we recall here a concept of an α-cut and a family of α-cuts and show that they relate to fuzzy sets in an $[0,1] \to [0,1]$ intuitive and transparent way. Let us re-visit the semantics of α-cuts: an α-cut of A embraces all elements of the fuzzy set whose degrees of belongingness (membership) to this fuzzy set are at least equal to α. In this sense, by selecting a sufficiently high value of α, we identify (tag) elements of A that belongs to it to a significant extent and thus could be sought as those highly representative of the concept conveyed by A. Those elements of $\mathbf{X}$ exhibiting lower values of the membership grades are suppressed so this allows us to selectively focus on the elements with the highest degrees of membership while dropping the others.

For α-cuts A_α the following properties hold

$$\begin{aligned}&\text{(a) } A_0 = X\\&\text{(b) If } \alpha \le \beta \text{ then } A_\alpha \supseteq A_\beta\end{aligned} \tag{2.46}$$

The first property shows that if we allow for the zero value of α, then all elements of $\mathbf{X}$ are included in this α-cut (0-cut, to be more specific). The second property underlines the monotonic character of the construct: higher values of the threshold imply that more elements are accepted in the resulting α-cuts. In other words, we may say that the level sets (α-cuts) A_α form a nested family of sets indexed by some parameter (α). If we consider the limit value of α, that is $\alpha = 1$, the corresponding α-cut is nonempty if and only if A is a normal fuzzy set.

It is also worth to remember that α-cuts, in contrast to fuzzy sets, are sets. We showed how for some given fuzzy set, its α-cut could be formed. An interesting question arises as to the construction that could be realized when moving into the opposite direction. Could we "reconstruct" a fuzzy set on a basis of an infinite family of sets? The answer to this problem is offered in what is known as the representation theorem for fuzzy sets (Klir and Yuan 1995).

Theorem Let $\{A_\alpha\}$ $\alpha \in [0,1]$ be a family of sets defined in **X** such that they satisfy the following properties

a) $A_0 = \mathbf{X}$
b) If $\alpha \le \beta$ then $A_\alpha \supseteq A_\beta$
c) For the sequence of threshold values $\alpha_1 \le \alpha_2 \le \ldots$ such that $\lim \alpha_n = \alpha$, we have $A_\alpha = \bigcap_{n=1}^{\infty} A_{\alpha_n}$

Then there exists a unique fuzzy set B defined in **X** such that $B_\alpha = A_\alpha$ for each $\alpha \in [0,1]$.

In other words, the representation theorem states that any fuzzy set A can be uniquely represented by an infinite family of its α-cuts. The following reconstruction expression shows how the corresponding α-cuts contribute to the formation of the corresponding fuzzy set

$$A = \bigcup_{\alpha > 0} \alpha A_\alpha \tag{2.47}$$

that is

$$A(x) = \sup_{\alpha \in [0,1]} [\alpha A_\alpha(x)] \tag{2.48}$$

where A_α denotes the corresponding α-cut.

The essence of this construct is that any fuzzy set can be uniquely represented by the corresponding family of nested sets (namely, ordered by the inclusion relation). The illustration of the concept of the α-cut and a way in which the representation of the corresponding fuzzy set becomes realized is shown in Figure 2.20.

More descriptively, we may say that fuzzy sets can be reconstructed by a family of sets. Apparently, we need a family of sets (intervals, in particular) to capture the essence of a single fuzzy set. The reconstruction scheme illustrated in Figure 2.20 is self-explanatory with this regard. In more descriptive terms, we may look at the expression offered by Eqs. (2.48)–(2.49) as a way of decomposing A into a series of layers (indexed sets) being calibrated by the values of the associated levels of α.

For the finite universe of discourse, dim (**X**) = n, we encounter a finite number of membership grades and subsequently a finite number of α-cuts. This finite family of α-cuts is then sufficient to fully "represent" or reconstruct the original fuzzy set.

Example 2.3 To illustrate the essence of α-cuts and the ensuing reconstruction, let us consider a fuzzy set with a finite number of membership grades, A = [0.8 1.0 0.2 0.5 0.1 0.0 0.0 0.7]. The corresponding α-cuts of A are equal to

$\alpha = 1.0$ $A_{1.0} = [0\ 1\ 0\ 0\ 0\ 0\ 0\ 0]$
$\alpha = 0.8$ $A_{0.8} = [1\ 1\ 0\ 0\ 0\ 0\ 0\ 0]$

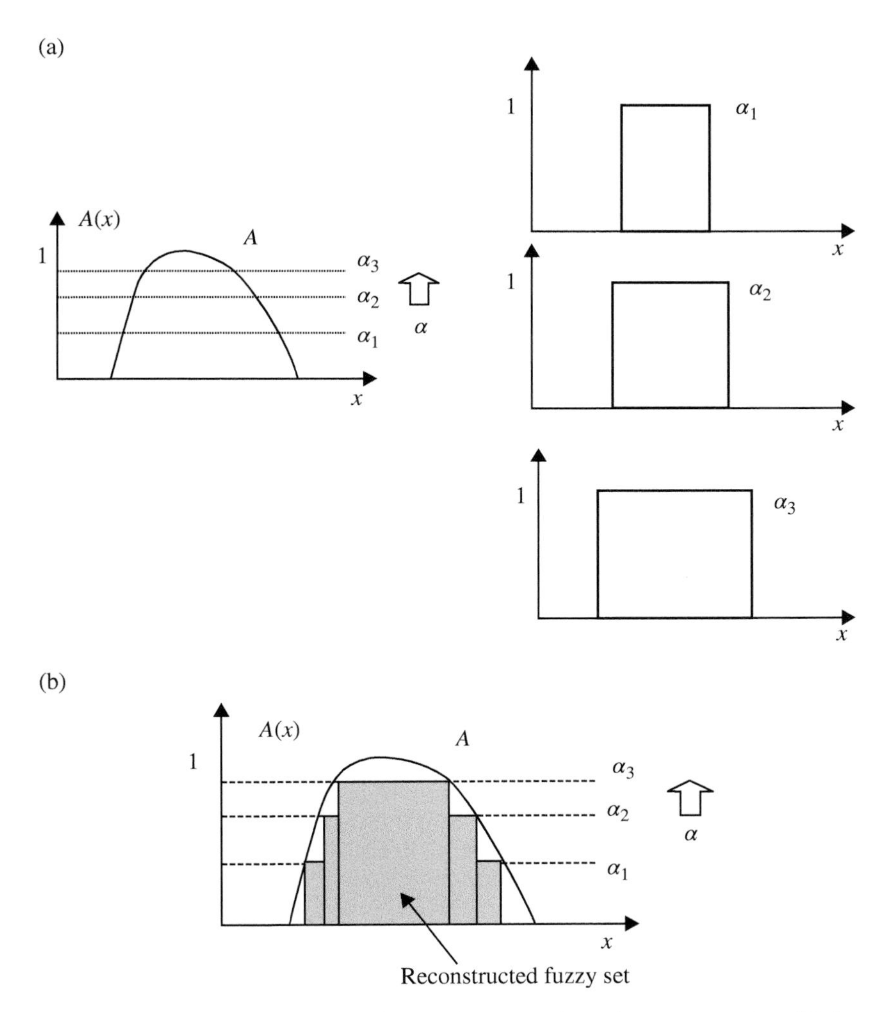

Figure 2.20 Fuzzy set A, examples of some of its α-cuts (a) and a representation of A through the corresponding family of sets (α-cuts) (b).

$\alpha = 0.7\ A_{0.7} = [1\ 1\ 0\ 0\ 0\ 0\ 0\ 1]$
$\alpha = 0.5\ A_{0.5} = [1\ 1\ 0\ 1\ 0\ 0\ 0\ 1]$
$\alpha = 0.2\ A_{0.2} = [1\ 1\ 1\ 1\ 0\ 0\ 0\ 1]$
$\alpha = 0.1\ A_{0.1} = [1\ 1\ 1\ 1\ 1\ 0\ 0\ 1]$

We clearly see the layered character of the consecutive α-cuts indexed by the sequence of the increasing values of α. Because of the finite number of membership grades, the reconstruction realized in terms of Eq. (2.47) returns the original fuzzy set (which is possible given the finite space over which the original fuzzy set has been defined) $A(x) = \max\ (1.0A_{1.0}(x),\ 0.8A_{0.8}(x),\ 0.7A_{0.7}(x),\ 0.5A_{0.5}(x),\ 0.2A_{0.2}(x),\ 0.1A_{0.1}(x))$.

Considering a finite number of α-cuts, an important question arises as to the choice of values of such thresholds (values of α) so that a fuzzy set could be approximated to the highest extent. This is a problem formulated and solved with the aid of population-based optimization in Pedrycz et al. (2009). Refer also to Bodjanova (2006).

2.14 Fuzzy Sets of Higher Type and Fuzzy Order

Fuzzy sets discussed so far can be regarded as generic constructs embracing elements belonging to a given concept to some degree. Two categories of generalizations are considered, namely type-2 and order-2 fuzzy sets.

2.14.1 Fuzzy Sets of Type –2

These fuzzy sets are defined with the use of parameters whose values are non-numeric but come in the form of information granules. They can be realized in a number of ways where non-numeric parameters of the fuzzy set can be defined in a variety of ways, see Figure 2.21. The quantification of parameters is dependent on the way in which formalism of information granules has been used. This, in turn, is implied by the applied aspect of the problem in which fuzzy sets are being used.

Evidently, there is a collection of possibilities. Interval-valued fuzzy sets are the simplest alternative; the degree of membership becomes an interval located in the unit interval. The characterization of membership degree could come in the form of a fuzzy set defined in the unit interval; in this case we are concerned with the construct, which is usually considered in the literature as a type-2 fuzzy set. The degree of membership can be described in the form of some probability density function; in this case one refers to them as so-called probabilistic sets. In general, by type-2 information granule we view an information granule whose parameters are information granules themselves.

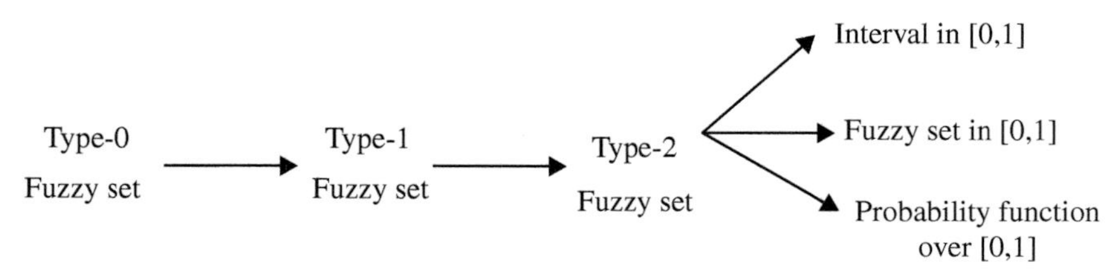

Figure 2.21 Hierarchy of type-2 fuzzy sets and a diversity of existing realizations of non-numeric values of membership grades.

2.14.2 Fuzzy Sets of Order-2

If a universe of discourse over which a fuzzy set is defined is composed of some fuzzy sets (referred to as reference fuzzy sets), the corresponding fuzzy set is called a fuzzy set of order-2. For example, a fuzzy set *comfortable* temperature defined in the space of real numbers (temperature) is a fuzzy set of order-1. Define a finite family of referential fuzzy sets A_1 = *cold*, A_2 = *moderate*, A_3 = *hot*, A_4 = *very hot*. A fuzzy set of order-2 is the one defined over these landmarks (referential fuzzy sets), [0.2 1.0 0.6 0.1], so the term *moderate* is preferred with the concept *comfortable* to the degree equal to 1, while the concept *very hot* is preferred to the degree 0.1.

It becomes apparent that type-2 and order-2 fuzzy sets are generalizations of type-1 fuzzy sets, namely the fuzzy sets discussed so far. Conceptually, they are legitimized given the plethora of applications and real-world environments. For instance, an emergence of type-2 fuzzy sets can be motivated by a variety of arguments:

i) variability in the perception of a given concept; this concerns situations when dealing with a group of observers, experts, or decision-makers. Type-2 fuzzy sets
ii) quality of the fuzzy model constructed with the aid of type-1 fuzzy sets can be quantified in terms of type-2 fuzzy sets of results produced by such models.

The constructs discussed here can be easily generalized to type-*n* and order-*n* fuzzy sets. This is done in a recursive way and the interpretation follows the same line of thought. For the sake of completeness of the overall taxonomy, *w* can refer to single-element information granules (fuzzy sets) as type-0 fuzzy sets. Evidently, numeric data are examples of such fuzzy sets (singletons).

Proceeding with further generalizations, one can encounter combined constructs in the form (type-*n*, order-*m*) fuzzy sets are considered. They are theoretically appealing however one has to be aware of their limited applicability (unless strongly motivated by the problem at hand).

2.15 Operations on Fuzzy Sets

Similar to set theory, we operate with fuzzy sets to obtain new fuzzy sets. The operations must possess properties to match intuition, to comply with the semantics of the intended operation, and to be flexible to fit application requirements. This chapter covers set operations beginning with early fuzzy set operations and continuing with their generalization, interpretations, formal requirements, and realizations. We emphasize complements, triangular norms, and triangular conorms as unifying, general constructs of the complement, intersection, and union operations. Combinations of fuzzy sets to provide aggregations are also essential when operating with fuzzy sets. Analysis of fundamental properties and characteristics of operations with fuzzy sets are discussed thoroughly.

It is instructive to start with the familiar operations of intersection, union, and complement encountered in set theory. They also exhibit some similarities with the operations commonly used in set theory. For instance, consider two sets $A = \{x \in \mathbf{R} \mid 1 \le x \le 3\}$ and $B = \{x \in \mathbf{R} \mid 2 \le x \le 4\}$, both being closed intervals in the real line. Their intersection is a set $A \cap B = \{x \in \mathbf{R} \mid 2 \le x \le 3\}$. Figure 2.22 illustrates the intersection operation represented in terms of the characteristic functions of A and B. Looking at the values of the characteristic function of $A \cap B$ that results when comparing the individual values of $A(x)$ and $B(x)$ at each $x \in \mathbf{R}$ we note that these are taken as the minimum between the values of $A(x)$ and $B(x)$.

In general, given the characteristic functions of A and B, the characteristic function of their intersection $A \cap B$ is computed in the following form

$$(A \cap B)(x) = \min\,[A(x), B(x)]\forall x \in X \tag{2.49}$$

where $(A \cap B)(x)$ denotes the characteristic function of the intersection $A \cap B$.

We now consider the union of sets A and B and express its characteristic function in terms of the respective characteristic functions of A and B. For example, if A and B are the same intervals as presented before, then $A \cup B = \{x \in \mathbf{R} \mid 1 \le x \le 4\}$. We note that the value of the characteristic function of the union is taken as the maximum of corresponding values of the characteristic functions $A(x)$ and $B(x)$ taken at each point of the universe of discourse, Figure 2.23.

Therefore, given the characteristic functions of A and B, we determine the characteristic function of the union to be computed as

$$(A \cup B)(x) = \max\,[A(x), B(x)]\ \forall x \in X \tag{2.50}$$

where $(A \cup B)(x)$ denotes the characteristic function of the intersection $A \cup B$.

Likewise, as Figure 2.24 suggests, the complement $\bar{A}$ of set A, expressed in terms of its characteristic function, is the complement of the characteristic function of A. For instance, if $A = \{x \in \mathbf{R} \mid 1 \le x \le 3\}$, which is the same interval as discussed before, then $\bar{A} = \{x \in R \mid 4 < x < 1\}$, Figure 2.24.

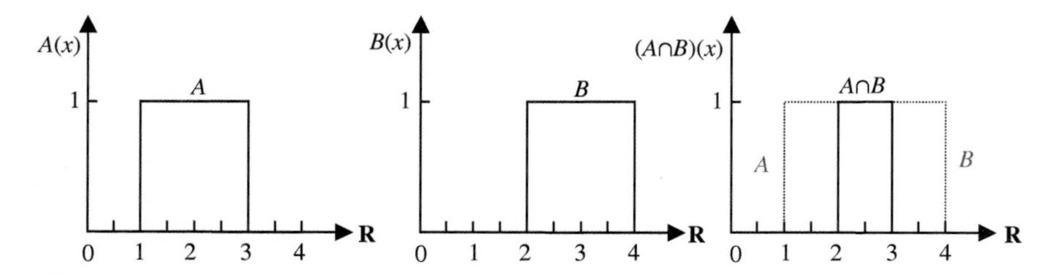

Figure 2.22 Intersection of sets represented in terms of their characteristic functions.

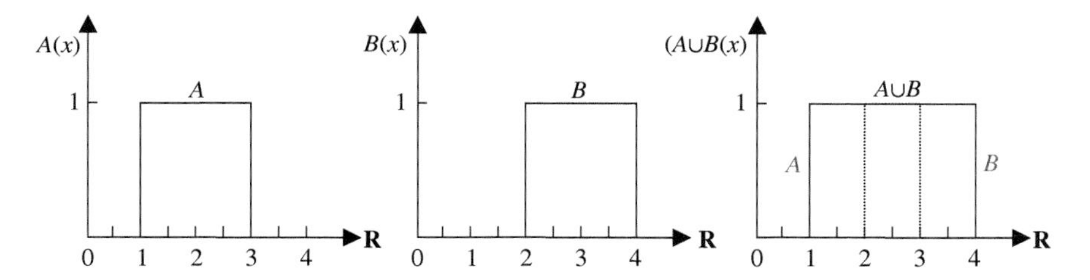

Figure 2.23 Union of two sets expressed in terms of their characteristic functions.

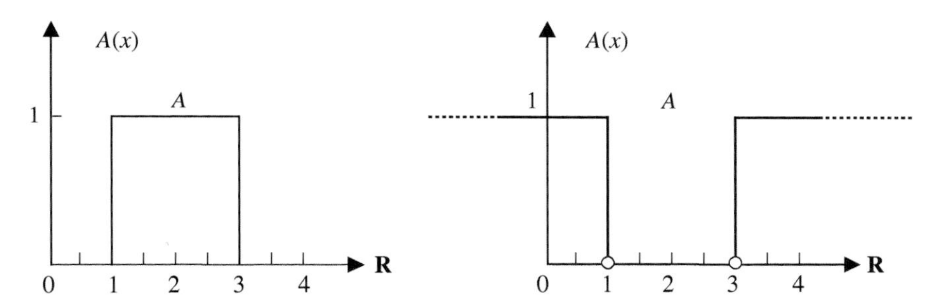

Figure 2.24 Complement of a set in terms of its characteristic function.

In general, the characteristic function of the complement of a set A is given in the form

$$\bar{A}(x) = 1 - A(x),\ \forall x \in \mathbf{X} \tag{2.51}$$

One may anticipate that since sets are particular instances of fuzzy sets, the operations of intersection, union, and complement as previously defined should apply equally well to fuzzy sets. Indeed, when we use membership functions in expressions Eqs. (2.49)–(2.51), these formulas serve as definitions of intersection, union, and complement of fuzzy sets. An illustration of these operations is included in Figure 2.25.

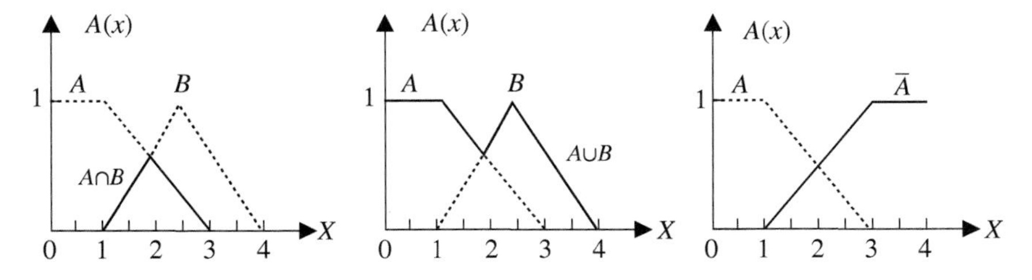

Figure 2.25 Operations on fuzzy sets realized with the use of min, max, and complement functions.

2.16 Triangular Norms and Triangular Conorms as Models of Operations on Fuzzy Sets

Operations on fuzzy sets concern manipulation of their membership functions. Therefore, they are domain dependent and different contexts may require their different realizations. For instance, since operations provide ways to combine information, they can be performed differently in image processing, control, and diagnostic systems, for example. When contemplating the realization of operations of intersection and union of fuzzy sets, we should require a satisfaction of the following intuitively appealing properties:

a) commutativity;
b) associativity;
c) monotonicity;
d) identity.

The last requirement of identity takes on a different form depending on the operation. More specifically, in the case of intersection, we anticipate that an intersection of any fuzzy set with the universe of discourse **X** should return this fuzzy set. For the union operations, the identity implies that the union of any fuzzy set and an empty fuzzy set returns the fuzzy set.

Thus, any binary operator $[0,1] \times [0,1] \rightarrow [0,1]$, which satisfies the collection of the requirements outlined before, can be regarded as a potential candidate to realize the intersection or union of fuzzy sets. Note also that identity acts as boundary conditions meaning that when confining to sets, the previously stated operations return the same results as encountered in set theory. In general, idempotency is not required, however, the realizations of union and intersection could be idempotent as this happens for the operations of minimum and maximum where $\min(a, a) = a$ and $\max(a, a) = a$.

In the theory of fuzzy sets, triangular norms offer a general class of operators of intersection and union. For instance, t-norms generalize intersection of fuzzy sets. Given a t-norm, a dual operator called a t-conorm (or s-norm) can be derived using the relationship $x \; s \; y = 1 - (1 - x) \; t \; (1 - y)$, $\forall \; x,y \in [0,1]$, the De Morgan law but t-conorm can also be described by an independent axiomatic system (Valverde and Ovchinnikov 2008). Triangular conorms provide generic models for the union of fuzzy sets.

A triangular norm, t-norm for brief, is a binary operation t: $[0,1] \times [0,1] \rightarrow [0,1]$ that satisfies the following properties

1) Commutativity: $a \; t \; b = b \; t \; a$
2) Associativity: $a \; t \; (b \; t \; c) = (a \; t \; b) \; t \; c$
3) Monotonicity: if $b \leq c$ then $a \; t \; b \leq a \; t \; c$
4) Boundary conditions: $a \; t \; 1 = a$
   $a \; t \; 0 = 0$

where $a,b,c \in [0,1]$.

Let us elaborate on the meaning of these requirements vis-à-vis the use of t-norms as models of operators of union and intersection of fuzzy sets. There is a one-to-one correspondence between the general requirements outlined in the previous section and the properties of t-norms. The first three reflect the general character of set operations. Boundary conditions stress the fact all t-norms attain the same values at boundaries of the unit square $[0,1] \times [0,1]$. Thus, for sets, any t-norm produces the same result that coincides with the one we could have expected in set theory when dealing with intersection of sets, that is $A \cap \mathbf{X} = A$, $A \cap \varnothing = \varnothing$. Some commonly encountered examples of t-norms include the following operations:

1) Minimum: $a\ t_m\ b = \min(a, b) = a \wedge b$
2) Product: $a\ t_p\ b = ab$
3) Lukasiewicz: $a\ t_l\ b = \max(a + b - 1, 0)$
4) Drastic product: $a\ t_d\ b = \begin{cases} a & \text{if } b = 1 \\ b & \text{if } a = 1 \\ 0 & \text{otherwise} \end{cases}$

In general, t-norms cannot be linearly ordered. One can demonstrate that the min (t_m) t-norm is the largest t-norm, while the drastic product is the smallest one. They form the lower and upper bounds of the t-norms in the following sense

$$a\ t_d\ b \le a\ t\ b \le a\ t_m\ b = \min(a, b) \tag{2.52}$$

Triangular conorms are functions s: $[0,1] \times [0,1] \to [0,1]$. that serve as generic realizations of the union operator on fuzzy sets. Similar to triangular norms, conorms provide the highly desirable modeling flexibility needed to construct fuzzy models. Triangular conorms can be viewed as dual operators to the t-norms and as such, explicitly defined with the use of De Morgan laws. We may characterize them in a fully independent fashion by offering the following definition.

A triangular conorm (s-norm) is a binary operation s: $[0,1] \times [0,1] \to [0,1]$ that satisfies the following requirements

1) Commutativity: $a\ s\ b = b\ s\ a$
2) Associativity: $a\ s\ (b\ s\ c) = (a\ s\ b)\ s\ c$
3) Monotonicity: if $b \le c$ then $a\ s\ b \le a\ s\ c$
4) Boundary conditions: $a\ s\ 0 = a$
   $a\ s\ 1 = 1$

$a,b,c \in [0,1]$.

One can show that s: $[0,1] \times [0,1] \to [0,1]$ is a t-conorm if and only if (iff) there exists a t-norm (dual t-norm) such that for $\forall\ a,b \in [0,1]$ we have

$$a\, s\, b = 1 - (1 - a)\, t\, (1 - b) \tag{2.53}$$

For the corresponding dual t-norm we have

$$a\, t\, b = 1 - (1 - a)\, s\, (1 - b) \tag{2.54}$$

The duality expressed by Eqs. (2.53) and (2.54) can be viewed as an alternative definition of *t*-conorms. This duality allows us to deduce the properties of t-conorms on the basis of the analogous properties of *t*-norms. Notice that after rewriting Eqs. (2.53) and (2.54), we obtain

$$(1 - a)\, t\, (1 - b) = 1 - a\, s\, b \tag{2.55}$$

$$(1 - a)\, s\, (1 - b) = 1 - a\, t\, b \tag{2.56}$$

These two relationships can be expressed symbolically as

$$\overline{A} \cap \overline{B} = \overline{A \cup B} \tag{2.57}$$

$$\overline{A} \cup \overline{B} = \overline{A \cap B} \tag{2.58}$$

that are nothing but the De Morgan laws.

The boundary conditions mean that all t-conorms behave similarly at the boundary of the unit square [0,1] × [0,1]. Thus, for sets, any *t*-conorm returns the same result as encountered in set theory.

A list of commonly used *t*-conorms includes the following examples

$$\text{Maximum: } a\, s_m\, b = \max(a, b) = a \vee b \tag{2.59}$$

$$\text{Probabilistic sum: } a\, s_p\, b = a + b - ab \tag{2.60}$$

$$\text{Lukasiewicz: } a\, s_l\, b = min\,(a + b, 1) \tag{2.61}$$

$$\text{Drastic sum: } a\, s_d\, b = \begin{cases} a & \text{if } b = 0 \\ b & \text{if } a = 0 \\ 1 & \text{otherwise} \end{cases} \tag{2.62}$$

2.17 Negations

Negation is a single-argument operation, which is a generalization of the complement operation encountered in set theory. More formally, by a negation we mean a function N: [0,1]→ [0,1] satisfying the following conditions:

Monotonicity: N is nonincreasing
Boundary conditions: $N(0) = 1$ and $N(1) = 0$

If the function N is continuous and decreasing, the negation is called *strict* (Fodor 1993). If, in addition, a strict negation is involutive, that is

$$N(N(x)) = x \text{ for all } x \in [0, 1] \tag{2.63}$$

it is called *strong*.

Two realizations of the negation operator are presented next

$$N(x) = \sqrt[w]{1 - x^w}, w > 0 \quad (2.64)$$

$$N(x) = \frac{1 - x}{1 + \lambda x}, \lambda > -1 \quad (2.65)$$

Interestingly, if in these expressions we set $w = 1$ or $\lambda = 0$, these realizations of the negation return the standard complement function, that is $N(x) = 1 - x$.

It is worth noting that the negation is a logic operation but not a model of antonyms encountered in natural language (Kim et al. 2000).

2.18 Fuzzy Relations

Relations represent and quantify associations between objects. They provide a fundamental vehicle to describe interactions and dependencies between variables, components, modules, and so on. Fuzzy relations generalize the concept of relations in the same manner as fuzzy sets generalize the fundamental idea of sets (Kandel and Yelowitz 1974; Naessens et al. 2002; Tsabadze 2008). Fuzzy relations are highly instrumental in problems of information retrieval, pattern classification, control, and decision-making. In particular, in decision-making, the notion of fuzzy relation is significant visibility. In what follows, we introduce the idea of fuzzy relations, present some illustrative examples, discuss the main properties of fuzzy relations and provide with some interpretation. The discussed properties exhibit interesting linkages with the essentials of decision-making where they come with some useful characterizations of the underlying decision processes and decision-makers involved there.

2.19 The Concept of Relations

Before proceeding with fuzzy relations, we provide a few introductory lines about relations. Relations capture associations between objects. For instance, consider the space of documents **X** and a space of keywords **Y** these documents contain. Now form a Cartesian product of **X** and **Y**, that is $\mathbf{X} \times \mathbf{Y}$. Recall that the Cartesian product of **X** and **Y**, denoted $\mathbf{X} \times \mathbf{Y}$, is the set of all pairs (x,y) such that $x \in \mathbf{X}$ and $y \in \mathbf{Y}$. We define a relation R as the set of pairs of documents and keywords, $R = \{(d_i, w_j) \mid d_i \in \mathbf{X} \text{ and } w_j \in \mathbf{Y}\}$. In terms of the characteristic function we express this as follows: $\mathbf{R}(d_i, w_j) = 1$ if keyword w_j is in document d_i, and $\mathbf{R}(d_i, w_j) = 0$ otherwise. In decision-making, situations and actions are related: top each situation (state of nature) we assign a collection of pertinent actions that are of interest.

More generally, a relation **R** defined over the Cartesian product of **X** and **Y**, is a collection of selected pairs (x,y) where $x \in \mathbf{X}$ and $y \in \mathbf{Y}$. Equivalently, it is a mapping.

$$R: \mathbf{X} \times \mathbf{Y} \rightarrow \{0,1\} \tag{2.66}$$

The characteristic function of **R** is such that if $\mathbf{R}(x,y) = 1$, then we say that the two elements x and y are related. If $\mathbf{R}(x,y) = 0$, we sat that these two elements (x and y) are unrelated. For example, suppose that $\mathbf{X} = \mathbf{Y} = \{2, 4, 6, 8\}$. The relation "equal to" formed over $\mathbf{X} \times \mathbf{X}$ is the set of pairs $\mathbf{R} = \{(x,y) \in \mathbf{X} \times \mathbf{X} \mid x = y\} = \{(2,2), (4,4), (6,6), (8,8)\}$, refer to Figure 2.26a. Its characteristic function is equal to

$$R(x,y) = \begin{cases} 1 & \text{if } x = y \\ 0 & \text{otherwise} \end{cases} \tag{2.67}$$

The plot of this characteristic function is included in Figure 2.26b.

Depending on the nature of the universe, which could be either finite or infinite, relations are represented in a tabular, matrix form, or described analytically. For instance, the set $\mathbf{X} = \{2, 4, 6, 8\}$ is finite and the relation "equal to" in $\mathbf{X} \times \mathbf{X}$ has a representation in the 4×4 matrix.

$$\boldsymbol{R} = \begin{bmatrix} 1 & 0 & 0 & 0 \\ 0 & 1 & 0 & 0 \\ 0 & 0 & 1 & 0 \\ 0 & 0 & 0 & 1 \end{bmatrix}$$

In general, if **X** and **Y** are finite, say Card(**X**) = n and Card(**Y**) = m, then **R** is a $n \times m$ matrix $\boldsymbol{R} = [r_{ij}]$ with the entries r_{ij} being equal to 1 if and only if $(x_i,y_j) \in \boldsymbol{R}$. Elementary geometry provide examples of relations on infinite universes such as $\mathbf{R} \times \mathbf{R} = \mathbf{R}^2$. In these cases, characteristic functions can, in general, be expressed analytically:

$$R(x,y) = \begin{cases} 1 & \text{if } |x| \le 1 \text{ and } |y| \le 1 \\ 0 & \text{otherwise} \end{cases} \quad \text{square}$$

$$R(x,y) = \begin{cases} 1 & \text{if } x^2 + y^2 = r^2 \\ 0 & \text{otherwise} \end{cases} \quad \text{circle}$$

Relations subsume functions but not vice versa; all functions are relations, but not all relations are functions. For instance, the relation "equal to" shown here is a function but the relations "square" and "circle" are not. A relation is a function if and only if for every x in **X** there is only a single element $y \in \mathbf{Y}$ such that $\mathbf{R}(x,y) = 1$. Therefore, functions are directional constructs, clearly implying a certain direction, for example from **X** to **Y**, say

$$f: X \rightarrow Y$$

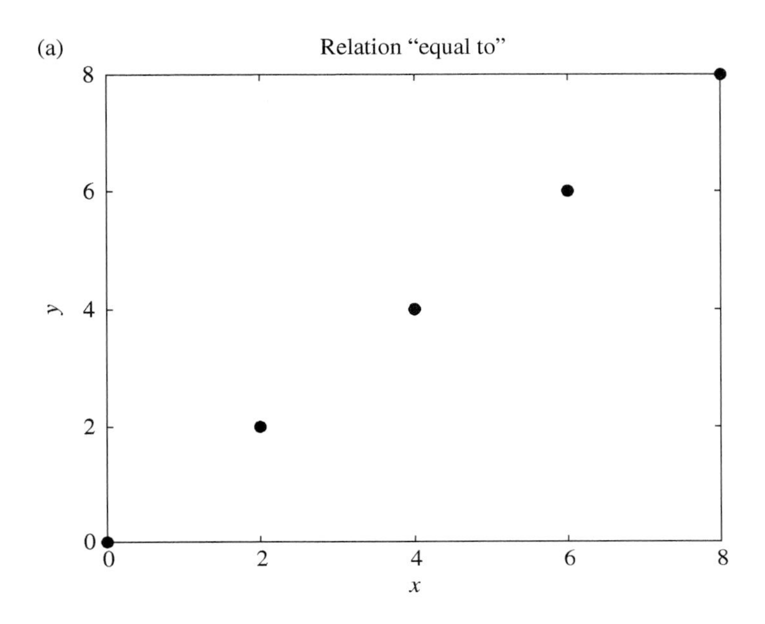

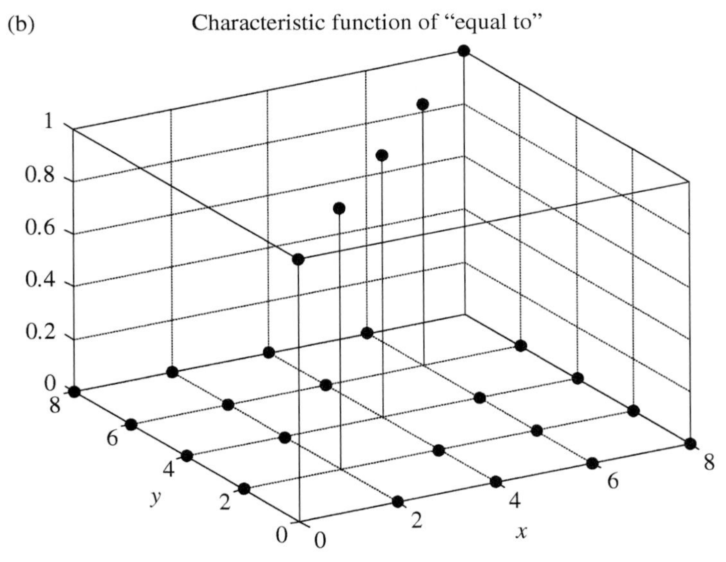

Figure 2.26 Relation "equal to" and its characteristic function.

If the mapping "f" is a function, there is no guarantee that the mapping f^{-1}: $\mathbf{Y} \to \mathbf{X}$ is also a function, except in some case when f^{-1} exists. In contrast, relations are direction free as there is no specific direction identified. Being more descriptive, they can be accessed from any direction. This makes a significant conceptual and computational difference.

When a space under discussion involves "n" universes as its coordinate, a n-ary relation is any subset of the Cartesian product of these universes,

$$\mathbf{R}: \mathbf{X}_1 \times \mathbf{X}_2 \times \ldots \times \mathbf{X}_n \to \{0,1\} \tag{2.68}$$

If $\mathbf{X}_1$, $\mathbf{X}_2$,...,$\mathbf{X}_n$ are finite and Card($\mathbf{X}_1$) = n_1 ... Card($\mathbf{X}_n$) = n_p, then $\mathbf{R}$ can be written as a ($n_1 \times \ldots \times n_p$) matrix $\mathbf{R} = [r_{ij..k}]$ with $r_{ij..k} = 1$ if and only if $(x_i, x_j, \ldots, x_k) \in \mathbf{R}$.

2.20 Fuzzy Relations

Fuzzy relations generalize the concept of relations by admitting the notion of partial association between elements of universes. Given two universes $\mathbf{X}$ and $\mathbf{Y}$, a fuzzy relation $\mathbf{R}$ is any fuzzy subset of the Cartesian product of $\mathbf{X}$ and $\mathbf{Y}$ (Zadeh 1971). Equivalently, a fuzzy relation on $\mathbf{X} \times \mathbf{Y}$ is a mapping.

$$\mathbf{R}: \mathbf{X} \times \mathbf{Y} \to [0,1] \tag{2.69}$$

The membership function of $\mathbf{R}$ for some pair (x,y), $\mathbf{R}(x,y) = 1$, denotes that the two elements x and y are fully related. On the other hand, $\mathbf{R}(x,y) = 0$ means that these elements are unrelated while the values in-between, $0 < \mathbf{R}(x,y) < 1$, underline a partial association. For instance, if d_{fs}, d_{nf}, d_{ns}, d_{gf} are documents whose subjects concern mainly fuzzy systems, neural fuzzy systems, neural systems, and genetic fuzzy systems, with keywords w_f, w_n, and w_g, respectively, then a relation $\mathbf{R}$ on $\mathbf{D} \times \mathbf{W}$, $\mathbf{D} = \{d_{fs}, d_{nf}, d_{ns}, d_{gf}\}$ and $\mathbf{W} = \{w_f, w_n, w_g\}$ can assume the matrix form with the following entries

$$\mathbf{R} = \begin{bmatrix} 1 & 0 & 0.6 \\ 0.8 & 1 & 0 \\ 0 & 1 & 0 \\ 0.8 & 0 & 1 \end{bmatrix}$$

Since the universes are discrete, $\mathbf{R}$ can be represented as a 4×3 matrix (four documents and three keywords) and entries, for example $\mathbf{R}(d_{fs}, w_f) = 1$ means that the document content d_{fs} is fully compatible with the keyword w_f whereas $\mathbf{R}(d_{fs}, w_n) = 0$ and $\mathbf{R}(d_{fs}, w_g) = 0.6$ indicates that d_{fs} does not mention neural systems, but does have genetic systems as part of its content, Figure 2.27a. As with relations, when $\mathbf{X}$ and $\mathbf{Y}$ are finite with Card($\mathbf{X}$) = n and Card($\mathbf{Y}$) = m, then $\mathbf{R}$ can be can be arranged into a certain n × m matrix $\mathbf{R} = [r_{ij}]$, with $r_{ij} \in [0,1]$ being the corresponding degrees of association between x_i and y_j.

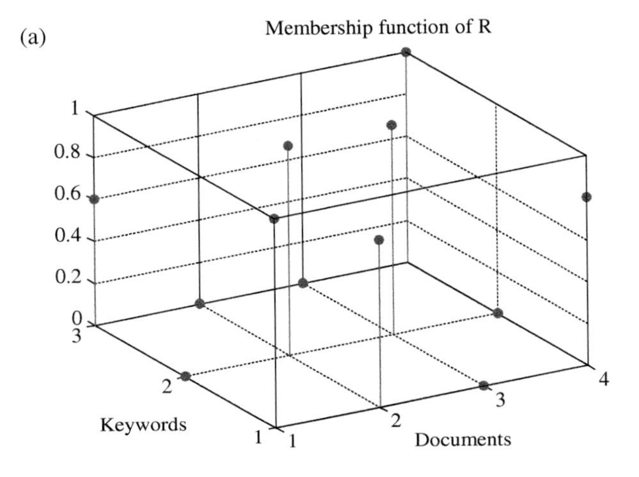

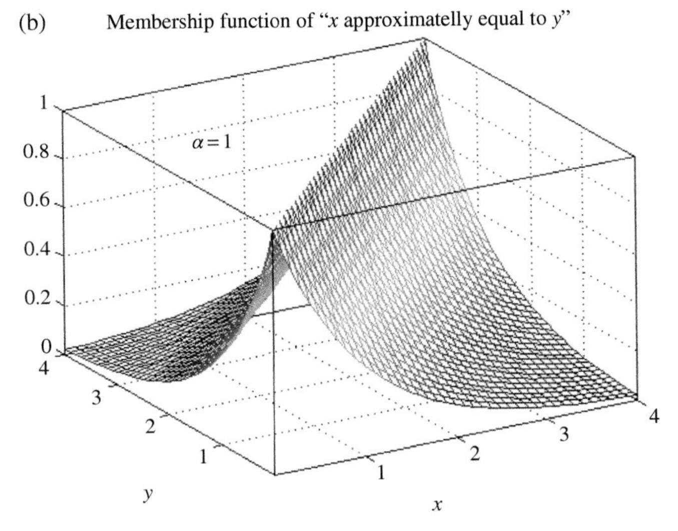

Figure 2.27 Membership functions of the relation **R** (a) and "*x* approximately equal to *y*" (b).

Fuzzy relations defined on some continuous spaces such as $\mathbf{R}^2$, say "much smaller than," "approximately equal," and "similar" could, for instance, be characterized by the following membership functions.

$$R_m(x, y) = \begin{cases} 1 - \exp(-\mid y - x \mid) & \text{if } x \leq y \\ 0 & \text{otherwise} \end{cases} \quad x \text{ much smaller than } y$$

$$R_e(x, y) = \exp\left\{\frac{-\mid x - y \mid}{\alpha}\right\}, \ \alpha > 0 \ x \text{ and } y \text{ approximately equal}$$

$$R_s(x,y) = \begin{cases} \exp[-(x-y)/\beta] & \text{if } |x-y| \le 5 \\ 0 & \text{if } |x-y| \ge 5 \end{cases}, \quad \beta > 0\, x \text{ and } y \text{ similar}$$

Figure 2.27b displays the membership function of the relation "x approximately equal to y" on $\mathbf{X} = \mathbf{Y} = [0,4]$ assuming that $\alpha = 1$.

2.21 Properties of the Fuzzy Relations

Fuzzy relations come with a number of properties, which capture the nature of the relationships conveyed by relations.

2.21.1 Domain and Codomain of Fuzzy Relations

The domain, dom **R**, of a fuzzy relation **R** defined in $\mathbf{X} \times \mathbf{Y}$ is a fuzzy set whose membership function is equal to

$$\text{dom}\,\mathbf{R}(x) = \sup_{y \in \mathbf{Y}} \mathbf{R}(x,y) \tag{2.70}$$

while its codomain, cod **R**, is a fuzzy set whose membership function is given as

$$\text{cod}\,\mathbf{R}(y) = \sup_{x \in \mathbf{X}} \mathbf{R}(x,y) \tag{2.71}$$

Considering finite universes of discourse, domain, and codomain can be viewed as the height of the rows and columns of the fuzzy relation matrix (Zadeh 1971).

2.21.2 Representation of Fuzzy Relations

Similar to the case of fuzzy sets, fuzzy relations can be represented by their α-cuts, that is

$$\mathbf{R} = \bigcup_{\alpha \in [0,1]} \alpha\, \mathbf{R}_\alpha \tag{2.72}$$

or, in terms of the membership function **R**(x,y) of R.

$$\mathbf{R}(x,y) = \sup_{\alpha \in [0,1]} \{\min[\alpha, \mathbf{R}(x,y)]\} \tag{2.73}$$

2.21.3 Equality of Fuzzy Relations

We say that two fuzzy relations **P** and **Q** defined in the same Cartesian product of spaces $\mathbf{X} \times \mathbf{Y}$ are equal if and only if their membership functions are identical, that is,

$$\mathbf{P}(x,y) = \mathbf{Q}(x,y) \;\; \forall (x,y) \in \mathbf{X} \times \mathbf{Y} \tag{2.74}$$

2.21.4 Inclusion of Fuzzy Relations

A fuzzy relation **P** is included in **Q**, denoted by $\mathbf{P} \subseteq \mathbf{Q}$, if and only if

$$\mathbf{P}(x,y) \le \mathbf{Q}(x,y) \quad \forall (x,y) \in \mathbf{X} \times \mathbf{Y} \tag{2.75}$$

Similar to what was presented in the case of relations, given n-fold Cartesian product of these universes we define the fuzzy relation in the form

$$\mathbf{R} : \mathbf{X}_1 \times \mathbf{X}_2 \times \ldots \times \mathbf{X}_n \rightarrow [0,1] \tag{2.76}$$

If the spaces $\mathbf{X}_1, \mathbf{X}_2, \ldots, \mathbf{X}_n$ are finite with Card($\mathbf{X}_1$) = n_1 ... Card($\mathbf{X}_n$) = n_n, then **R** can be expressed as an n-fold ($n_1 \times \ldots \times n_p$) matrix $\mathbf{R} = [r_{ij..k}]$ with $r_{ij..k} \in [0,1]$ being the degree of association assigned to the n-tuple $(x_i, x_j, \ldots, x_k) \in \mathbf{X}_1 \times \mathbf{X}_2 \times \ldots \times \mathbf{X}_n$. If $\mathbf{X}_1$, $\mathbf{X}_2 \ldots \mathbf{X}_n$ are infinite, then the membership function of **R** is a certain function of many variables. The concepts of equality and inclusion of fuzzy relations could be easily extended for relations defined in multidimensional spaces.

2.21.5 Operations on Fuzzy Relations

The basic operations on fuzzy relations, say union, intersection, and complement, conceptually follow the corresponding operations on fuzzy sets once fuzzy relations are fuzzy sets formed on multidimensional spaces. For illustrative purposes, the definitions of union, intersection, and complement that follow involve two-argument fuzzy relations. Without any loss of generality, we can focus on binary fuzzy relations **P**, **Q**, **R** defined in $\mathbf{X} \times \mathbf{Y}$. As in the case of fuzzy sets, all definitions are defined pointwise.

2.21.6 Union of Fuzzy Relations

The union **R** of two fuzzy relations **P** and **Q** defined in $\mathbf{X} \times \mathbf{Y}$, $\mathbf{R} = \mathbf{P} \cup \mathbf{Q}$ is defined with the use of the following membership function.

$$\mathbf{R}(x,y) = \mathbf{P}(x,y)\, s\, \mathbf{Q}(x,y) \quad \forall (x,y) \in \mathbf{X} \times \mathbf{Y} \tag{2.77}$$

recall that "s" stands for some t-conorm.

2.21.7 Intersection of Fuzzy Relations

The intersection **R** of fuzzy relations **P** and **Q** defined in $\mathbf{X} \times \mathbf{Y}$, $\mathbf{R} = \mathbf{P} \cap \mathbf{Q}$ is defined in the following form,

$$\mathbf{R}(x,y) = \mathbf{P}(x,y)\, t\, \mathbf{Q}(x,y) \quad \forall (x,y) \in \mathbf{X} \times \mathbf{Y} \tag{2.78}$$

2.21.8 Complement of Fuzzy Relations

The complement $\bar{\mathbf{R}}$ of the fuzzy relation $\mathbf{R}$ is defined by the membership function.

$$\bar{\mathbf{R}}(x,y) = 1 - \mathbf{R}(x,y) \quad \forall (x,y) \in \mathbf{X} \times \mathbf{Y} \tag{2.79}$$

2.21.9 Transposition of Fuzzy Relations

Given a fuzzy relation $\mathbf{R}$, its transpose, denoted by $\mathbf{R}^{\mathrm{T}}$, is a fuzzy relation on $\mathbf{Y} \times \mathbf{X}$ such that the following relationship holds

$$\mathbf{R}^{T}(y,x) = \mathbf{P}(x,y) \ \forall (x,y) \in \mathbf{X} \times \mathbf{Y} \tag{2.80}$$

If $\mathbf{R}$ is a relation defined in some finite space, then $\mathbf{R}^{\mathrm{T}}$ is the transpose of the corresponding $n \times m$ matrix representation of $\mathbf{R}$. Therefore, the form of $\mathbf{R}^{\mathrm{T}}$ is a $m \times n$ matrix whose columns are now the rows of $\mathbf{R}$.

The following properties are direct consequences of the definitions provided before

$$\left(\mathbf{R}^{\mathrm{T}}\right)^{\mathrm{T}} = \mathbf{R} \tag{2.81}$$

$$\left(\bar{\mathbf{R}}\right)^{\mathrm{T}} = \overline{\left(\mathbf{R}^{\mathrm{T}}\right)} \tag{2.82}$$

2.21.10 Cartesian Product of Fuzzy Relations

Given fuzzy sets $A_1, A_2, \ldots, A_n$ defined in universes $\mathbf{X}_1, \mathbf{X}_2, \ldots, \mathbf{X}_n$, respectively, their Cartesian product $A_1 \times A_2 \times \ldots \times A_n$ is a fuzzy relation $\mathbf{R}$ on $\mathbf{X}_1 \times \mathbf{X}_2 \times \ldots \times \mathbf{X}_n$ with the following membership function.

$$\mathbf{R}(x_1,x_2,\ldots,x_n) = \min\{A_1(x_1),A_2(x_2),\ldots,A_n(x_n)\} \ \forall x_1 \in \mathbf{X}_1, \ \forall x_2 \in \mathbf{X}_2,\ldots,\forall x_n \in \mathbf{X}_n \tag{2.83}$$

In general, we can generalize the concept of this Cartesian product by using some t-norms

$$\mathbf{R}(x_1,x_2,\ldots,x_n) = A_1(x_1)\, t A_2(x_2)\, t \ldots t A_n(x_n) \quad \forall x_1 \in \mathbf{X}_1, \ \forall x_2 \in \mathbf{X}_2,\ldots,\forall x_n \in \mathbf{X}_n \tag{2.84}$$

2.21.11 Projection of Fuzzy Relations

In contrast to the concept of the Cartesian product, the idea of projection is to construct fuzzy relations on some subspaces of the original relation. Projection reduces the dimensionality of the original space over which the original fuzzy relation as defined.

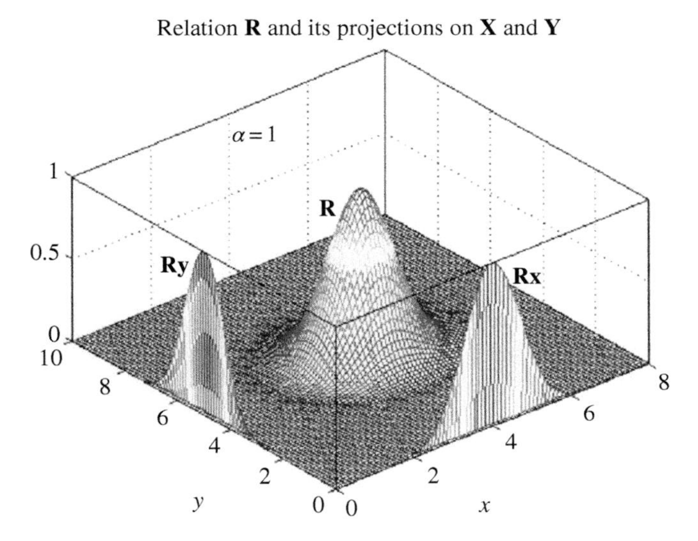

Figure 2.28 Fuzzy relation **R** along with its projections on **X** and **Y**.

Given that **R** is a fuzzy relation defined in $\mathbf{X}_1 \times \mathbf{X}_2 \times \ldots \times \mathbf{X}_n$, its projection on $\mathbf{X} = \mathbf{X}_i \times \mathbf{X}_j \times \ldots \times \mathbf{X}_k$, where $I = \{i,j,\ldots,k\}$ is a subsequence of the set of indexes $\mathbf{N} = \{1,2,\ldots,n\}$, is a fuzzy relation $\mathbf{R_X}$ with the membership function (Zadeh 1971).

$$\mathbf{R_X}\left(x_i,x_j,\ldots,x_k\right) = \text{Proj}_\mathbf{X}\mathbf{R}(x_1,x_2,\ldots,x_n) = \sup_{x_t,x_u,\ldots,x_v} \mathbf{R}(x_1,x_2,\ldots,x_n) \tag{2.85}$$

where $\mathbf{J} = \{t, u,\ldots,v\}$ is a subsequence of **N** such that $\mathbf{I} \cup \mathbf{J} = \mathbf{N}$ and $\mathbf{I} \cap \mathbf{J} = \emptyset$. In other words, **J** is the complement of I with respect to N. Notice that expression (2.85) is computed for all values of $(x_1, x_2, \ldots,x_n) \in \mathbf{X}_i \times \mathbf{X}_j \times \ldots \times \mathbf{X}_k$.

For instance, Figure 2.28 shows the projections $\mathbf{R_X}$ and $\mathbf{R_Y}$ of a certain Gaussian binary fuzzy relation **R** defined in $\mathbf{X} \times \mathbf{Y}$ with $\mathbf{X} = [0, 8]$ and $\mathbf{Y} = [0, 10]$, whose membership function is equal to $\mathbf{R}(x,y) = \exp\{-\alpha[(x\text{-}4)^2 + (y\text{-}5)^2]\}$. In this case, the projections are formed as

$$\mathbf{R_X}(x) = \text{Proj}_\mathbf{X}\mathbf{R}(x,y) = \sup_y \mathbf{R}(x,y) \tag{2.86}$$

$$\mathbf{R_Y}(y) = \text{Proj}_\mathbf{Y}\mathbf{R}(x,y) = \sup_x \mathbf{R}(x,y) \tag{2.87}$$

To find projections of the fuzzy relations defined in some finite spaces, the maximum operation replaces the sup operation occurring in the definition provided before. For example, for the fuzzy relation $\mathbf{R}: \mathbf{X} \times \mathbf{Y} \rightarrow [0,1]$ with $\mathbf{X} = \{1,2,3\}$ and $\mathbf{Y} = \{1,2,3,4,5\}$.

$$\mathbf{R}(x,y) = \begin{bmatrix} 1.0 & 0.6 & 0.8 & 0.5 & 0.2 \\ 0.6 & 0.8 & 1.0 & 0.2 & 0.9 \\ 0.8 & 0.6 & 0.8 & 0.3 & 0.9 \end{bmatrix}$$

the three elements of the projection $\mathbf{R_X}$ are taken as the maximum computed for each of the three rows of $\mathbf{R}$

$$\mathbf{R_X} = [\max(1,0,0.6,0.8,0.5,0.2)\,\max(0.6,0.8,1.0,0.2,0.9)\,\max(0.8,0.6,0.8,0.3,0.9)]$$
$$= [1.0\ 1.0\ 0.9]$$

Similarly, the five elements of $\mathbf{R_Y}$ are taken as the maximum among the entries of the five columns of $\mathbf{R}$. Figure 2.29 shows $\mathbf{R}$ and its projections $\mathbf{R_X}$ and $\mathbf{R_Y}$.

$$R_Y = [1.0 \quad 0.8 \quad 1.0 \quad 0.5 \quad 0.9]$$

Note that domain and codomain of the fuzzy relation are examples of its projections.

2.21.12 Cylindrical Extension

The cylindrical extension increases the number of coordinates of the Cartesian product over which the fuzzy relation is formed. In this sense, cylindrical extension is an operation that is complementary to the already discussed projection operation (Zadeh 1971).

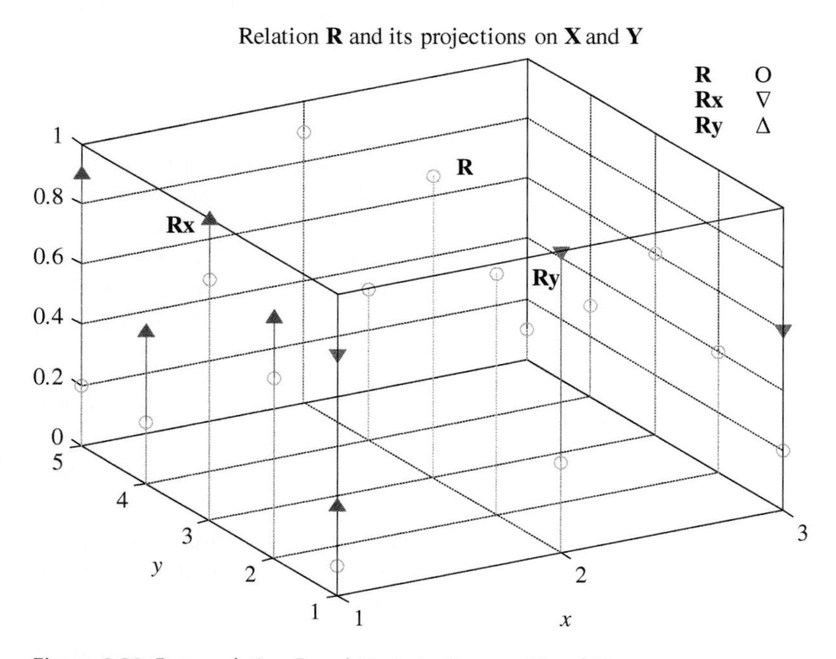

Figure 2.29 Fuzzy relation **R** and its projections on **X** and **Y**.

The cylindrical extension on $\mathbf{X} \times \mathbf{Y}$ of a fuzzy set of $\mathbf{X}$ is a fuzzy relation cyl$A$ whose membership function is equal to

$$\text{cyl}A(x,y) = A(x),\ \forall x \in \mathbf{X}, \forall y \in \mathbf{Y} \tag{2.88}$$

If the fuzzy relation is viewed as a two-dimensional matrix, the operation of cylindrical extension forms identical columns indexed by the successive values of $y \in \mathbf{Y}$. The main intent of cylindrical extensions is to achieve compatibility of spaces over which fuzzy sets and fuzzy relations are formed. For instance, let A be a fuzzy set of $\mathbf{X}$ and $\mathbf{R}$ a fuzzy relation on $\mathbf{X} \times \mathbf{Y}$. Suppose we attempt to compute union and intersection of A and $\mathbf{R}$. Because the universes over which A and $\mathbf{R}$ are defined are different, we cannot carry out any set-based operations on A and $\mathbf{R}$. The cylindrical extension of A, denoted by cyl$A$ provides the compatibility required. Then the operations such as (cyl$A$) $\cup\mathbf{R}$ and (cyl$A$) $\cap\mathbf{R}$ make sense. The concept of cylindrical extension can be easily generalized to multi-dimensional cases.

2.21.13 Reconstruction of Fuzzy Relations

Projections do not retain complete information conveyed by the original fuzzy relation. This means that in general one cannot faithfully reconstruct a relation from its projections. In other words, projections $\text{Proj}_{\mathbf{X}}\mathbf{R}$ and $\text{Proj}_{\mathbf{Y}}\mathbf{R}$ of some fuzzy relation $\mathbf{R}$, do not necessarily lead to the original fuzzy relation $\mathbf{R}$. In general, the reconstruction of a relation via the Cartesian product of its projections is a relation that includes the original relation, that is

$$\text{Proj}_{\mathbf{X}}\mathbf{R} \times \text{Proj}_{\mathbf{Y}}\mathbf{R} \supseteq \mathbf{R} \tag{2.89}$$

If, however, in the relationship in (2.89) the equality holds, then we call the relation $\mathbf{R}$ noninteractive.

2.21.14 Binary Fuzzy Relations

A binary fuzzy relation $\mathbf{R}$ on $\mathbf{X} \times \mathbf{X}$ is defined as follows:

$$\mathbf{R} : \mathbf{X} \times \mathbf{X} \rightarrow [0,1] \tag{2.90}$$

There are several important features of binary fuzzy relations:

a) Reflexivity: $\mathbf{R}(x,x) = 1\ \forall x \in \mathbf{X}$, Figure 2.30a. When $\mathbf{X}$ is finite $\mathbf{R} \supseteq \mathbf{I}$ where $\mathbf{I}$ is an identity matrix, $\mathbf{I}(x,y) = 1$ if $x = y$ and $\mathbf{I}(x,y) = 0$ otherwise. Reflexivity can be relaxed by admitting a concept of so-called ε-reflexivity, $\varepsilon \in [0,1]$. This means $\mathbf{R}(x,x) \geq \varepsilon$. When $\mathbf{R}(x,x) = 0$ the fuzzy relation is irreflexive. A fuzzy relation is locally reflexive if, for any $x,y \in \mathbf{X}$, $\max\{\mathbf{R}(x,y), \mathbf{R}(y,x)\} \leq \mathbf{R}(x,x)$.
b) Symmetry: $\mathbf{R}(x,y) = \mathbf{R}(y,x)\ \forall (x,y) \in \mathbf{X} \times \mathbf{X}$, Figure 2.30b. For finite $\mathbf{X}$, the matrix representing $\mathbf{R}$ has entries distributed symmetrically along the main diagonal. Clearly, if $\mathbf{R}$ is symmetric, then $\mathbf{R}^{\mathrm{T}} = \mathbf{R}$.

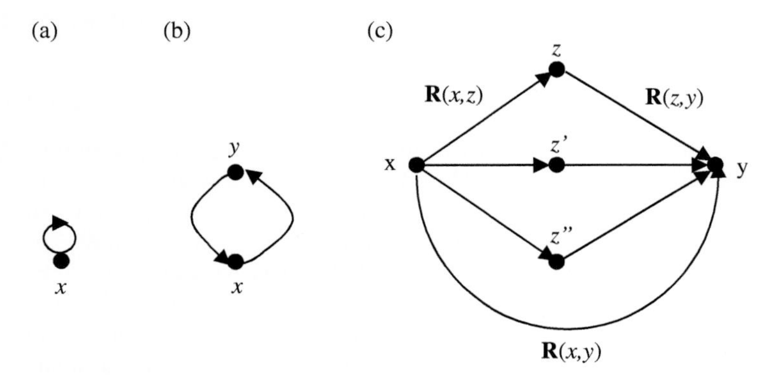

Figure 2.30 Main characteristics of binary fuzzy relations; see the details in the text.

c) Transitivity: $\sup_{z \in X}\{\mathbf{R}(x,z)t\mathbf{R}(z,y)\} \leq \mathbf{R}(x,y) \forall x,y,z \in \mathbf{X}$. In particular, if this relationship holds for $t = \min$, then the relation is called sup-min transitive. Looking at the levels of associations $\mathbf{R}(x,z)$ and $\mathbf{R}(z,y)$ occurring between x, and z, and z and y, the property of transitivity reflects the maximal strength among all possible links arranged in series (such as ($\mathbf{R}(x,z)$ and $\mathbf{R}(z,y)$)) that does not exceed the strength of the direct link $\mathbf{R}(x,z)$, Figure 2.30c.

2.21.15 Transitive Closure

Given a binary fuzzy relation in a finite universe $\mathbf{X}$, there exists a unique fuzzy relation $\overleftrightarrow{\mathbf{R}}$ on $\mathbf{X}$, called transitive closure of $\mathbf{R}$, that contains $\mathbf{R}$ and itself is included in any transitive fuzzy relation on $\mathbf{X}$ that contains $\mathbf{R}$. Therefore, if $\mathbf{R}$ is defined on a finite universe of cardinality n, the transitive closure is given by

$$\text{trans}(\mathbf{R}) = \overleftrightarrow{\mathbf{R}} = \mathbf{R} \cup \mathbf{R}^2 \cup \cdots \cup \mathbf{R}^n \tag{2.91}$$

where, by definition,

$$\mathbf{R}^2 = \mathbf{R} \circ \mathbf{R} \cdot \cdots \cdot \mathbf{R}^p = \mathbf{R} \circ \mathbf{R}^{p-1} \tag{2.92}$$

$$\mathbf{R} \circ \mathbf{R}(x,y) = \max_z \{\mathbf{R}(x,z)t\mathbf{R}(z,y)\} \tag{2.93}$$

Notice that the composition $\mathbf{R} \circ \mathbf{R}$ can be computed similarly as encountered in matrix algebra by replacing the ordinary multiplication by some t-norm and the sum by the max operations. In other words, if $r_{ij}^2 = [\mathbf{R}^2]_{ij} = [\mathbf{R} \circ \mathbf{R}]_{ij}$ then

$$r_{ij}^2 = \max_k (r_{ik} t\, r_{kj}) \tag{2.94}$$

If $\mathbf{R}$ is reflexive, then

$$I \subseteq \mathbf{R} \subseteq \mathbf{R}^2 \subseteq \cdots \subseteq \mathbf{R}^{n-1} = \mathbf{R}^n \tag{2.95}$$

The transitive closure of the fuzzy relation **R** can be found by computing the successive k max-t products of **R** until $\mathbf{R}^k = \mathbf{R}^{k-1}$, a procedure whose complexity is $O(n^3 \log_2 n)$ in time and $O(n^2)$ in space (Naessens et al. 2002; De Baets and Meyer 2003). Refer also to (Wallace et al. 2006).

2.21.16 Equivalence and Similarity Relations

Equivalence relations are relations that are reflexive, symmetric, and transitive (Foulloy and Benoit 2006). Suppose that one of the arguments of $\mathbf{R}(x,y)$, "x" for example, has been fixed. Thus, all elements related to x constitute a set called as an equivalence class of **R** with respect to "x," denoted by

$$A_x = \{y \in \mathbf{Y} \mid \mathbf{R}(x,y) = 1\} \tag{2.96}$$

The family of all equivalence classes of **R**, denoted **X**/**R**, is a partition of **X**. In other words, **X**/**R** is a family of pairwise disjoint nonempty subsets of **X** whose union is **X**. Equivalence relations can be viewed as a generalization of the equality relations in the sense that members of an equivalence class can be considered equivalent to each other under the relation **R**.

Similarity relations are fuzzy relations that are reflexive, symmetric, and transitive. Like any fuzzy relation, a similarity relation can be represented by a nested family of its α-cuts, $\mathbf{R}_\alpha$. Each α-cut constitutes an equivalence relation and forms a partition of **X**. Therefore, each similarity relation is associated with a set P(**R**) of partitions of **X**,

$$\mathrm{P}(\mathbf{R}) = \{\mathbf{X}/\mathbf{R}_\alpha \mid \alpha \in [0,1]\} \tag{2.97}$$

Partitions are nested in the sense that, if $\alpha > \beta$, then the partition $\mathbf{X}/\mathbf{R}_\alpha$ is finer than the partition $\mathbf{X}/\mathbf{R}_\beta$. For example, consider the relation defined in $\mathbf{X} = \{a, b, c, d, e\}$ in the following way

$$R = \begin{bmatrix} 1.0 & 0.8 & 0 & 0 & 0 \\ 0.8 & 1.0 & 0 & 0 & 0 \\ 0 & 0 & 1.0 & 0.9 & 0.5 \\ 0 & 0 & 0.9 & 1.0 & 0.5 \\ 0 & 0 & 0.5 & 0.5 & 1.0 \end{bmatrix}$$

One can easily verify that **R** is a symmetric matrix, has values of 1 at its main diagonal, and is max-min transitive. Therefore, **R** is a similarity relation. The levels of refinement of the similarity relation **R** can be represented in the form of partition tree in which each node corresponds to a fuzzy relation on **X** whose degrees of association between the elements are greater than or equal to the threshold value α.

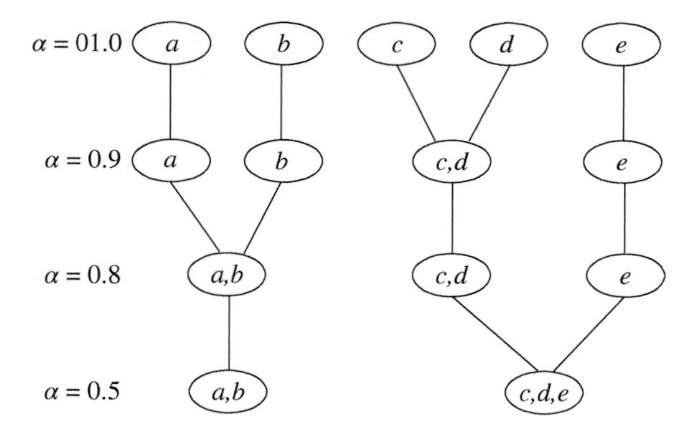

Figure 2.31 Partition tree induced by binary fuzzy relation **R**.

For instance, we have the following fuzzy relations for $\alpha = 0.5$, 0.8, and 0.9, respectively:

$$\mathbf{R}_{0.5} = \begin{bmatrix} 1 & 1 & 0 & 0 & 0 \\ 1 & 1 & 0 & 0 & 0 \\ 0 & 0 & 1 & 1 & 1 \\ 0 & 0 & 1 & 1 & 1 \\ 0 & 0 & 1 & 1 & 1 \end{bmatrix}, \mathbf{R}_{0.8} = \begin{bmatrix} 1 & 1 & 0 & 0 & 0 \\ 1 & 1 & 0 & 0 & 0 \\ 0 & 0 & 1 & 1 & 0 \\ 0 & 0 & 1 & 1 & 0 \\ 0 & 0 & 0 & 0 & 1 \end{bmatrix}, \mathbf{R}_{0.9} = \begin{bmatrix} 1 & 0 & 0 & 0 & 0 \\ 0 & 1 & 0 & 0 & 0 \\ 0 & 0 & 1.0 & 1 & 0 \\ 0 & 0 & 1 & 1.0 & 0 \\ 0 & 0 & 0 & 0 & 1 \end{bmatrix}$$

Notice that $\mathbf{R} = \bigcup_{\alpha \in \Lambda} \alpha \mathbf{R}_\alpha$ where $\cup = \max$ and $\Lambda = \{0.5, 0.8, 0.9, 1.0\}$ is the level set of **R**. Also, notice that the greater the value of α, the finer the classes are, as shown in Figure 2.31.

2.21.17 Compatibility and Proximity Relations

Compatibility relations are reflexive and symmetric relations. Associated with any compatibility relation are sets called compatibility classes. A compatibility class is a subset A of a universe **X** such that $\mathbf{R}(x,y) = 1$ for all $x,y \in A$.

Proximity relations are reflexive and symmetric fuzzy relations. Let A be a subset of a universe **X**. Thus, A is a ε-proximity class of **R** if $\mathbf{R}(x,y) \geq \varepsilon$ for all $x,y \in A$. For instance, the relation **R** on $\mathbf{X} = \{1, 2, 3, 4, 5\}$

$$\mathbf{R} = \begin{bmatrix} 1.0 & 0.7 & 0.0 & 0.0 & 0.6 \\ 0.7 & 1.0 & 0.6 & 0.0 & 0.0 \\ 0.0 & 0.6 & 1.0 & 0.7 & 0.4 \\ 0.0 & 0.0 & 0.7 & 1.0 & 0.5 \\ 0.6 & 0.0 & 0.4 & 0.5 & 1.0 \end{bmatrix}$$

has the unity in its main diagonal and is symmetric. Therefore, **R** is a proximity relation. Compatibility classes and α-compatibility classes do not necessarily induce partitions of **X** (Klir and Yuan 1995).

Proximity is an important concept in pattern recognition being used in contexts such as visual images as under these circumstances human subjectivity leads to some useful information that could be represented in the form of proximity relations.

2.22 Conclusions

Fuzzy sets provide a conceptual and operational framework to deal with granular information. Particular cases of degenerate membership functions (singletons) represent numeric information. Fuzzy sets come with a well-defined semantics whose formal description is conveyed in the form of membership functions. The variety of available membership functions offers a great deal of flexibility of capturing the meaning of the information granule. We showed the relationships between fuzzy sets and sets by stressing that sets are subsumed as particular cases of sets. The reconstruction of fuzzy sets through a finite or infinite family of α-cuts emphasizes the computational linkages between fuzzy sets and sets and indicates that the set-theoretic methods can be effectively utilized when processing fuzzy sets. Fuzzy relations and their important classes bring an interesting view at the characterization of dependencies captured by membership grades.

Exercises

2.1 There is an interesting problem posed by Borel (1950) that could be now conveniently handled in the setting of fuzzy sets:

One seed does not constitute a pile nor two or three. From the other side, everybody will agree that 100 million seeds constitute a pile. What therefore is the appropriate limit?
Given this description, suggest a membership function of the concept discussed here. What type of membership function would you consider in this problem? Why?

2.2 Consider two situations: (i) the number of expected people to ride on a bus on a certain day; (ii) the number of people that could ride in a bus at any one time. Both situations describe an uncertain scenario. Which of these two situations involves randomness? Which one involves fuzziness? What is the nature of fuzziness: similarity, possibility, or preference?

2.3 We are interested in describing the state of an environment by quantifying temperature as *very cold, cold, comfortable, warm,* and *hot.* Choose an appropriate universe of discourse. Represent state values using (i) sets and (ii) fuzzy sets.

2.4 Suppose that allowed speed values in a city street range between 0 and 60 km/hour. Describe the speed values such as low, medium, and high using sets and fuzzy sets. Would this description be adequate also for highways? Justify the answer.

2.5 Given is the fuzzy set A with the membership function

$$A(x) = \begin{cases} x - 4 & \text{if } 4 \le x \le 5 \\ -x + 6 & \text{if } 5 < x \le 6 \\ 0 & \text{otherwise} \end{cases}$$

a) Plot the membership function and identify its shape.
b) What type of linguistic label (semantics) could be associated with the concept conveyed by A?

2.6 Consider the fuzzy set A with the following membership function

$$A(x) = \begin{cases} x - 4/2 & \text{if } 4 \le x \le 5 \\ -x + 6/2 & \text{if } 5 < x \le 6 \\ 0 & \text{otherwise} \end{cases}$$

a) Plot this membership function
b) Is A normal? Does A have a core? What is the height of this fuzzy set?
c) Find the support of A. Is A a convex fuzzy set?

2.7 Assume a fuzzy sets A whose membership functions is defined in the following form

$$A(x) = \begin{cases} x - 4 & \text{if } 4 \le x \le 5 \\ 1 & \text{if } 5 < x \le 6 \\ -x + 7 & \text{if } 6 < x \le 7 \\ 0 & \text{otherwise} \end{cases}$$

a) Sketch the graph of the membership function.
b) Find an analytic expression for its α-cuts.
c) Is A a convex fuzzy set?

2.8 Demonstrate that if a fuzzy set is convex, then all its α-cuts are convex.

2.9 Consider the following fuzzy sets defined in the finite universe of discourse $\mathbf{X} = \{1, 2, 3. \ldots\ldots, 10\}$

$A = (0, 0, 0, 0, 0.4, 0.6, 0.8, 1, 0.8, 0.6)$
$B = (0, 0, 0, 0, 0.4, 0.5, 0.6, 1, 0.6, 0.4)$
$C = (0, 1, 0.2, 0.3, 0.4, 0.5, 0.6, 1, 0.5, 0)$

a) Is $A \subseteq B$? $B \subseteq A$?
b) Is $C \subseteq A$? $C \subseteq B$?
c) Quantify the findings obtained in (a) and (b).

2.10 The concept of sensitivity of a fuzzy set A defined in the space of real numbers is expressed by taking an absolute value of the derivative of the membership function $\left|\frac{dA}{dx}\right|$ (consider arguments of the membership function for which the derivative exists). Discuss several classes of membership functions

– triangular with parameters a, m, and b
– piecewise linear shown in the figure here

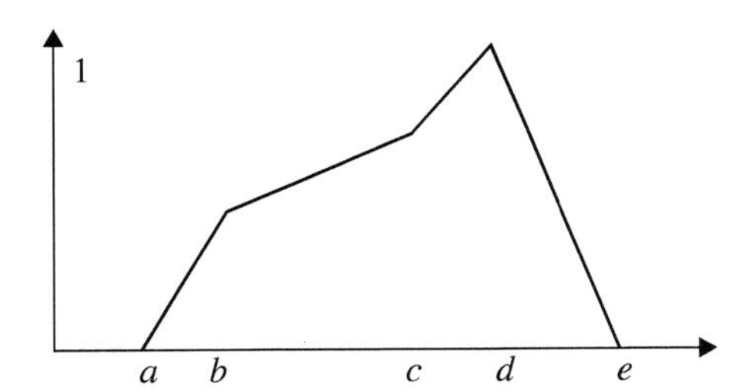

– parabolic centered around 0 and with spread a.

Determine their sensitivity and interpret the obtained results. What could be the possible implications of these results on the selection of a certain class of membership function?

2.11 Consider a membership function defined over [0, 1] and described as $A(x) = \exp(-0.03x)$.

Determine the following:

a) energy measure of fuzziness of A, $E(A)$; use $e(u) = u^2$
b) entropy measure of fuzziness of A, $H(A)$; use $h(u)$ as a parabolic function over [0,1]
c) specificity of A. Use $\int_0^1 \text{sp}(A_\alpha)d\alpha$ with $\text{sp}(A_\alpha) = 1 - \frac{|b_\alpha - a_\alpha|}{range}$, where a_α and b_α are the bounds of the α-cut of A. The value of *range* is 3.

2.12 A finite collection of alternatives $\mathbf{X} = \{x_1, x_2, \ldots, x_9\}$ has been evaluated with respect to two criteria giving rise to the following fuzzy sets

$A = [0.8\ 0.2\ 0.5\ 0.7\ 1.0\ 0.9\ 0.5\ 0.6\ 0.1]$
$B = [0.5\ 0.3\ 0.7\ 0.5\ 0.4\ 0.0\ 0.6\ 0.7\ 0.8]$

The same alternatives were assessed with respect to some global criterion producing the fuzzy set with the membership function

$C = [0.2\ 0.8\ 0.6\ 0.5\ 0.8\ 1.0\ 1.0\ 0.9\ 0.2]$

The resulting fuzzy set C can be regarded as a logic aggregation of A and B, that *is* $g(A, B)$.

Experiment with various t-norms and t-conorms as here:

t-norms: min, product, Lukasiewicz *and*-connective
t-conorm: max, probabilistic sum, Lukasiewicz *or*-connective

Determine their performance and suggest the best logic operator for this aggregation. As the performance index use a Hamming distance between C and $g(A, B)$ where "g" stands for the selected logic operator (t-norm or t-conorm). Discuss the obtained results.

2.13 Suppose that fuzzy sets A and B defined in $\mathbf{X} = \{x_1, x_2, x_3\}$ are represented as vectors whose components are the membership degrees of x_1, x_2, and x_3 in A and B. Plot A and B in the unit cube for each of the following cases

a) $A = (1, 0, 0)$ and $B = (0, 1, 1)$,
b) $A = (0, 1, 0)$ and $B = (1, 0, 1)$,
c) $A = (0, 0, 1)$ and $B = (1, 1, 0)$,
d) $A = (0.5, 0.5, 0.5)$ and $B = (0.5, 0.5, 0.5)$.

2.14 Let $R_\alpha = \{(x,y) \in \mathbf{X} \times \mathbf{Y} \mid R(x,y) \geq \alpha\}$ be the α-cut of the fuzzy relation R. Show that any fuzzy relation $R: \mathbf{X} \times \mathbf{Y} \rightarrow [0,1]$ can be represented in the following canonical form

$$R = \bigcup_{\alpha \in (0,1]} \alpha R_\alpha$$

where $\cup$ denotes standard union operation, and αR_α is a subnormal fuzzy set whose membership function is α if $(x,y) \in R_\alpha$ and zero otherwise.

2.15 How can the algorithm to compute the transitive closure of a fuzzy relation be used to verify if a fuzzy relation is transitive or not?

2.16 Show that if R is a similarity relation, then each of its α-cut R_α is an equivalence relation.

2.17 Verify that the transitive closure of a fuzzy proximity relation is a similarity relation.

2.18 A tolerance relation R in $\mathbf{X} \times \mathbf{Y}$ is a reflexive and symmetric ordinary relation. Show that if R is a proximity relation, then for any $0 < \alpha \leq 1$, R_α is a tolerance relation.

References

Alefeld, G. and Herzberger, J. (1983). *Introduction to Interval Computations.* New York: Academic Press.

Bargiela, A. and Pedrycz, W. (2003). *Granular Computing: An Introduction.* Dordecht: Kluwer Academic Publishers.

Bargiela, A. and Pedrycz, W. (2005). Granular mappings. *IEEE Transactions on Systems, Man, and Cybernetics-Part A: Systems and Humans* 35 (2): 292–297.

Bargiela, A. and Pedrycz, W. (2008). Toward a theory of granular computing for human-centered information processing. *IEEE Transactions on Fuzzy Systems* 16 (2): 320–330.

Bodjanova, S. (2006). Median alpha-levels of a fuzzy number. *Fuzzy Sets and Systems* 157 (7): 879–891.

Borel, E. (1950). *Probabilité e Certitude.* Paris: Press Universite de France.

De Baets, B. and Meyer, H. (2003). On the existence and construction of T-transitive closures. *Information Sciences* 152 (1): 167–179.

De Luca, A. and Termini, S. (1972). A definition of nonprobabilistic entropy in the setting of fuzzy sets. *Information and Control* 20 (3): 301–312.

Dubois, D. and Prade, H. (1979). Outline of fuzzy set theory: an introduction. In: *Advances in Fuzzy Set Theory and Applications* (eds. M.M. Gupta, R.K. Ragade and R.R. Yager), 27–39. North-Holland: Amsterdam.

Dubois, D. and Prade, H. (1997). The three semantics of fuzzy sets. *Fuzzy Sets and Systems* 90 (2): 141–150.

Dubois, D. and Prade, H. (1998). An introduction to fuzzy sets. *Clinica Chimica Acta* 70 (1): 3–29.

Fodor, J. (1993). A new look at fuzzy connectives. *Fuzzy Sets and Systems* 57 (2): 141–148.

Foulloy, L. and Benoit, E. (2006). Building a class of fuzzy equivalence relations. *Fuzzy Sets and Systems* 157 (11): 1417–1437.

Kandel, A. and Yelowitz, L. (1974). Fuzzy chains. *IEEE Transactions Systems, Man, and Cybernetics* SMC-4 (5): 472–475.

Kim, C.S., Kim, D.S., and Park, J.S. (2000). A new fuzzy resolution principle based on the antonym. *Fuzzy Sets and Systems* 13 (2): 299–307.

Klir, G. and Yuan, B. (1995). *Fuzzy Sets and Fuzzy Logic: Theory and Applications.* Upper Saddle River: Prentice-Hall.

Kosko, B. (1992). *Neural Networks and Fuzzy Systems.* Englewood Cliffs: Prentice-Hall International.

Moore, R. (1966). *Interval Analysis.* Englewood Cliffs: Prentice Hall.

Moore, R., Kearfott, R.B., and Cloud, M.J. (2009). *Introduction to Interval Analysis.* Philadelphia: SIAM.

Naessens, H., Meyer, H., and De Baets, B. (2002). Algorithms for the computation of T-transitive closures. *IEEE Transactions on Fuzzy Systems* 10 (4): 541–551.

Nguyen, H. and Walker, E. (1999). *A First Course in Fuzzy Logic*. Boca Raton: Chapman Hall, CRC Press.

Pawlak, Z. (1982). Rough sets. *International Journal of Information and Computer Science* 11 (15): 341–356.

Pedrycz, W. (1998). Shadowed sets: representing and processing fuzzy sets. *IEEE Transactions on Systems, Man, and Cybernetics. Part B, Cybernetics* 28: 103–109.

Pedrycz, W. (2005). From granular computing to computational intelligence and human-centric systems. *IEEE Connections* 3 (2): 6–11.

Pedrycz, A., Dong, F., and Hirota, K. (2009). Finite α cut-based approximation of fuzzy sets and its evolutionary optimization. *Fuzzy Sets and Systems* 160 (24): 3550–3564.

Tsabadze, T. (2008). The reduction of binary fuzzy relations and its applications. *Information Sciences* 178 (2): 562–572.

Valverde, L. and Ovchinnikov, S. (2008). Representations of T-similarity relations. *Fuzzy Sets and Systems* 159 (17): 2211–2220.

Wallace, M., Acrithis, Y., and Kollias, S. (2006). Computationally efficient sup-t transitive closure for sparse fuzzy binary relations. *Fuzzy Sets and Systems* 157 (3): 341–372.

Yager, R. (1983). Entropy and specificity in a mathematical theory of evidence. *International Journal of General Systems* 9 (1): 249–260.

Zadeh, L.A. (1965). Fuzzy sets. *Information and Control* 8 (3): 33–353.

Zadeh, L.A. (1971). Similarity relations and fuzzy orderings. *Information Sciences* 3 (2): 177–200.

Zadeh, L.A. (1975). The concept of linguistic variables and its application to approximate reasoning I, II, III. *Information Sciences* 8 (3): 199–249.

Zadeh, L.A. (1978). Fuzzy sets as a basis for a theory of possibility. *Fuzzy Sets and Systems* 1 (1): 3–28.

Zadeh, L.A. (1997). Toward a theory of fuzzy information granulation and its centrality in human reasoning and fuzzy logic. *Fuzzy Sets and Systems* 90 (2): 111–127.

Zadeh, L.A. (1999). From computing with numbers to computing with words: from manipulation of measurements to manipulation of perceptions. *IEEE Transactions on Circuits and Systems* 45 (1): 105–119.

Zadeh, L.A. (2005). Toward a generalized theory of uncertainty (GTU) – an outline. *Information Sciences* 172: 1–40.

3

Design and Processing Aspects of Fuzzy Sets

In this chapter, we focus on the fundamentals of fuzzy sets by concentrating on the three main processing issues; (i) design of fuzzy sets (membership functions), (ii) logic operations and aggregation of fuzzy sets, and (iii) transformations (mappings) of fuzzy sets including fuzzy arithmetic. Those are essentials, which make the framework of fuzzy sets fully operational when supporting a wide range of applications.

3.1 The Development of Fuzzy Sets: Elicitation of Membership Functions

The problem of elicitation and interpretation of fuzzy sets (their membership functions) is of significant relevance from the conceptual, algorithmic, and application-oriented standpoints (Klir and Yuan 1995; Nguyen and Walker 1999). In the existing literature, we encounter a great number of methods that support the construction of membership functions. In general, we distinguish here between *user-driven* and *data-driven* approaches with a number of techniques that share some features specific to both data- and user-driven techniques, and hence are located somewhere in between. The determination of membership functions has been a debatable issue for a long time, almost since the very inception of fuzzy sets. In contrast to interval analysis and set theory where the estimation of bounds of the interval constructs did not attract a great deal of attention and seemed to be somewhat taken for granted, an estimation of membership degrees (and membership functions, in general) became essential and over time has led us to sound, well justified, and algorithmically appealing estimation techniques (Civanlar and Trussell 1986; Dombi 1990; Turksen 1991; Chen and Wang 1999; Medaglia et al. 2002).

Multicriteria Decision-Making under Conditions of Uncertainty: A Fuzzy Set Perspective, First Edition. Petr Ekel, Witold Pedrycz, and Joel Pereira, Jr.

3.1.1 Semantics of Fuzzy Sets: Some General Observations

Fuzzy sets are constructs that come with a well-defined meaning. They capture the semantics of the framework they intend to operate within. Fuzzy sets are the building conceptual blocks (generic constructs) that are used in problem description, modeling, control, and pattern classification tasks. Before discussing specific techniques of membership function estimation, it is worth casting the overall presentation in a certain context by emphasizing the aspect of the use of a finite number of fuzzy sets leading to some essential vocabulary reflective of the underlying domain knowledge. In particular, we are concerned with the related semantics, calibration capabilities of membership functions, and the locality of fuzzy sets.

The limited capacity of a short term memory, as identified by Miller (1956), suggests that we could easily and comfortably handle and process 7 ± 2 items. This implies that the number of fuzzy sets to be considered as meaningful conceptual entities should be kept at the same level. The observation sounds reasonable – quite commonly in practice we witness situations in which this assumption holds. For instance, when describing linguistically quantified variables, say error or change of error, quantify temperature (warm, hot, cold, etc.) we may use seven generic concepts (descriptors) labeling them as positive *large*, positive *medium*, positive *small*, *around* zero, negative *small*, negative *medium*, and negative *large*. When characterizing speed, we may talk about its quite intuitive descriptors such as *low*, *medium*, and *high* speed. In the description of an approximation error, we may typically use the concept of a *small* error around a point of linearization (in all these examples, the terms are indicated in italics to emphasize the granular character of the constructs and the role being played there by fuzzy sets). While embracing very different tasks, these descriptors exhibit a striking similarity. All of them are information granules, not numbers. We can stress that the descriptive power of numbers is very much limited and numbers themselves are not used to abstract concepts. In general, the use of an excessive number of terms does not offer any advantage. On the contrary: it remarkably clutters our description of the phenomenon and hampers further effective usage of such concepts we intend to establish to capture the essence of the domain knowledge. With the increase in the number of fuzzy sets, their semantics and interpretation capabilities also become negatively impacted. Fuzzy sets may be built into a hierarchy of terms (descriptors) but at each level of this hierarchy (when moving down toward higher specificity that is an increasing level of detail), the number of fuzzy sets is kept relatively low.

While fuzzy sets capture the semantics of the concepts, they may require some calibration depending on the specification of the problem at hand. This flexibility of fuzzy sets should not be treated as any shortcoming but rather viewed as a certain and fully exploited advantage. For instance, a term *low* temperature comes with a clear meaning yet it requires a certain calibration depending upon the environment and the context it was put into. The concept of *low* temperature is used in different climate zones and is of relevance in any communication between people, yet for each individual in the community the meaning of the term is different thereby requiring some calibration. This could be realized, for example, by shifting the membership function along the universe of discourse of temperature, affecting the universe of discourse by some translation, dilation, and the like. As a communication means, linguistic terms are fully legitimate and as such they appear in different settings. They require some refinement so that their meaning is fully understood and shared by the community of the users.

When discussing the methods aimed at the determination of membership functions or membership grades, it is worthwhile underlining the existence of the two main categories of approaches being reflective of the origin of the numeric values of membership. The first one is reflective of the domain knowledge and opinions of experts. In the second one, we consider experimental data whose global characteristics become reflected in the form and parameters of the membership functions. In the first group, we can refer to the pairwise comparison (for instance, Saaty's approach, as discussed later in this chapter) as one of the quite visible and representative examples while fuzzy clustering is usually presented as a typical example of the data-driven method of membership function estimation. In what follows, we elaborate on several representative methods, which will help us appreciate the level and flexibility of fuzzy sets.

3.1.2 Fuzzy Set as a Descriptor of Feasible Solutions

The aim of the method is to relate membership function to the level of feasibility of individual elements of a family of solutions associated with the problem at hand. Let us consider a certain function $F(x)$ defined in L, that is F: $L \rightarrow \mathbf{R}^+$, where $L \subset \mathbf{R}$. Our intent is to determine its maximum, namely $x^0 = \arg\max_{x \in L} F(x)$. on a basis of the values of $F(x)$, we can form a fuzzy set A describing a collection of feasible solutions that could be labeled as optimal. Being more specific, we use the fuzzy set to represent an extent (degree) to which some specific values of "x" could be sought as potential (optimal) solutions to the problem. Taking this into consideration, we relate the membership function of A with the corresponding value of $F(x)$ cast in the context of the boundary values assumed by "F." For instance, the membership function of A could be expressed in the following form:

$$A(x) = \frac{F(x) - \min\limits_{x \in L} F(x)}{\max\limits_{x \in L} F(x) - \min\limits_{x \in L} F(x)} \tag{3.1}$$

The boundary conditions are intuitively associated with values of $\min_{x \in L} F(x)$ and $\max_{x \in L} F(x)$. For other values of "x" where F attains is maximal value, $A(x)$ is equal to 1 and around this point, the membership values are reduced when "x" is likely to be a solution to the problem $F(x) < \max_{x \in L} F(x)$. The form of the membership function depends upon the character of the function under consideration.

If the fuzzy set is used to quantify the quality (performance) of the solution to the minimization problem, then the resulting membership function reads as follows

$$A(x) = 1 - \frac{F(x) - \min\limits_{x \in L} F(x)}{\max\limits_{x \in L} F(x) - \min\limits_{x \in L} F(x)} \tag{3.2}$$

If the function of interest assumes values in **R**, then the two formulas are modified by including the absolute values of the differences, that is

$$A(x) = \frac{\left| F(x) - \min\limits_{x \in L} F(x) \right|}{\left| \max\limits_{x \in L} F(x) - \min\limits_{x \in L} F(x) \right|} \text{ and } A(x) = 1 - \frac{\left| F(x) - \min\limits_{x \in L} F(x) \right|}{\left| \max\limits_{x \in L} F(x) - \min\limits_{x \in L} F(x) \right|}.$$

Linearization, its quality, and description of such quality fall under the same banner as the optimization problem. We show how the membership function could be formed in this case. When linearizing a function around some predetermine point, a quality of the linearization scheme can be quantified in a form of some fuzzy set. Its membership function attains one for all these points where the linearization error is equal to zero (in particular, this holds at the point around which the linearization is carried out). The following example illustrates this idea.

Example 3.1 We are interested in the linearization of the function $y = g(x) = x^2$ around $x_0 = 1$ and assessing the quality of this linearization in the range [0, 4]. The linearization formula reads as $y - y_0 = g'(x_0)(x - x_0)$ where $y_0 = g(x_0)$ and $g'(x_0)$ is the derivative of $g(x)$ at x_0. Given the form of the function under consideration, its linearized version comes in the form $(2x_0)(x - x_0) = 2(x-1)$. We define the quality of this linearization by taking the absolute value of the difference between the original function and its linearization, $f(x) = |g(x) - 2(x-1)| = |x^2 - 2(x-1)|$. As the fuzzy set A describes the quality of linearization, its membership function has to take into consideration the following expression

$$A(x) = 1 - \frac{\left| F(x) - \min\limits_{x \in L} F(x) \right|}{\left| \max\limits_{x \in L} F(x) - \min\limits_{x \in L} F(x) \right|} \tag{3.3}$$

where $\max_{x \in L} F(x) = F(4) = 10$ and $\min_{x \in L} F(x) = 0.0$. When at some z, $F(z) = \min_{x \in L} F(z)$, this means that $A(z) = 1$, which in the sequel indicates that the linearization at this point is perfect; no linearization error has been generated. We note that the higher quality of approximation is achieved for the arguments.

3.1.3 Fuzzy Set as a Descriptor of the Notion of Typicality

Fuzzy sets address an issue of gradual *typicality* of elements to a given concept whose essence is being captured by the fuzzy set. They stress the fact that there are elements that fully satisfy the concept (are typical for it) and there are various elements that are allowed only with partial membership degrees. The form of the membership function is reflective of the semantics of the concept. Its details could be conveniently captured by adjusting the parameters of the membership function or choosing its form depending upon available experimental data. For instance, consider a fuzzy set of circles. Formally, an ellipsoid includes a circular shape as its very special example that satisfies the condition of equal axes, that is $a = b$, see Figure 3.1. What if we have $a = b + \varepsilon$ where ε is a very small positive number? Could this figure be sought as a circle? It is very likely so. Perhaps not a circle in a straight mathematical sense, which we may note by assigning the membership grade that is very close to 1, say 0.97. Our perception of the concept, which comes with some level of tolerance to imprecision, does not allow us to tell apart this figure from the ideal circle.

It is intuitively appealing to see that higher differences between the values of the axes "a" and "b" result in lower values of the membership function. The definition of the fuzzy set of circle could be formed in a number of ways. Prior to the definition or even before a visualization of the shape of the membership

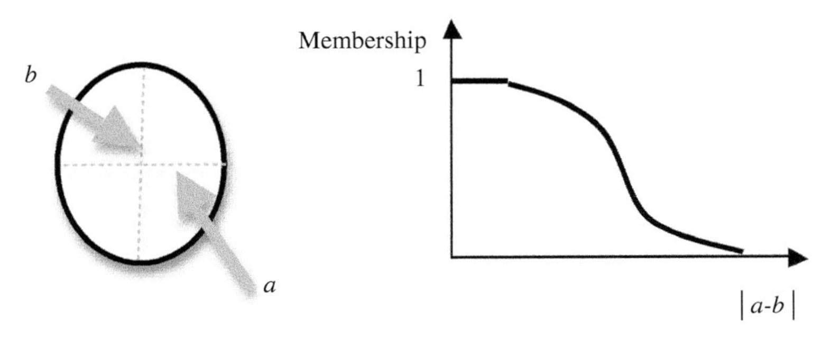

Figure 3.1 Perception of geometry of ellipsoids and quantification of their membership grades to the concept of "fuzzy circles."

function, it is important to formulate a universe of discourse over which it is to be defined. There are several sound alternatives worth considering:

a) for each pair of values of the axes (a and b), collect an experimental assessment of membership of the ellipsoids to the category of circles. Here, the membership function is defined over a Cartesian space of the spaces of lengths of axes of the ellipsoids. While selecting a form of the membership we require that it assumes values at $a = b$ and becomes gradually reduced when the arguments start getting more different.
b) we can define an absolute distance between "a" and "b," $|a-b|$ and form a fuzzy set over this space **X**; $\mathbf{X} = \{x|x = |a-b|\}$ $\mathbf{X} \subset \mathbf{R}_+$. This semantic constraints translate into the condition of $A(0) = 1$. For higher values of x we may consider monotonically decreasing values of A.
c) we can envision ratios of a and b $x = a/b$ and construct a fuzzy set over the space of $\mathbf{R}_+$ such that $\mathbf{X} = \{x|x = a/b\}$. Here we require that $A(1) = 1$. We also anticipate lower values of membership grades when moving to the left and to the right from $x = 1$. Note that the membership function could be asymmetric so we allow for different membership values for the same length of the sides, say $a = 6$, $b = 5$ and $a = 6$ and $b = 5$ (the effect could be quite apparent due to the occurrence of visual effects when perceiving geometric phenomena). The previous model of **X** as outlined in (a) cannot capture this effect.

Once the form of the membership function has been defined, it could be further adjusted by modifying the values of its parameters on a basis of some experimental findings. They come in the form of ordered triples or pairs, say (a, b, μ), $(a/b, \mu)$ or $(|a-b|, \mu)$ depending on the previously accepted definition of the universe of discourse. The membership values μ are those available from the expert offering an assessment of the likeness of the corresponding geometric figure. Note that the resulting membership functions become formulated in different universes of discourse.

3.1.4 Vertical and Horizontal Schemes of Membership Function Estimation

The vertical and horizontal modes of membership estimation are two standard approaches used in the determination of fuzzy sets. They reflect distinct ways of looking at fuzzy sets whose membership functions at some finite number of points are quantified by experts. In the horizontal approach, we identify a collection of elements in the universe of discourse **X** and request that an expert answers the following question

$$\text{– does x belong to concept } A? \tag{3.4}$$

The answers are expected to come in a binary (yes-no) format. The concept A defined in $\mathbf{X}$ could be any linguistic notion, say *high* speed, *low* temperature, and so on. Given "n" experts whose answers for a given point of $\mathbf{X}$ form a mix of yes-no replies, we count the number of "yes" answers and compute the ratio of the positive answers (p) versus the total number of replies(n), that is p/n. This ratio (likelihood) is treated as a membership degree of the concept at the given point of the universe of discourse. When all experts accept that the element belongs to the concept, then its membership degree is equal to 1. Higher disagreement between the experts (quite divided opinions) results in lower membership degrees. The concept A defined in $\mathbf{X}$ requires collecting results for some other elements of $\mathbf{X}$ and determining the corresponding ratios as outlined in Figure 3.2 (observe a series of estimates that are determined for selected elements of $\mathbf{X}$; note also that the elements of $\mathbf{X}$ need not to be evenly distributed).

If replies follow some, for example, binomial, distribution then we can y determine a confidence interval of the individual membership grade. The standard deviation of the estimate of the ration of the positive answers associated with the point x, denoted here by σ, is given in the form

$$\sigma = \sqrt{\frac{p/n(1-p/n)}{n}} \tag{3.5}$$

The associated confidence interval, which describes a range of membership values, is then determined as

$$[p-\sigma, p+\sigma] \tag{3.6}$$

In essence, when the confidence intervals are taken into consideration, the membership estimates become intervals of possible membership values and this leads to the concept of so-called interval-valued fuzzy sets. By assessing the width of the estimates, we could control the execution of the experiment: when the ranges are too long, one could re-design the experiment and monitor closely the consistency of the responses collected in the experiment.

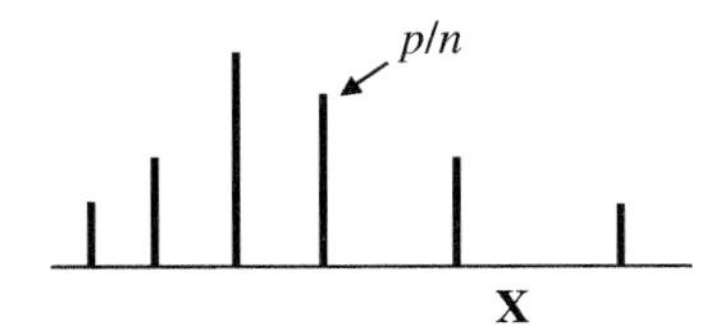

Figure 3.2 A horizontal method of the estimation of the membership function.

Example 3.2 Let us consider responses of 10 experts who came up with the following assessment of the concept *high* interest rate (%) with the number of "yes" responses collected as follows:

$x(\%)$	2	3	5	8	10
no. of "yes" replies	0	2	4	7	10

Following these responses, the membership function and its confidence values σ producing confidence intervals are given here:

$x(\%)$	2	3	5	8	10
$A(x)$ (*high* interest rate)	0.0	0.2	0.4	0.7	1.0
σ	0.0	0.126	0.155	0.144	0.0

The advantage of the method comes with its simplicity as the technique relies explicitly upon a direct counting of responses. The concept is also intuitively appealing. The probabilistic nature of the replies helps us construct confidence intervals that are essential to the assessment of the specificity of the membership quantification. A certain drawback is related with the local character of the construct: as the estimates of the membership function are completed separately for each element of the universe of discourse, they could exhibit a lack of continuity when moving from certain point to its neighbor. This concern is particularly valid in the case when **X** is a subset of real numbers.

The vertical mode of membership estimation is concerned with the estimation of the membership function by focusing on the determination of the successive α-cuts. The experiment focuses on the unit interval of membership grades. The experts involved in the experiment are asked the questions of the following form:

– what are the elements of **X** that belong to fuzzy set A at degree not lower than α? (3.7)

where α is a certain level (threshold) of membership grades in [0, 1]. The essence of the method is illustrated in Figure 3.3. Note that the satisfaction of the inclusion constraint is obvious: we envision that for higher values of α, the expert is going to provide more limited subsets of **X**; the vertical approach leads to the fuzzy set by combining the estimates of the corresponding α-cuts. Given the nature of this method, we are referring to the collection of random sets as these estimates appear in the successive stages of the estimation process.

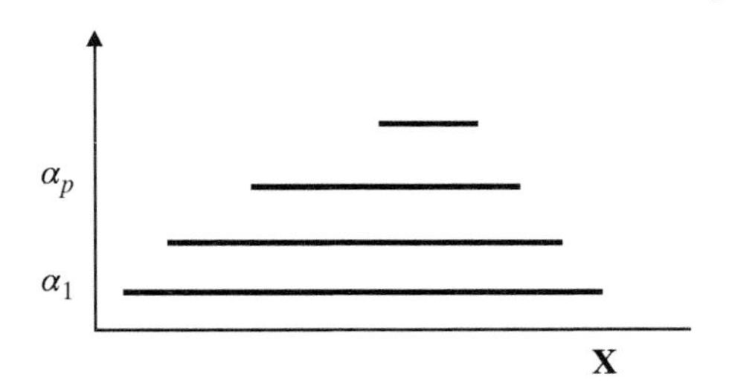

Figure 3.3 A vertical approach of membership estimation through the reconstruction of a fuzzy set through its estimated α-cuts.

These elements are identified by the expert as those forming the corresponding α-cuts of A. By repeating the process for several selected values of α we end up with the α-cuts and using them we reconstruct the fuzzy set. The simplicity of the method is its genuine advantage. As in the horizontal method of membership estimation, a possible lack of continuity is a certain disadvantage one has to be aware of. Here, the selection of suitable levels of α needs to be carefully investigated. Similarly, an order at which different levels of α are used in the experiment could impact the estimate of the membership function. The discussion on the optimization of a series of α-cuts (which might be of relevance in the context of the estimation of membership functions) is given in (Pedrycz et al. 2009).

3.1.5 Saaty's Priority Approach of Pairwise Membership Function Estimation

The priority approach introduced by Saaty (Saaty 1980, 1986a) forms another interesting alternative used to estimate the membership function that help alleviate the limitations that are associated with the horizontal and vertical schemes of membership function estimation. To explain the essence of the method, let us consider a collection of elements $X_1, X_2, \ldots, X_n$ (those could be, for instance, some alternatives whose allocation to a certain fuzzy set is sought) for which given are membership grades $A(X_1)$, $A(>X_2)$, …, $A(X_n)$. Let us organize them into a so-called reciprocal matrix of the following form

$$\mathbf{R} = [\mathbf{R}(X_i, X_j)] = \begin{bmatrix} \frac{A(X_1)}{A(X_1)} & \frac{A(X_1)}{A(X_2)} & \cdots & \frac{A(X_1)}{A(X_n)} \\ \frac{A(X_2)}{A(X_1)} & \frac{A(X_2)}{A(X_2)} & \cdots & \frac{A(X_2)}{A(X_n)} \\ \cdots & \cdots & & \\ \frac{A(X_n)}{A(X_1)} & \frac{A(X_n)}{A(X_2)} & \cdots & \frac{A(X_n)}{A(X_n)} \end{bmatrix} = \begin{bmatrix} 1 & \frac{A(X_1)}{A(X_2)} & \cdots & \frac{A(X_1)}{A(X_n)} \\ \frac{A(X_2)}{A(X_1)} & 1 & \cdots & \frac{A(X_2)}{A(X_n)} \\ & & \cdots & \\ \frac{A(X_n)}{A(X_1)} & \frac{A(X_n)}{A(X_2)} & \cdots & 1 \end{bmatrix} \tag{3.8}$$

Noticeably, the diagonal values of **R** are equal to 1. The entries that are symmetrically positioned with respect to the diagonal satisfy the condition of reciprocality that is $\mathbf{R}(X_i,X_j) = 1/\mathbf{R}(X_j, X_i)$. We will be referring to this form of reciprocality as *multiplicative reciprocality* as opposed to so-called *additive reciprocality* for which $\mathbf{R}(X_i,X_j) + \mathbf{R}(X_j,X_i) = 1$. Furthermore an important transitivity property holds that is $\mathbf{R}(X_i,X_k)\ \mathbf{R}(X_k,X_j) = \mathbf{R}(X_i,X_j)$ for all indexes *i*, *j*, and *k*. This property holds because of the way in which the matrix has been constructed. By plugging in the corresponding ratios, one obtains $\mathbf{R}(X_i,X_k)\mathbf{R}(X_k,X_j) = \frac{A(X_i)A(X_k)}{A(X_k)A(X_j)} = \frac{A(X_i)}{A(X_j)} = \mathbf{R}(X_i,X_j)$. Let us now multiply the matrix by the vector of the membership grades $A = [A(X_1)\ A(X_2)\ \ldots\ A(X_n)]^T$. For the i-th row of *R* (that is the i-th entry of the resulting vector of results) we obtain

$$[\mathbf{R}A]_i = \begin{bmatrix} \frac{A(X_i)}{A(X_1)} & \frac{A(X_i)}{A(X_2)} \cdots \frac{A(X_i)}{A(X_n)} \end{bmatrix} \begin{bmatrix} A(X_1) \\ A(X_2) \\ \cdots \\ A(X_n) \end{bmatrix}, i = 1,2,\ldots,n \tag{3.9}$$

Thus the *i*-th element of the vector is equal to $nA(X_i)$. Overall once completing the calculations for all "*i*," this leads us to the expression $\mathbf{R}A = nA$. In other words, we conclude that *A* is the eigenvector of *R* associated with the largest eigenvalue of $\mathbf{R}$, which is equal to *n*. In this scenario, we have assumed that the membership values $A(x_i)$ are given and then showed what form of results could they lead to. In practice, the membership grades are not given and have to be looked for.

The starting point of the estimation process are entries of the reciprocal matrix that are obtained through collecting results of pairwise evaluations offered by an expert, designer, or user (depending on the character of the task at hand). Prior to making any assessment, the expert is provided with a finite scale with values spread in between 1 and 7. Some other alternatives of the scales such as those involving five or nine levels could be sought as well. If X_i is strongly preferred over X_j when being considered in the context of the fuzzy set whose membership function we would like to estimate, then this judgment is expressed by assigning high values of the available scale, say 6 or 7. If we still sense that X_i is preferred over X_j, yet the strength of this preference is lower in comparison with the previous case, then this is quantified using some intermediate values of the scale, say 3 or 4. If no difference is sensed, the values close to 1 are the preferred choice, say 2 or 1. The value of 1 indicates that X_i and X_j are equally preferred. The general quantification of preferences positioned on the scale of 1–7 can be described as in Table 3.1 (Saaty 1986b).

On the other hand, if X_j is preferred over X_i, the corresponding entry assumes values below 1. Given the reciprocal nature of the assessment, once the preference of X_i over X_j has been quantified, the inverse of this number is plugged into

Table 3.1 Scale of intensities of relative importance.

Intensity of relative importance	Description
1	Equal importance
3	Moderate importance of one element over another
5	Essential or strong importance
7	Demonstrated importance
9	Extreme importance
2, 4, 6, 8	Intermediate values between the two adjacent judgments

the entry of the matrix that is located at the (*j, i*)-th coordinate. As indicated earlier, the elements on the main diagonal are equal to 1. Next the maximal eigenvalue is computed along with its corresponding eigenvector. The normalized version of the eigenvector is then the membership function of the fuzzy set we considered when doing all pairwise assessments of the elements of its universe of discourse. The effort to complete pairwise evaluations is far more manageable in comparison to any experimental overhead we need when assigning membership grades to all elements (alternatives) of the universe in a single step. Practically, the pairwise comparison helps the expert focus only on two elements once at a time thus reducing uncertainty and hesitation while leading to the higher level of consistency. The assessments are not free of bias and could exhibit some inconsistent evaluations. In particular, we cannot expect that the transitivity requirement could be fully satisfied. Fortunately, the lack of consistency could be quantified and monitored. The largest eigenvalue computed for R is always greater than the dimensionality of the reciprocal matrix (recall that in reciprocal matrices the elements positioned symmetrically along the main diagonal are inverse of each other), $\lambda_{\max} > n$ where the equality $\lambda_{\max} = n$ occurs only if the results are fully consistent. The ratio

$$\nu = (\lambda_{\max} - n)/(n-1). \tag{3.10}$$

can be regarded as a sound index of inconsistency of the data; the higher its value, the less consistent are the collected experimental results. This expression can be sought as the indicator of the quality of the pairwise assessments provided by the expert. If the value of ν is too high exceeding a certain superimposed threshold, the experiment may need to be repeated. Typically if ν is less than 0.1 the assessment is sought to be consistent while higher values of ν call for the re-examination of the experimental data and a re-run of the experiment. To quantify how much the experimental data deviate from the transitivity requirement, we calculate the absolute differences between the corresponding experimentally obtained entries of the reciprocal matrix, namely $\mathbf{R}(X_i, X_k)$ and $\mathbf{R}(X_i, X_j)\mathbf{R}(X_j, X_k)$. The sum is expressed in the form

$$V(i,k) = \sum_{j=1}^{n} | \mathbf{R}(X_i,X_j)\mathbf{R}(X_j,X_k) - \mathbf{R}(X_i,X_k) | \qquad (3.11)$$

serves as a useful indicator of the lack of transitivity of the experimental data for the given pair of elements (i, k). If required, we may repeat the experiment if this sum takes high values. The overall sum $\sum_{i,k}^{n} V(i,k)$ becomes then a global evaluation of the lack of transitivity of the experimental assessment.

Example 3.3 Let us estimate the membership function of the concept *hot* temperature for the space of temperatures consisting of 10, 20, 30, 30, 45° C. The scale in which the pairs of these elements are evaluated consists of five levels (say, 1, 2, ..., 5). The experimental results of the pairwise comparison are collected in the reciprocal matrix **R**,

$$\mathbf{R} = \begin{bmatrix} 1 & 1/2 & 1/4 & 1/5 \\ 2 & 1 & 1/3 & 1/4 \\ 4 & 3 & 1 & 1/3 \\ 5 & 4 & 3 & 1 \end{bmatrix} \qquad (3.12)$$

Calculating the maximal eigenvalue, we obtain $\lambda_{\max} = 4.114$, which is slightly higher than the dimension ($n = 4$) of the reciprocal matrix. The corresponding eigenvector is equal to [0.122 0.195 0.438 0.869] which after normalization gives rise to the membership function of *hot* temperature to be equal to [0.14 0.22 0.50 1.00]. The value of the inconsistency index is equal to (4.114–4)/3 = 0.038 and is far lower than the threshold of 0.1.

Example 3.4 Now let us consider some modified version of the previously discussed reciprocal matrix with the following entries

$$R = \begin{bmatrix} 1 & 1/2 & 1/4 & 1/5 \\ 2 & 1 & 1/3 & 4 \\ 4 & 3 & 1 & 1/3 \\ 5 & 1/4 & 3 & 1 \end{bmatrix} \qquad (3.13)$$

Now the maximal eigenvalue is far higher than the dimensionality of the problem, $\lambda_{max} = 5.426$. In this case, given the high value of the inconsistency index, $\nu = (5.426 - 4)/3 = 0.475$, there is no point in computing the corresponding eigenvector. To fix the problem we could compute the lack of transitivity for the triples of indexes (i, j, k) and in this way highlight these assessments that tend to be highly inconsistent. These are the candidates whose evaluation has to be revised.

The method of pairwise comparison has been generalized in many different ways by allowing for estimates being expressed as fuzzy sets (van Laarhoven and Pedrycz 1983). One can refer to a number of applications in which the technique of pairwise comparison has been directly applied (Kulak and Kahraman 2005).

3.1.6 Fuzzy Sets as Granular Representatives of Numeric Data – The Principle of Justifiable Granularity

The principle of justifiable granularity (Pedrycz 2013) delivers a comprehensive conceptual and algorithmic setting to develop information granules. The principle is general in the sense it shows a way of forming information granules without being restricted to certain formalism in which granules are formalized. Information granules are built by considering available experimental evidence.

Let us start with a simple scenario using which we illustrate the key components of the principle and its underlying motivation.

Consider a collection of one-dimensional numeric real-number data of interest (for which an information granule is to be formed) $\mathbf{X} = \{x_1, x_2, \ldots, x_N\}$. Denote the largest and the smallest element in $\mathbf{X}$ by x_{min} and x_{max}, respectively. On a basis of this experimental evidence $\mathbf{X}$, we form an interval information granule A so that it satisfies the requirements of coverage and specificity. The first requirement implies that the information granule is justifiable, namely it embraces (covers) as many elements of $\mathbf{X}$ as possible and can be sought as a sound representative. The quest for meeting the requirement of the well-defined semantics is quantified in terms of high specificity of A. In other words, for given $\mathbf{X}$, the interval A has to satisfy the requirement of high coverage and specificity; these two concepts have been already discussed in the previous chapter. In other words, the construction of $A = [a, b]$ leads to the optimization of its bounds a and b so that at the same time we maximize the coverage and specificity. It is known that these requirements are in conflict: the increase in the coverage values leads to lower values of specificity. To transform the two-objective optimization problem in a scalar version of the optimization, we consider the performance index built as a product of the coverage and specificity

$$V(a, b) = \text{cov}(A) * \text{sp}(A) \tag{3.14}$$

and determine the solution (a_{opt}, b_{opt}) such that $V(a, b)$ becomes maximized.

The ensuing approach can be established as a two-phase algorithm. In phase one, we proceed with a formation of a numeric representative of **X**, say a mean, median, or a modal value (denoted here by r) that can be regarded as a rough initial representative of **X**. In the second phase, we independently determine the lower bound (i) and the upper bound (ii) of the interval by maximizing the product of the coverage and specificity as specified by the optimization criterion. This simplifies the process of building the granule as we encounter two separate optimization tasks:

$$\begin{aligned} a_{\text{opt}} &= \arg \text{Max}_a V(a) \;\; V(a) = \text{cov}([a,r])*\text{sp}([a,r]) \\ b_{\text{opt}} &= \arg \text{Max}_a V(b) \;\; V(b) = \text{cov}([r,b])*\text{sp}([r,b]) \end{aligned} \tag{3.15}$$

We calculate $\text{cov}([r, b]) = \text{card}\{x_k \mid x_k \in [r, b]\}/N$. The specificity model has to be provided in advance. Its simplest version is expressed as a linearly decreasing function, $\text{sp}([r, b]) = 1 - |b-r|/(x_{\text{max}}-r)$. By sweeping through possible values of b positioned within the range $[r, x_{\text{max}}]$, we observe that the coverage is a stairwise increasing function whereas the specificity decreases linearly, see Figure 3.4. The maximum of the product can be easily determined.

The determination of the optimal value of the lower bound of the interval a is completed in the same way as before. We determine the coverage by counting the data located to the left from the numeric representative r, namely $\text{cov}([a, r]) = \text{card}\{x_k \mid x_k \in [a, r]\}/N$ and compute the specificity as $\text{sp}([a, r]) = 1 - |a-r|/(r-x_{\text{min}})$.

Some additional flexibility can be added to the optimized performance index by adjusting the impact of the specificity in the construction of the information granule. This is done by bringing a weight factor ξ as follows

$$V(a,b) = \text{cov}(A)*\text{sp}(A)^{\xi} \tag{3.16}$$

Note that the values of ξ lower than 1 discount the influence of the specificity; in the limit case this impact is eliminated when $\xi = 0$. The value of ξ set to 1

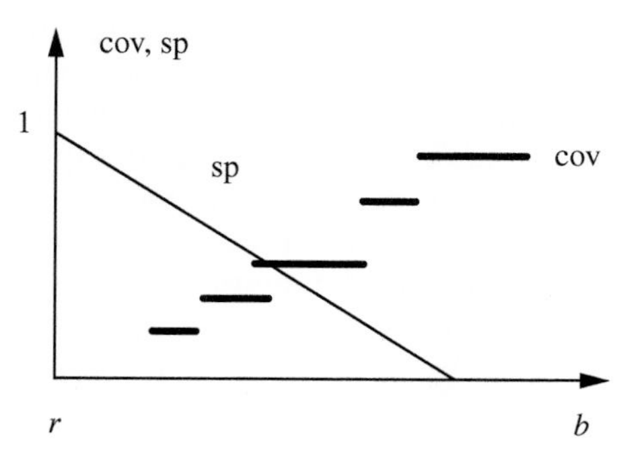

Figure 3.4 Example plots of coverage and specificity (linear model) regarded as a function of b.

returns the original performance index, whereas the values of ξ greater than 1 stress the importance of specificity by producing results that are more specific.

In case the data are governed by some given probability function $p(x)$, the coverage is computed as an integral $\text{cov}([r,b]) = \int_r^b p(x)dx$ and the specificity expressed in the form $\text{sp}([r,b]) = 1 - |b-r|/(r - x_{\max}) \cdot \dfrac{b}{x_{\max}}$.

As an illustrative example, consider the data governed by the Gaussian probability density function $p(x)$ with a zero mean and some standard deviation σ; $x_{\max}$ is set as 3σ. The corresponding plots of the product of coverage and specificity are displayed in Figure 3.5. The obtained functions are smooth and exhibit clearly visible maximal values. For $\sigma = 1$, b_{opt} is equal to 1.16 while for $\sigma = 2$, the optimal location of the upper bound is 2.31, $b_{\text{opt}} = 2.31$. The values of b are moved toward higher values to reflect higher dispersion of the available data.

In the case of n-dimensional multivariable data, $\mathbf{X} = \{\boldsymbol{x}_1, \boldsymbol{x}_2, \ldots, \boldsymbol{x}_N\}$, the principle is realized in a similar manner. For convenience, we assume that the data are normalized to [0, 1] meaning that each coordinate of the normalized $\boldsymbol{x}_k$ assumes values positioned in [0, 1]. The numeric representative r is determined first and then the information granule is built around it. The coverage is expressed in the form of the following count

$$\text{Cov}(A) = \text{card}\left\{x_k \; \|x_k - r\|^2 \leq n\rho^2\right\} \tag{3.17}$$

Note that the geometry of the resulting information granule is implied by the form of the distance function ||. || used in Eq. (3.17). For the Euclidean distance, the granule is a disc. For the Tchebyschev one, we end up with hyper rectangular shapes (hypercube). The specificity is expressed as $\text{sp}(A) = 1-\rho$. For these two

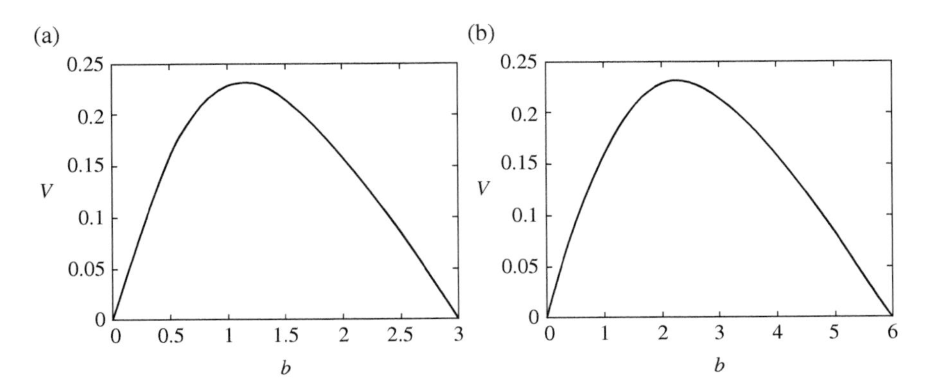

Figure 3.5 $V(b)$ as a function b: (a) $\sigma = 1.0$, (b) $\sigma = 2.0$.

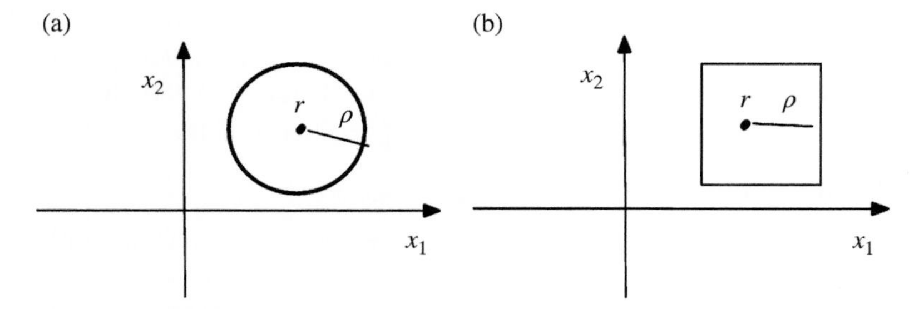

Figure 3.6 Development of information granules in the two-dimensional case when using two distance functions: (a) Euclidean distance, and (b) Tchebyschev distance.

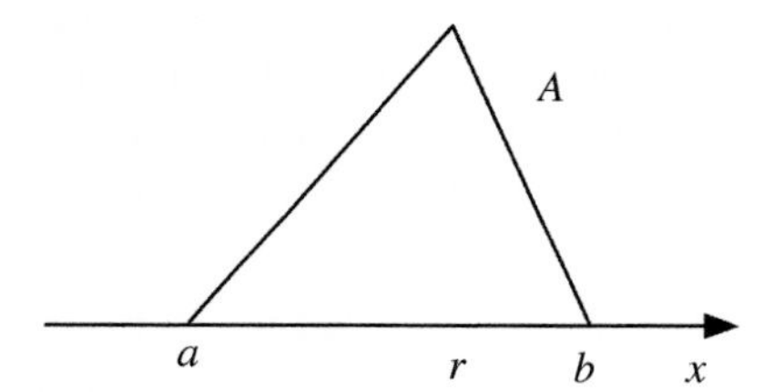

Figure 3.7 Triangular membership function with adjustable (optimized) bounds a and b.

distance functions, the corresponding plots for a two-dimensional case are illustrated in Figure 3.6.

So far, we have presented the development of the principle of justifiable granularity when building an interval information granule.

When constructing an information granule in the form of a fuzzy set, the implementation of the principle has to be modified. Considering some predetermined form of the membership function, say a triangular one, the parameters of this fuzzy set (lower and upper bounds, a and b) are optimized. See Figure 3.7.

The coverage is replaced by a σ-count by summing up the membership grades of the data in A (in what follows we are concerned with the determination of the upper bound of the membership function, namely b)

$$\text{cov}(A) = \sum_{x_k : xk > r} A(x_k) \tag{3.18}$$

The coverage computed to determine the lower bound (i) is expressed in the form

$$\text{cov}(A) = \sum_{x_k : xk < r} A(x_k) \tag{3.19}$$

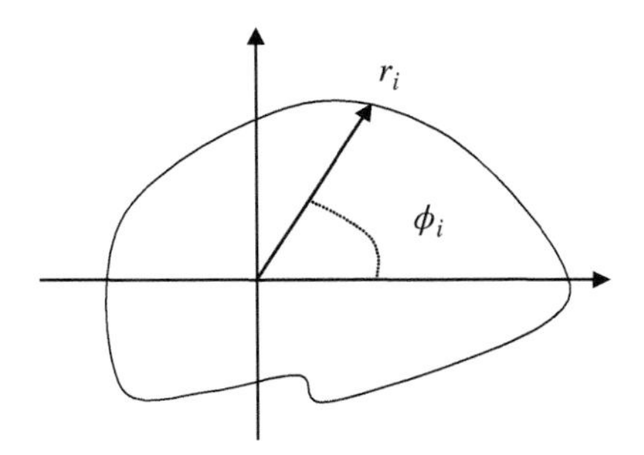

Figure 3.8 Example of a figure that can be presented as fuzzy circle with its radius described by a fuzzy set.

The specificity of the fuzzy set was discussed in Chapter 2; recall that it is computed as an integral of the specificity values of the α-cuts of A.

Example 3.5 Let us consider a geometrical figure that resembles a fuzzy circle, Figure 3.8. The coordinates of the central point are given as (x_0, y_0). Let us represent the figure as a fuzzy circle, namely a circle whose radius is a fuzzy set (fuzzy number).

The membership of the fuzzy radius is determined on a basis of numeric values of the radii obtained for several successive discrete values of the angle ϕ_i thus giving rise to the values of the corresponding distance $r_1, r_2, \ldots, r_n$. Next, the determination of the fuzzy set of a radius (fuzzy circle) is realized following the optimization scheme governed by Eq. (3.17).

3.1.7 From Type-0 to Type-1 Information Granules

A collection of experimental data – numerical data, namely type-0 information granules, leads to a single information granule of elevated type in comparison to the available experimental evidence we have started with. This comes as a general regularity of elevation of type of information granularity (Pedrycz 2013): data of type-1 transform into type-2, data of type-2 into type-3, and so on. Hence, we talk about type-2 fuzzy sets, granular intervals, imprecise probabilities. Let us recall that type-2 information granule is a granule whose parameters are information granules rather than numeric entities. Figure 3.9 illustrates a hierarchy of information granules built successively by using the principle of justifiable granularity.

The principle of justifiable granularity applies to various formalisms of information granules and this makes this approach substantially general.

Several important variants of the principle are discussed next where its generic version becomes augmented by available domain knowledge.

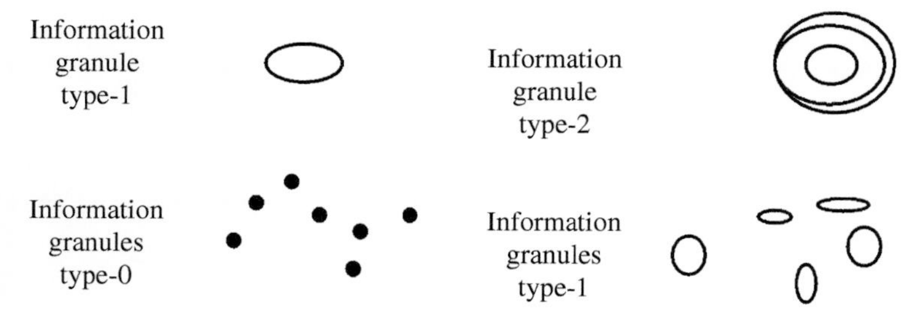

Figure 3.9 Aggregation of experimental evidence through the principle of justifiable granularity: an elevation of type of information granularity.

3.2 Weighted Data

The data can come in a weighted format meaning that each data point x_k is associated with a weight w_k assuming values in [0, 1] and quantifying the relevance (importance) of the data. The higher the value of the weight w_k, the higher the importance of x_k becomes. Apparently, this situation generalized the previously discussed in which all w_k's can be treated as equal to 1. The computing the coverage is modified to accommodate the varying values of the weight. When forming an interval information granule, we consider the sum of the weights leading to the coverage expressed in the form (we are concerned with the optimization of the upper bound of the interval $[a, b]$).

$$\text{cov}([r,b]) = \sum_{x_k: xk > r} w_k \tag{3.20}$$

When building a fuzzy set, we additionally accommodate the values of the corresponding membership grades thus computing the coverage in the form (again the computations are concerned with the optimization of the upper bound of the support of A).

$$\text{cov}([r,b]) = \sum_{x_k: xk < r} w_k \tag{3.21}$$

Note that the minimum operation leads to the conservative way of determining the contribution of x_k to the computing the coverage.

The definition of specificity and its computing is kept unchanged.

The approach presented here can be referred to the filter-based (or context-based) principle of justifiable granularity. The weights associated with the data play a role of a filter delivering some auxiliary information about the data for which an information granule is being constructed.

3.3 Inhibitory Data

In a number of problems, especially in classification tasks, where we usually encounter data (patterns) belonging to several classes, an information granule is built for the data belonging to a given class. In terms of coverage, the objective is to embrace (cover) as much experimental evidence behind the given class, but at the same time an inclusion of data of inhibitory character (those coming from other classes) has to be penalized. This leads us to the modification of the coverage to accommodate the data of the inhibitory character. Consider the interval information granule and focus on the optimization of the upper bound. As usual, the numeric representative is determined by taking a weighted average of the excitatory data (r). Along with the excitatory data to be represented (x_k, w_k), the inhibitory data come in the form of the pairs (z_k, v_k). The weights w_k and v_k assume the values in the unit interval. The computing of the coverage has to take into consideration the discounting nature of the inhibitory data, namely

$$\text{cov}([r,b]) = \max\left(0, \sum_{x_k: x_k \geq r} w_k - \gamma \sum_{x_k: z_k \in [r,b]} v_k\right) \tag{3.22}$$

The alternative version of the coverage can be expressed as follows

$$\text{cov}([r,b]) = \sum_{\substack{x_k: x_k \geq r \\ z_k \in [r,b]}} [\max(0, w_k - \gamma v_k)] \tag{3.23}$$

As seen here, the inhibitory data reduce to the reduction of the coverage; the non-negative parameter γ is used to control an impact coming from the inhibitory data. The specificity of the information granule is computed in the same way as done previously.

We consider data governed by the normal probability function $N(0, 1)$ ($p_1(x)$). The inhibitory data are also governed by the normal distribution N (2, 2) ($p_2(x)$). The plots of these pdfs are shown in Figure 3.10. In virtue of the given pdfs, the coverage is easily computed as follows (here γ =1), $\text{cov} = \max\left(0, \int_0^b p_1(x)dx - \int_0^b p_2(x)dx\right)$. We are interested in forming an optimal interval [0, b] with the upper bound being optimized.

The optimal value of b is determined by maximizing the product of this coverage and specificity (which in this case is taken as $1-b/4$. V = b_{opt} = arg $\max_b V(b)$.

The plots of the maximized performance index V versus values of b are displayed in Figure 3.11. The maximum is clearly visible as it is achieved for b = 1.05. For comparison, when γ = 0 (so no inhibitory data are taken into

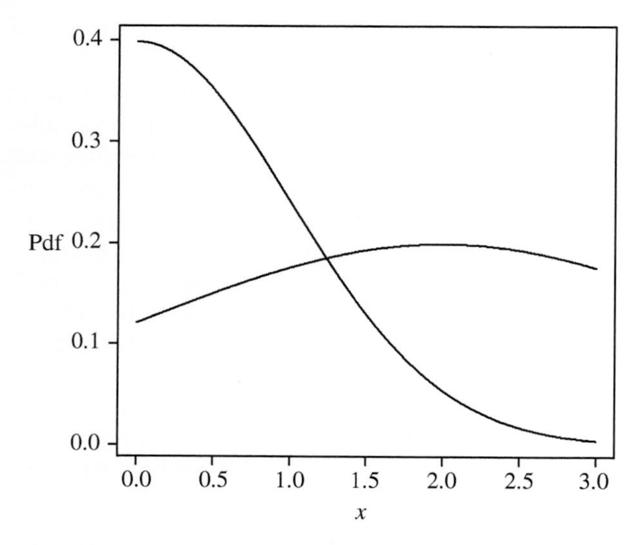

Figure 3.10 Plots of probability density function of the data for which the principle of justifiable granularity has been applied. Also shown are the inhibitory data (governed by p_2).

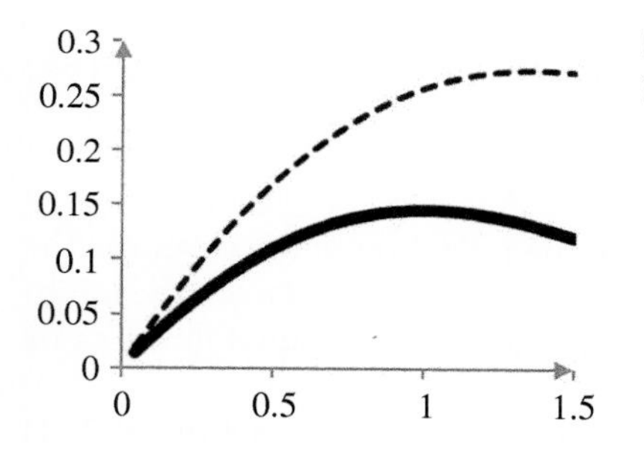

Figure 3.11 Plots of $V(b)$ for $\gamma = 1$ (solid line) and $\gamma = 0$ (dotted line).

consideration), the optimal value of b becomes higher and equal to 1.35 (which is not surprising as we are not penalizing by the inhibitory data). In this case, the corresponding plot of V is also shown in Figure 3.11 (dotted curve).

3.3.1 Design of Fuzzy Sets Through Fuzzy Clustering: From Data to Their Granular Abstraction

Fuzzy sets can be formed on a basis of numeric data through their clustering (groupings). The groups of data give rise to membership functions that convey a global more abstract view at the available data. With this regard Fuzzy

C-Means (FCM, for brief) is one of the commonly used mechanisms of fuzzy clustering (Bezdek 1981; Pedrycz 2005).

Let us review its formulation, develop the algorithm and highlight the main properties of the fuzzy clusters. Given a collection of n-dimensional data set $\{\mathbf{x}_k\}$, $k = 1,2, \ldots, N$, the task of determining its structure – a collection of c clusters, is expressed as a minimization of the following objective function (performance index) Q being regarded as a sum of the squared distances

$$Q = \sum_{i=1}^{c}\sum_{k=1}^{N} u_{ik}^{m} \|\mathbf{x}_k - \mathbf{v}_i\|^2 \tag{3.24}$$

where $\mathbf{v}_1, \mathbf{v}_2, \ldots, \mathbf{v}_c$ are n-dimensional prototypes of the clusters and $U = [u_{ik}]$ stands for a partition matrix expressing a way of allocation of the data to the corresponding clusters; u_{ik} is the membership degree of data $\mathbf{x}_k$ in the i-th cluster. The distance between the data $\mathbf{z}_k$ and prototype $\mathbf{v}_i$ is denoted by $||.||$. The fuzzification coefficient m (>1.0) expresses the impact of the membership grades on the individual clusters.

A partition matrix U satisfies two important properties

$$\text{(a)}\ \ 0 < \sum_{k=1}^{N} u_{ik} < N, \quad i = 1,2,\ldots,c \tag{3.25}$$

$$(b)\ \ \sum_{i=1}^{c} u_{ik} = 1, \quad k = 1,2,\ldots,N \tag{3.26}$$

Let us denote by $\mathbf{U}$ a family of matrices satisfying these two requirements (a)–(b). The first requirement states that each cluster has to be nonempty and different from the entire set. The second requirement states that the sum of the membership grades should be confined to 1.

The minimization of Q completed with respect to $U \in \mathbf{U}$ and the prototypes $\mathbf{v}_\mathrm{i}$ of $V = \{\mathbf{v}_1, \mathbf{v}_2, \ldots, \mathbf{v}_c\}$ of the clusters. More explicitly, we write it down as follows

$$\min Q \text{ with respect to } U \in \mathbf{U}, \mathbf{v}_1, \mathbf{v}_2, \ldots, \mathbf{v}_c \in \mathbf{R}^n \tag{3.27}$$

From the optimization standpoint, there are two individual optimization tasks to be carried out separately for the partition matrix and the prototypes. The first one concerns the minimization with respect to the constraints given the requirement in the form Eq. (3.26), which holds for each data point $\mathbf{x}_k$. The use of Lagrange multipliers transforms the problem into its constraint-free version. The augmented objective function V formulated for each data point, $k = 1, 2, \ldots, N$, reads as

$$V = \sum_{i=1}^{c} u_{ik}^{m} d_{ik}^{2} + \lambda\left(\sum_{i=1}^{c} u_{ik} - 1\right) \tag{3.28}$$

where $d_{ik}^2 = ||\boldsymbol{x}_k - \boldsymbol{v}_i||^2$. Proceeding with the necessary conditions for the minimum of V for $k = 1, 2, \ldots, N$, one has

$$\frac{\partial V}{\partial u_{st}} = 0 \quad \frac{\partial V}{\partial \lambda} = 0 \tag{3.29}$$

$s = 1, 2 \ldots, c$, $t = 1, 2, \ldots, N$. Now we calculate the derivative of V with respect to the elements of the partition matrix in the following way

$$\frac{\partial V}{\partial u_{st}} = m u_{st}^{m-1} d_{st}^2 + \lambda \tag{3.30}$$

From Eq. (3.20) and the use of Eq. (3.21) we calculate the membership grade u_{st} to be equal to

$$u_{st} = -\left(\frac{\lambda}{m}\right)^{\frac{1}{m-1}} d_{st}^{\frac{2}{m-1}} \tag{3.31}$$

Given the normalization condition $\sum_{j=1}^{c} u_{jt} = 1$ and plugging it into Eq. (3.22), one has

$$-\left(\frac{\lambda}{m}\right)^{\frac{1}{m-1}} \sum_{j=1}^{c} d_{jt}^{\frac{2}{m-1}} = 1 \tag{3.32}$$

We complete some re-arrangements of expression Eq. (3.32) by isolating the term including the Lagrange multiplier

$$-\left(\frac{\lambda}{m}\right)^{\frac{1}{m-1}} = \frac{1}{\sum_{j=1}^{c} d_{jt}^{\frac{2}{m-1}}} \tag{3.33}$$

Inserting this expression into Eq. (3.22), we obtain the successive entries of the partition matrix

$$u_{st} = \frac{1}{\sum_{j=1}^{c} \left(\frac{d_{st}^2}{d_{jt}^2}\right)^{\frac{1}{m-1}}} \tag{3.34}$$

The optimization of the prototypes $\mathbf{v}_i$ is carried out assuming the Euclidean distance between the data and the prototypes that is $||\boldsymbol{x}_k - \boldsymbol{v}_i||^2 = \sum_{j=1}^{n} (x_{kj} - v_{ij})^2$. The objective function reads now as follows $Q = \sum_{i=1}^{c} \sum_{k=1}^{N} u_{ik}^m \sum_{j=1}^{n} (x_{kj} - v_{ij})^2$ and its gradient with respect to $\mathbf{v}_i$, $\nabla_{\mathbf{v}_i} Q$ made equal to zero yields the system of linear equations

$$\sum_{k=1}^{N} u_{ik}^{m}(x_{kt} - v_{st}) = 0, s = 1,2,\ldots,c\, t = 1,2,\ldots,n \tag{3.35}$$

Thus

$$v_{st} = \frac{\sum_{k=1}^{N} u_{ik}^{m} x_{kt}}{\sum_{k=1}^{N} u_{ik}^{m}} \tag{3.36}$$

One should emphasize that the use of some other distance different from the Euclidean brings some computational complexity and the formula for the prototype cannot be presented in the concise manner as given previously.

Overall, the FCM clustering is completed through a sequence of iterations where we start from some random allocation of data (a certain randomly initialized partition matrix) and carry out the following updates by adjusting the values of the partition matrix and the prototypes. Iteration is repeated until a certain termination criterion has been satisfied. Typically, the termination condition is quantified by looking at the changes in the membership values of the successive partition matrices. Denote by $U(t)$ and $U(t+1)$ the two partition matrices produced in two consecutive iterations of the algorithm. If the distance $||U(t+1)-U(t)||$ is less than a small predefined threshold ε, then we terminate the algorithm. Typically, one considers the Tchebyschev distance between the partition matrices meaning that the termination criterion reads as follows

$$\max_{i,k} \mid u_{ik}(t+1) - u_{ik}(t) \mid \; \le \varepsilon \tag{3.37}$$

The key components of the FCM and a quantification of their impact on the form of the produced results are summarized in Table 3.2.

The fuzzification coefficient exhibits a direct impact on the geometry of fuzzy sets generated by the algorithm. Typically, the value of m is assumed to be equal to 2.0. Lower values of m (that are closer to 1) yield membership functions that start resembling characteristic functions of sets; most of the membership values become localized around 1 or 0. The increase of the fuzzification coefficient ($m = 3, 4$, etc.) produces "spiky" membership functions with the membership grades equal to 1 at the prototypes and a fast decline of the values when moving away from the prototypes. Several illustrative examples of the membership functions are included in Figure 3.12. Here the prototypes are equal to 1, 3.5, and 5 while the fuzzification coefficient assumes values of 1.2 (a), 2.0 (b), and 3.5 (c). In addition to the varying shape of the membership functions, observe that the requirement put on the sum of membership grades imposed on the fuzzy sets yields some rippling effect: the membership functions are not unimodal but may exhibit some ripples whose intensity depends on the distribution of the prototypes and the values of the fuzzification coefficient. The intensity of the rippling effect is also affected by the values of m and increases with the higher values of m.

Table 3.2 The Main Features of the FCM Clustering Algorithm.

Feature of the FCM algorithm	Representation and optimization aspects
Number of clusters (c)	structure in the data set and the number of fuzzy sets estimated by the method; the increase in the number of clusters produces lower values of the objective function, however, given the semantics of fuzzy sets one should maintain this number quite low (5–9 information granules)
Objective function Q	Develops the structure aimed at the minimization of Q; iterative process supports the determination of the local minimum of Q
Distance function $\|\|.\|\|$	Reflects (or imposes) a geometry of the clusters one is looking for; essential design parameter affecting the shape of the membership functions
Fuzzification coefficient (m)	Implies a certain shape of membership functions present in the partition matrix; essential design parameter. Low values of "m" (being close to 1.0) induce characteristic function. The values higher than 2.0 yield spiky membership functions
Termination criterion	Distance between partition matrices in two successive iterations; the algorithm terminated once the distance goes below some assumed positive threshold (ε); that is $\|\| U(iter+1) - U(iter)\|\| < \varepsilon$

The membership functions offer an interesting feature of evaluating an extent to which a certain data point is shared between different clusters and in this sense become difficult to allocate to a single cluster (fuzzy set). Let us introduce the following index, which can serve as a certain separation measure

$$\varphi(u_1, u_2, \ldots, u_c) = 1 - c^c \prod_{i=1}^{c} u_i \tag{3.38}$$

where $u_1, u_2, \ldots, u_c$ are the membership degrees for some data point. If only one of membership degrees, say $u_i = 1$, and the remaining are equal to zero, then the separation index attains its maximum equal to 1. On the other extreme, when the data point is shared by all clusters to the same degree being equal to $1/c$, then the value of this index is reduced to zero. This means that there is no separation between the clusters as reported for this specific point.

It is worth emphasizing that the FCM algorithm is a highly representative method of membership estimation that profoundly dwells on the use of experimental data. In contrast to some other techniques presented so far that are also data-driven, FCM can easily cope with multivariable experimental data.

In case m tends to 1, the partition matrix comes with the entries equal to 0 or 1 so the results come in the form of the Boolean partition matrix. The algorithm becomes then the well-known K-means clustering.

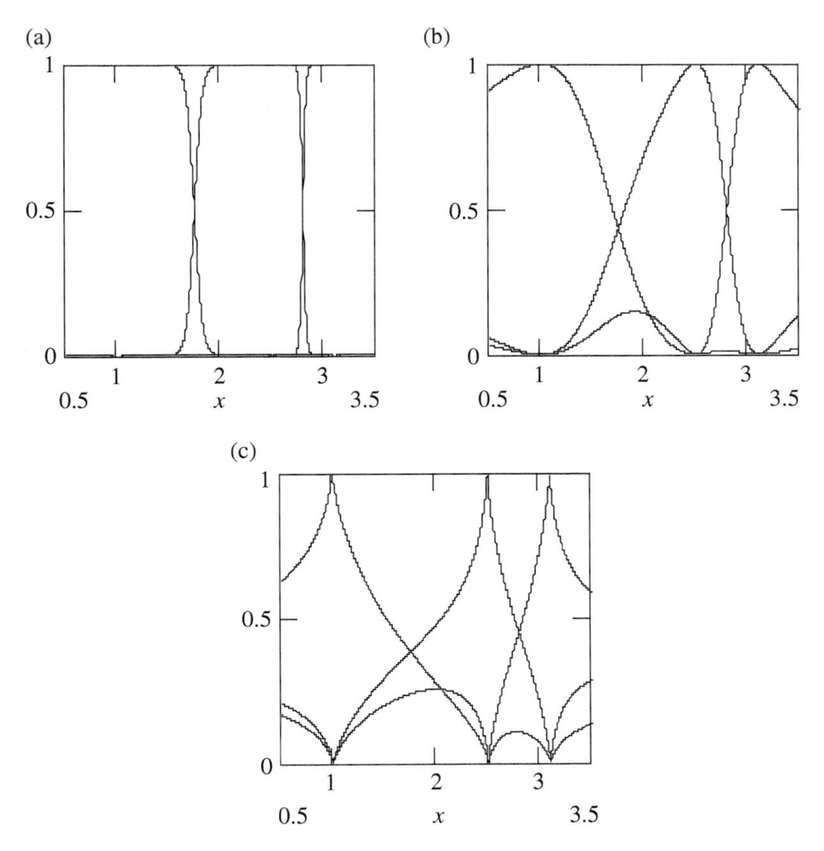

Figure 3.12 Examples of membership functions of fuzzy sets; see the detailed description in the text. The fuzzification coefficient assumes values of 1.2 (a), 2.0 (b), and 3.5 (c).

From the perspective of information granules built through clustering and fuzzy clustering, let us note that the partition matrix U can be sought as a collection of membership functions of the information granules occupying successive rows of the matrix. In other words, fuzzy sets A_1, A_2, …, A_c with their membership values assumed for the data $\mathbf{x}_1$, $\mathbf{x}_2$, …, $\mathbf{x}_N$ form the corresponding rows of U that is

$$U = \begin{bmatrix} A_1 \\ A_2 \\ \dots \\ A_c \end{bmatrix} = \begin{bmatrix} \mathbf{u}_1 \\ \mathbf{u}_2 \\ \dots \\ \mathbf{u}_c \end{bmatrix}$$

where $A_i(\mathbf{x}_k) = u_{ik}$ and $\mathbf{u}_i$ stands for a vector of membership degrees of data localized in the i-th cluster.

The constructed partition matrix offers also another insight into the revealed structure at the level of pairs of data yielding a so-called proximity matrix and a linkage matrix of clusters.

Proximity matrix The proximity matrix $P = [p_{kl}]$, $k, l = 1,2, \ldots, N$ implied by the partition matrix comes with the entries defined in the form

$$p_{kj} = \sum_{i=1}^{c} \min(p_{ik}, p_{ij}) \tag{3.39}$$

p_{kj} characterizes a level of closeness between pairs of individual data. In light of formula Eq. (3.39), proximity of some pairs of data (k,j) is equal to one if and only if these two points are identical. If the membership grades of such points are closer to each other, the corresponding entry of the proximity matrix attains higher values. Note that p_{kj} can be sought as an approximation of a value of some kernel function K at the pairs of the corresponding pair of points, $p_{kj} \approx K(\mathbf{x}_k, \mathbf{x}_j)$.

Linkage matrix The $c{*}c$ dimensional linkage matrix $L = [l_{ij}]$ offers a global characterization of association between two clusters i and j determined (averaged) over all data,

$$l_{ij} = 1 - \frac{1}{N} \sum_{k=1}^{N} | u_{ik} \equiv u_{jk} | \tag{3.40}$$

where expression Eq. (3.40) uses the Hamming distance between the corresponding rows of the partition matrix. Having this in mind, the formula reads as

$$l_{ij} = 1 - \frac{1}{N} \|\mathbf{u}_i - \mathbf{u}_j\| \tag{3.41}$$

If the two corresponding rows (i and j) of the partition matrix are getting close to each other, then the clusters (information granules) become strongly linked (associated).

3.4 Quality of Clustering Results

In evaluating the quality of the clustering results, there are two main categories of quality indicators that might be classified as internal and external ones. The internal ones, which include a large number of alternatives, are referred to as cluster validity indexes. In one way or another, they tend to reflect a general nature of constructed clusters involving their compactness and separation abilities. The external measures of quality are used to assess the obtained clusters vis-à-vis the abilities of the clusters to cope with the inherent features of the data. The resulting clusters are evaluated with respect to the abilities to represent the data themselves or the potential of the clusters in the construction of predictors or classifiers.

3.4.1 Cluster Validity Indexes

There is a significant collection of indexes used to quantify the performance of the obtained clusters (so-called cluster validity indexes) (Davies and Bouldin, 1979; Milligan and Cooper 1985; Rousseeuw 1987; Krzanowski and Lai 1988; Vendramin et al. 2010; Arbelaitz et al. 2013). The main criteria that are used by the validity indexes concern the quality of clusters expressed in terms of compactness and separation of the resulting information granules (Davies and Bouldin 1979). The compactness can be quantified by the sum of distances between the data belonging to the same cluster whereas the separation is expressed by determining a sum of distances between the prototypes of the clusters. The lower the sum of the distances in the compactness measure and the higher the differences between the prototypes, the better the solution. For instance, the Xie-Beni (XB) index is reflective of this characterization of these properties by taking into account the following ratio,

$$V = \frac{\sum_{i=1}^{c}\sum_{k=1}^{N} u_{ik}^{m}\|\mathbf{x}_k - \mathbf{v}_i\|^2}{N\min_{\substack{i,j\\ i\neq j}} \|\mathbf{v}_i - \mathbf{v}_j\|} \tag{3.42}$$

Note that the nominator is a descriptor of the compactness while the denominator captures the separation aspect. The lower the value of V, the better the quality of the clusters. The index that follows is developed along the same line of thought with an exception that in the nominator one has the distance of the prototypes from the total mean $\boldsymbol{v}^{\sim}$ of the entire data set

$$V = \frac{\sum_{i=1}^{c}\sum_{k=1}^{N} u_{ik}^{m}\|\mathbf{x}_k - \mathbf{v}_i\|^2 + \frac{1}{c}\sum_{i=1}^{c}\|\mathbf{v}_i - \mathbf{v}^{\sim}\|^2}{N\min_{\substack{i,j\\ i\neq j}} \|\mathbf{v}_i - \mathbf{v}_j\|} \tag{3.43}$$

Some other alternative is expressed in the following form

$$V = \sum_{i=1}^{c}\sum_{k=1}^{N} u_{ik}^{m}\|\mathbf{x}_k - \mathbf{v}_i\|^2 + \frac{1}{c}\sum_{i=1}^{c}\|\mathbf{v}_i - \mathbf{v}^{\sim}\|^2 \tag{3.44}$$

As noted, there are a number of other alternatives; however, they quantify the criteria of compactness and separation in different ways.

3.4.2 Classification Error

The classification error is one of the obvious alternatives with this regard: we anticipate that clusters are homogeneous with respect to classes of patterns. In an ideal situation, a cluster should be composed of patterns belonging only to a single class. The more heterogeneous the cluster is, the lower its quality in terms of the classification error criterion. The larger the number of clusters is, the higher their likelihood becomes homogeneous. Evidently, this implies that the number of clusters should not be lower than the number of classes. If the topology of the data forming individual classes is more complex, we envision that the number of required classes to keep them homogeneous has to be higher in comparison with situations where the geometry of classes is quite simple (typically spherical geometry of highly disjoint classes, which is easily captured through the Euclidean distance function).

3.4.3 Reconstruction Error

Clusters tend to reveal an inherent structure in the data. To link this aspect of the clusters with the quality of the clustering result, we determine reconstruction abilities of the clusters. Given a structure in the data described by a collection of prototypes $\mathbf{v}_1, \mathbf{v}_2, \ldots, \mathbf{v}_c$, we first represent any data $\mathbf{x}$ with their help. For the K-Means, this representation returns a Boolean vector $[0\ 0 \ldots 1\ 0 \ldots 0]$ where a nonzero entry of the vector indicates the prototype that is the closest to $\mathbf{x}$. For the FCM, the representation of $\mathbf{x}$ comes as a vector of membership grades $\mathbf{u}(\mathbf{x}) = [u_1\ u_2 \ldots u_c]$ computed in a "usual" way. This step is referred to as a granulation, namely a process of representing $\mathbf{x}$ in terms of information granules (clusters). In the sequel, a degranulation phase returns a reconstructed datum $\mathbf{x}$ expressed as

$$\hat{\mathbf{x}} = \frac{\sum_{i=1}^{c} u_i^m(\mathbf{x})\mathbf{v}_i}{\sum_{i=1}^{c} u_i^m(\mathbf{x})} \tag{3.45}$$

The formula Eq. (3.45) results directly from the minimization of the following performance index with the weighted Euclidean distance used originally in the FCM algorithm, and being minimized with respect to the reconstructed data while the reconstruction error is expressed as the following sum

$$V = \sum_{k=1}^{N} ||\mathbf{x}_k - \hat{\mathbf{x}}_k||^2 \tag{3.46}$$

(with $||.||^2$ standing for the same weighted Euclidean distance as being used in the clustering algorithm).

3.5 From Numeric Data to Granular Data

Information granules are constructed on a basis of numeric representatives (prototypes) produced through clustering or fuzzy clustering (Gacek and Pedrycz 2015). They can be also selected randomly. Denote them by $\mathbf{v}_1, \mathbf{v}_2, \ldots, \mathbf{v}_c$.

Depending on the nature of the available data, two general design scenarios are considered.

3.5.1 Unlabeled Data

The patterns are located in the n-dimensional space of real numbers. Some preliminary processing is completed by selecting a subset of data (say, through clustering or some random mechanism) and building information granules around them. Such information granules could be intervals or set-based constructs or probabilities (estimates of probability density functions). The principle of justifiable granularity arises here as a viable alternative to be considered in this setting.

3.5.2 Labeled Data

In this scenario, as before we involve the principle of justifiable granularity and in the construction of the granule, a mechanism of supervision is invoked. The obtained granule is endowed with its content described by the number of patterns belonging to different classes.

Proceeding with the formation of information granules carried out in unsupervised mode, using the principle of justifiable granularity, we construct information granules $\mathbf{V}_1, \mathbf{V}_2, \ldots, \mathbf{V}_M$ that are further regarded as granular data. As usual, the two characteristics of granules are considered here.

We determine the coverage

$$\text{cov}(V_i) = \{\mathbf{x}_k \mid \|\mathbf{x}_k - \mathbf{v}_i\| \leq \rho_i\} \tag{3.47}$$

where $\rho_i \in [0, 1]$ is the size (diameter) of the information granule. The specificity of the granule is expressed in the form

$$\text{Sp}(V_i) = 1 - \rho_i \tag{3.48}$$

Proceeding with the details, the coverage involves the distance $||.||$, which can be specified in various ways. Here, we recall the two commonly encountered examples such as the Euclidean and Tchebyschev distances. This leads to the detailed formulas of the coverage

$$\text{cov}(V_i) = \left\{ \mathbf{x}_k \left| \frac{1}{n}\sum_{j=1}^{n} \frac{\left(x_{kj} - v_{ij}\right)^2}{\sigma_j^2} \leq n\rho_i^2 \right. \right\} \tag{3.49}$$

$$\text{cov}(V_i) = \{\mathbf{x}_k \mid \max_{j=1,2,\ldots,n} |x_{kj} - v_{ij}| \leq \rho_i\} \tag{3.50}$$

The size of the information granule ρ_i is optimized by maximizing the product of the coverage and specificity producing an optimal size of the granule

$$\rho_i = \arg\max_{\rho\in[0,1]}[\text{cov}(\mathbf{V}_i)*\text{sp}(\mathbf{V}_i)] \tag{3.51}$$

The impact of the specificity criterion can be weighted by bringing a nonzero weight coefficient β

$$\rho_i = \arg\max_{\rho\in[0,1]}\left[\text{cov}(\mathbf{V}_i)*\text{sp}^{\beta}(\mathbf{V}_i)\right] \tag{3.52}$$

The higher the value of β, the more visible impact of the specificity on the constructed information granule.

As a result, we form M information granules characterized by the prototypes and the corresponding sizes, namely $\mathbf{V}_1 = (\mathbf{v}_1, \rho_1)$, $\mathbf{V}_2 = (\mathbf{v}_2, \rho_2)$, ..., $\mathbf{V}_M = (\mathbf{v}_M, \rho_M)$.

In the supervised mode of the development of information granules, one has to take into consideration class information. It is very likely that patterns falling within the realm of information granules may belong to different classes. The class content of information granule can be regarded as a probability vector $\mathbf{p}$ with the entries $[p_1\, p_2 .. p_c] = \left[\frac{n_1}{n_1+n_2+\ldots+n_c}\ \frac{n_2}{n_1+n_2+\ldots+n_c}\ \cdots\ \frac{n_c}{n_1+n_2+\ldots+n_c}\right]$. Here $n_1, n_2, \ldots, n_c$ are the number of patterns embraced by the information granule and belonging to the corresponding class of patterns. Alternatively, one can consider a class membership vector being a normalized version of $\mathbf{p}$ in which all coordinates are divided by the highest entry of this vector yielding $[p_1/\max(p_1, p_2, \ldots, p_c)\ p_2/\max(p_1, p_2, \ldots, p_c) \ldots.\ p_c/\max(p_1, p_2, \ldots, p_c)]$.

The class content plays an integral component in the description of information granule. With this regard, an overall scalar characterization of the heterogeneity of information granule comes in the form of entropy function (Duda et al. 2001).

$$H(p) = -\frac{1}{c}\sum_{i=1}^{c} p_i \log_2 p_i \tag{3.53}$$

When constructing an information granule, the optimization criterion is augmented by the entropy component, which is incorporated in the already existing product of coverage and specificity, namely

$$\rho_i = \arg\max_{\rho\in[0,1]}[\text{cov}(\mathbf{V}_i)*\text{sp}(\mathbf{V}_i)*(1-H(\mathbf{p}))] \tag{3.54}$$

One can rewrite the coverage expression as follows

$$\text{cov}'(\mathbf{V}_i) = \text{cov}(\mathbf{V}_i)*(1-H(\mathbf{p})) \tag{3.55}$$

which underlines a fact that the coverage cov' is calibrated (more precisely, discounted) by the entropy of the data embraced by the information granule. Note that if entropy becomes higher, it impacts the resulting coverage cov' more visibly, which becomes lower than the original coverage *cov*. Only if the entropy is equal to zero (so we have data belonging only to a single class), is no reduction in the coverage values reported.

As a result of the use of the principle of justifiable granularity modified as shown previously, we obtain M granular data in the following triple ($\mathbf{v}_i$, ρ_i, $\mathbf{p}_i$), $i = 1,2, \ldots, M$ described in terms of the numeric prototypes, the size ρ_i and the class content $\mathbf{p}_i$. The Boolean manifestation of the same information granule is provided as ($\mathbf{v}_i$, ρ_i, $\mathbf{I}_i$) where $\mathbf{I}_i$ is a Boolean vector $\mathbf{I}_i = [0\ 0 \ldots 0\ 1\ 0 \ldots .0]$ with a single nonzero entry whose coordinate (class index) corresponds to an index of the class for which the entry of $\mathbf{p}_i$ is the highest.

3.5.3 Fuzzy Equalization as a Way of Building Fuzzy Sets Supported by Experimental Evidence

The underlying principle of this approach is based on an observation that while fuzzy sets are reflective of the perception of systems or phenomena, quite often there is some experimental evidence in the form of data whose nature could be captured in a more synthetic manner through the underlying probability function or probability density function $p(x)$. The essence of fuzzy equalization comes with the conjecture that given a collection of fuzzy sets $\{A_1, A_2, \ldots, A_c\}$ that are used to granulate some variable (say, inflation, profit, etc.) can be formed in such a way that each fuzzy set in this family comes with the same level of experimental evidence. To put it more formally, we consider that the integral (or sum) of the form

$$\int_X A_i(x)p(x)dx \tag{3.56}$$

assumes the same value for all fuzzy sets A_i, $i = 1, 2, \ldots, c$.

In other words, we require that the expected value expressed by Eq. (3.30) and computed for each fuzzy set is approximately the same, $\int_X A_i(x)p(x)dx = g$ where "g" is some constant. The essence of this construct is illustrated in Figure 3.13. This way of developing fuzzy sets is in agreement with our intuition: the less experimental evidence we have, the broader (less specific) the corresponding fuzzy set should be.

The underlying optimization task is concerned with the determination of the parameters of the fuzzy sets (assuming that their form has been already specified) so that Eq. (3.26) becomes satisfied. For triangular fuzzy sets with half overlap between neighboring fuzzy sets, this optimization requires an adjustment of the vector of the modal values of the fuzzy sets.

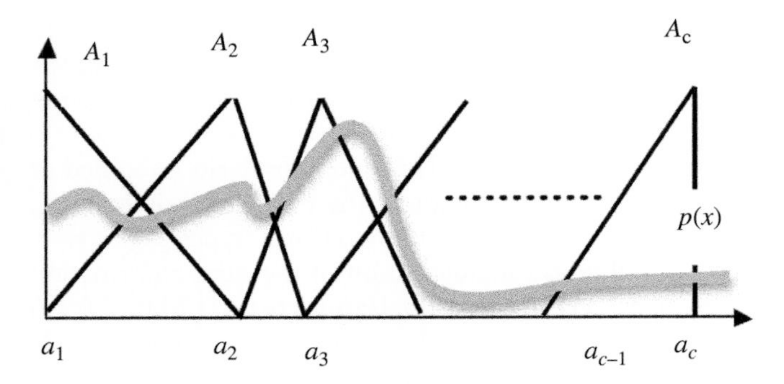

Figure 3.13 A collection of fuzzy sets complying with the equalization rule; note increased specificity of fuzzy sets in the regions of high density ($p(x)$) of experimental data.

One can note that while fuzzy sets and probability are two orthogonal concepts, there are a number of methods of membership function estimation that invoke some probabilistic information (Dishkant 1981; Civanlar and Trussell 1986; Hong and Lee 1996; Masson and Denoeux 2006).

3.5.4 Several Design Guidelines for the Formation of Fuzzy Sets

The considerations presented previously give rise to a number of general guidelines supporting the development of fuzzy sets:

a) highly visible and well-defined semantics of information granules. No matter what the determination technique is, one has to become cognizant of the semantics of the resulting fuzzy sets. Fuzzy sets are interpretable information granules of a well-defined meaning and this aspect needs to be fully captured. Given this, the number of information granules has to be kept quite small with their number being restricted to 7 ± 2 fuzzy sets.
b) There are several fundamental views at fuzzy sets and depending upon them, we could consider the use of various estimation techniques (e.g. by accepting the horizontal or the vertical view at fuzzy sets and adopting a pertinent technique).
c) Fuzzy sets are context-sensitive constructs and as such require careful calibration. This feature of fuzzy sets should be treated as their genuine advantage. The semantics of fuzzy sets can be adjusted through shifting fuzzy sets or/and adjusting their membership functions. The nonlinear transformation we introduced here helps complete an effective adjustment of the membership functions making use of some "standard" membership functions. The calibration mechanisms being used in the design of the membership function are reflective of human-centricity of fuzzy sets.

d) We have delineated between the two major categories of approaches supporting the design of membership functions, that is data-driven and expert (user)-based. There are very different in the sense of the origin of the supporting evidence. Fuzzy clustering is a fundamental mechanism of the development of fuzzy sets. It is important in the sense the method is equally suitable for one-dimensional and multivariable cases. The expert or simply user-based methods of membership estimation are important in the sense they offer some systematic and coherent mechanisms of elicitation of membership grades. With the regard to the consistency of the elicited membership grades, the pairwise estimation technique is of particular interest by providing well-quantified mechanisms of the assessment of the consistency of the produced membership grades. The estimation procedures underline some need of further development of higher type of constructs such as fuzzy sets of type-2 or higher and fuzzy sets of higher order that may be ultimately associated with constructs such as type-2 fuzzy sets or interval-valued fuzzy sets (this particular construct is visible when dealing with the horizontal method of membership estimation that comes with the associated confidence intervals).
e) The user-driven membership estimation uses the statistics of data yet in an *implicit* manner. The granular term, fuzzy sets, comes into existence once there is some experimental evidence behind them (otherwise there is no point forming such fuzzy sets).
f) the development of fuzzy sets can be carried out in a stepwise manner. For instance, a certain fuzzy set can be further refined, if required in the problem at hand. This could lead to several more specific fuzzy sets that are associated with the fuzzy set formed at the higher level. Being aware of the complexity of the granular descriptors, we should resist temptation of forming an excessive number of fuzzy sets at a single level as such fuzzy sets could be easily lacking any sound interpretation.

3.6 Aggregation Operations

Several fuzzy sets can be combined (aggregated), thus leading to a single fuzzy set forming the result of such an aggregation operation. For instance, when we compute intersection and union of fuzzy sets, the result is a fuzzy set whose membership function captures information conveyed by the original fuzzy sets. This fact suggests a general view of aggregation of fuzzy sets as a certain transformations performed on their membership functions. In general, we encounter a wealth of aggregation operations (Dubois and Prade 1980, 1985).

Formally, an aggregation operation is a n-ary function g: $[0, 1]^n \rightarrow [0, 1]$ satisfying the following requirements:

$$\text{Monotonicity} \quad g(x_1, x_2, \ldots, x_n) > g(y_1, y_2, \ldots, y_n) \text{ if } x_i > y_j \tag{3.57}$$

$$\text{Boundary conditions } g(0, 0, \ldots., 0) = 0 \text{ and } g(1, 1, \ldots.., 1) = 1 \tag{3.58}$$

An element $e \in [0, 1]$ is called a neutral element of the aggregation operation "g" and an element $l \in [0, 1]$ is called an annihilator (absorbing element) of the aggregation operation "g" if for each i = 1, 2, ..., n, $n \geq 2$ and for all $x_1, x_2, \ldots, x_{i-1}, x_{i+1}, \ldots., x_n \in [0, 1]$ we have

$$g(x_1, x_2, \ldots, x_{i-1}, e, x_{i+1} \ldots., x_n) = g(x_1, x_2, \ldots, x_{i-1}, x_{i+1} \ldots., x_n) \tag{3.59}$$

$$g(x_1, x_2, \ldots, x_{i-1}, l, x_{i+1} \ldots., x_n) = l \tag{3.60}$$

Since triangular norms and conorms are monotonic, associative, and satisfy the boundary conditions, they provide wide a class of associative aggregation operations whose neutral elements are equal to 1 and 0, respectively. We are, however, not restricted to those as the only available alternatives.

3.6.1 Averaging Operations

In addition to monotonicity and the satisfaction of the boundary conditions, averaging operations are idempotent and commutative. They can be described in terms of the generalized mean (Dyckhoff and Pedrycz 1984).

$$g(x_1, x_2, \ldots., x_n) = \sqrt[p]{\frac{1}{n}\sum_{i=1}^{n}(x_i)^p}, \quad p \in \mathbf{R}, p \neq 0 \tag{3.61}$$

Interestingly, generalized mean subsumes some well-known cases of well-known averages including

$p = 1$	$g(x_1, x_2, \ldots., x_n) = \frac{1}{n}\sum_{i=1}^{n} x_i$	arithmetic mean
$p \rightarrow 0$	$g(x_1, x_2, \ldots., x_n) = \sqrt[n]{\prod_{i=1}^{n} x_i}$	geometric mean
$p = -1$	$g(x_1, x_2, \ldots., x_n) = \frac{n}{\sum_{i=1}^{n} 1/x_i}$	harmonic mean
$p \rightarrow -\infty$	$g(x_1, x_2, \ldots., x_n) = \min(x_1, x_2, \ldots, x_n)$	minimum
$p \rightarrow \infty$	$g(x_1, x_2, \ldots., x_n) = \max(x_1, x_2, \ldots, x_n)$	maximum

The following inequalities hold:

$$\min(x_1, x_2, \ldots., x_n) \leq g(x_1, x_2, \ldots, x_n) \leq \max(x_1, x_2, \ldots., x_n) \tag{3.62}$$

Therefore, generalized means ranges over the values not being covered by triangular norms and conorms.

3.7 Transformations of Fuzzy Sets

Transformations of elements (points) through functions are omnipresent. An immediate generalization of such point transformations involves set transformations between spaces. Mappings of fuzzy sets between universes constitute another generalization of mapping sets between spaces. Thus, point transformations can be expanded to cover transformations involving fuzzy sets. Transformations of this nature can be realized using either functions or relations. In both cases, these transformations constitute an essential component of various pursuits including system modeling and control applications, pattern recognition, and information retrieval, to name a few representative areas. This chapter introduces two important mechanisms to transform fuzzy sets; namely, the extension principle and the calculus of fuzzy relations. We elaborate on their essential properties, present algorithmic aspects, and discuss various interpretations of the resulting constructs.

3.7.1 The Extension Principle

The extension principle is a fundamental construct that enables extensions of point operations to operations involving sets and fuzzy sets. Intuitively, the idea is as follows: given a function (mapping) from some domain **X** to codomain (range) **Y**, the extension principle offers a mechanism to transform a fuzzy set defined in **X** to some fuzzy set defined in **Y**.

Let f: $\mathbf{X} \rightarrow \mathbf{Y}$ be a function. Given any $x \in \mathbf{X}$, $y = f(x)$ denotes the image of x under f, namely, the point transformation of x under f, refer to Figure 3.14. This is the straightforward idea that the customary notion of any function conveys. The pointwise transformations can be naturally extended to handle transformations of sets.

Let $P(\mathbf{X})$ and $P(\mathbf{Y})$ be the power sets of **X** and **Y** and $A \in P(\mathbf{X})$ a set. The image of A under f can be determined by realizing point transformations $y = f(x)$ for all $x \in A$. In this sense, the image of A under f is some set B that arises in the following form:

$$B = f(A) = \{y \in \mathbf{Y} \mid y = f(x), \forall x \in A\} \tag{3.63}$$

Since A and B are sets, they can be expressed in terms of their characteristic functions as follows

$$B(y) = f(A)(y) = \sup_{x \in A}[A(x)] \tag{3.64}$$

as displayed in Figure 3.15. Notice that this mechanism provides a way to extend the notion of functions regarded as point transformations to the notion of set functions. Once viewed in terms of characteristic functions, it becomes natural to extend this notion to fuzzy sets as follows.

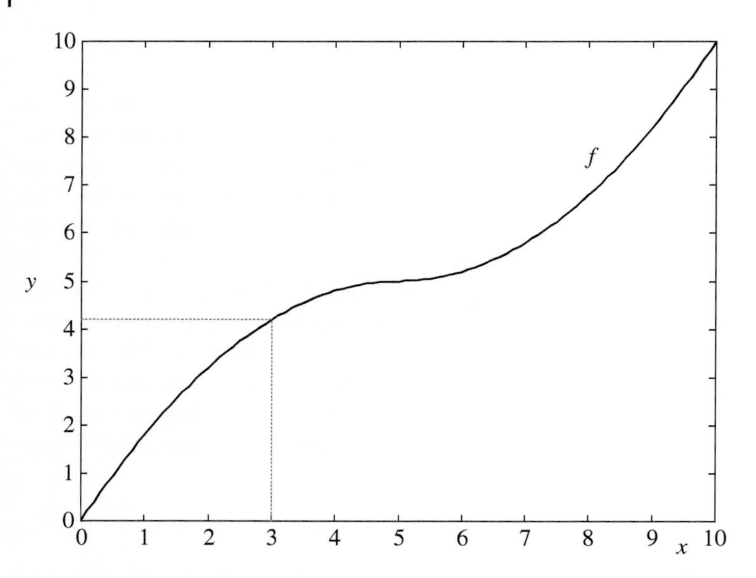

Figure 3.14 An example of function "f" along with its point transformation.

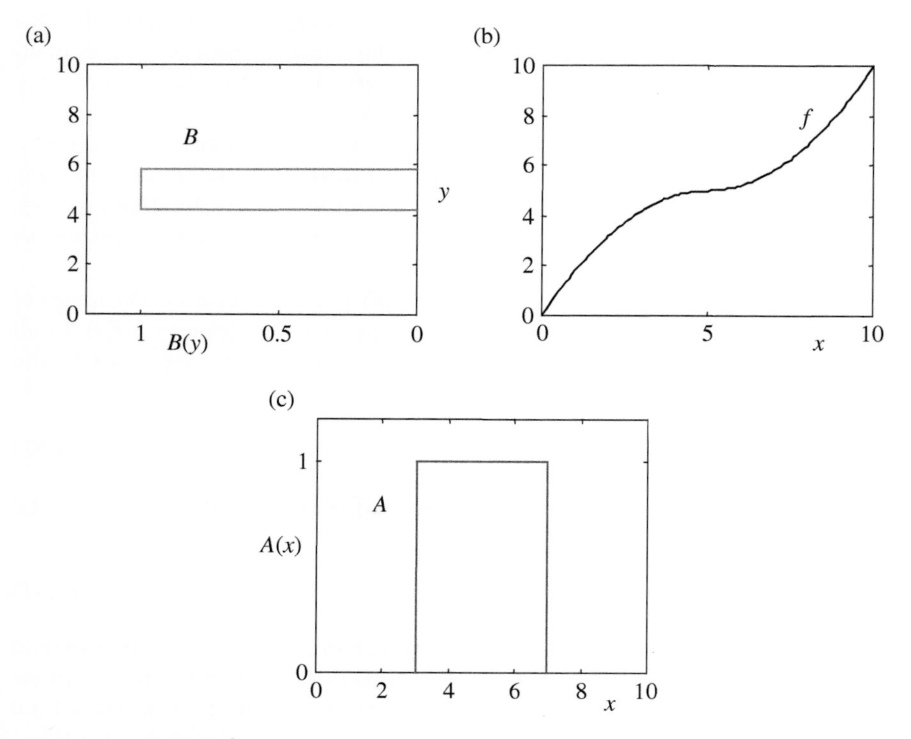

Figure 3.15 Set transformation.

Let $F(\mathbf{X})$ and $F(\mathbf{Y})$ denote the families of all fuzzy sets defined in $\mathbf{X}$ and $\mathbf{Y}$, respectively and $f: \mathbf{X} \rightarrow \mathbf{Y}$ be a function. Function "f" induces a mapping $f: F(\mathbf{X}) \rightarrow \mathrm{F}(\mathbf{Y})$ such that if A is a fuzzy set in $\mathbf{X}$, then its image under F is a fuzzy set $B = f(A)$ whose membership function is expressed as (Klir and Yuan 1995; Pedrycz and Gomide 1998)

$$B(y) = \sup_{x:y = f(x)} A(x) \tag{3.65}$$

For finite universes, consider $\mathbf{X} = \{-3, -2, -1, 0, 1, 2, 3\}$ and $y = f(x) = x^2$. Given the fuzzy set $A = \{0/-3, 0.1/-2, 0.3/-1, 1/0, 0.2/1, 0/2, 0/3\}$ defined in $\mathbf{X}$, the image $B = f(A)$ is a fuzzy set in $\mathbf{Y} = \{y \mid y = x^2\} = \{0, 1, 4, 9\}$ whose membership function is, see Figure 3.16, $B = \{1/0, \max(0.2, 0.3)/1, \max(0, 0.1)/4, 0/9\} = \{1/0, 0.3/1, 0.1/4, 0/9\}$.

The extension principle generalizes to functions of many variables as follows. Let $\mathbf{X}_i$ $i = 1, \ldots, n$ and $\mathbf{Y}$ be universes and $\mathbf{X} = \mathbf{X}_1 \times \mathbf{X}_2 \times \ldots \times \mathbf{X}_n$. Consider fuzzy sets A_i on $\mathbf{X}_i$, $i = 1, \ldots n$, and a function $y = f(\mathbf{x})$ with $\mathbf{x} = (x_1, x_2, \ldots, x_n)^{\mathrm{T}}$ a point of $\mathbf{X}$.

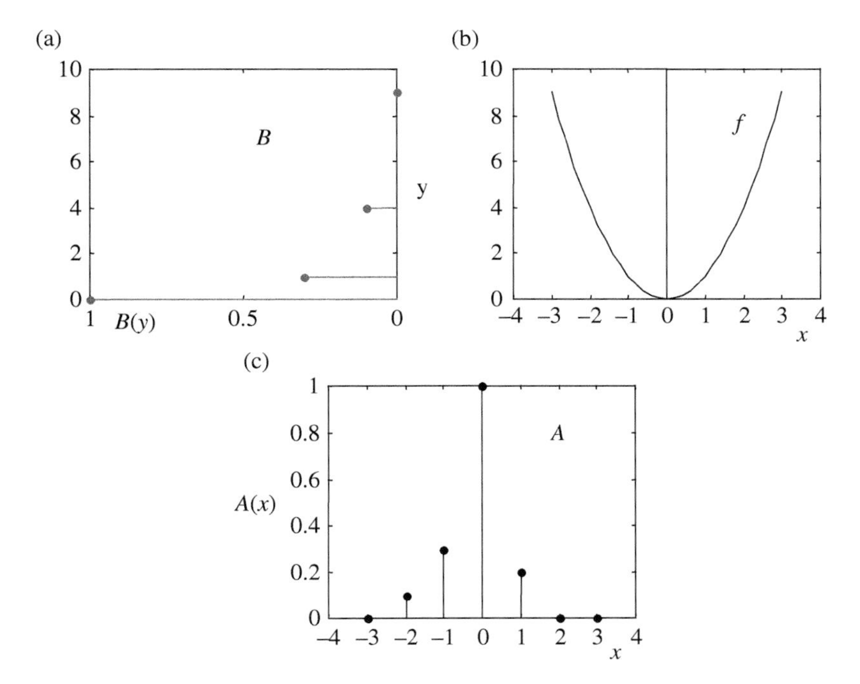

Figure 3.16 Extension principle applied in the case of a certain many-to-one mapping and finite universes.

Fuzzy sets $A_1, A_2, \ldots, A_n$ can be transformed through "f" to give a fuzzy set $B = f(A_1, A_2, \ldots, A_n)$ in $\mathbf{Y}$ with the membership function

$$B(y) = \sup_{\mathbf{x}|y=f(\mathbf{x})} \{\min[A_1(x_1), A_2(x_2), \ldots, A_n(x_n)]\} \tag{3.66}$$

In Eq. (3.40), the min operation is a certain choice coming from the family of triangular norms. Any t-norm can be adopted because each component x_i occurs concurrently in $\mathbf{x}$.

3.7.2 Fuzzy Numbers and Fuzzy Arithmetic

We call a membership function $A : \mathbf{X} \rightarrow [0, 1]$ is upper semicontinuous if the set $\{x \in \mathbf{X} |\, A(x) > \alpha\}$ is closed, that is, the α-cuts are closed intervals and, therefore, convex sets. If the universe $\mathbf{X}$ is the set $\mathbf{R}$ of real numbers and membership function is normal, $A(x) = 1\ \forall x \in [b, c]$, then $A(x)$ is a model of a fuzzy interval, with monotone increasing function f_A: $[a, b) \rightarrow [0, 1]$, monotone decreasing function g_A: $(c, d] \rightarrow [0, 1]$, and null otherwise. Fuzzy intervals $A(x)$ have the following canonical form.

$$A(x) = \begin{cases} f_A(x) & \text{if } x \in [a, b) \\ 1 & \text{if } x \in [b, c] \\ g_A(x) & \text{if } x \in (c, d] \\ 0 & \text{otherwise} \end{cases} \tag{3.67}$$

where $a \le b \le c \le d$, see Figure 3.17a.

When $b = c$, $A(x) = 1$ for exactly one element of $\mathbf{X}$, and the fuzzy quantity is called a fuzzy number, Figure 3.17b.

In general, the functions f_A and g_A are semicontinuous from the right and left, respectively. From the practical point of view, fuzzy intervals and numbers are mappings from the real line $\mathbf{R}$ to the unit interval that satisfy a series of

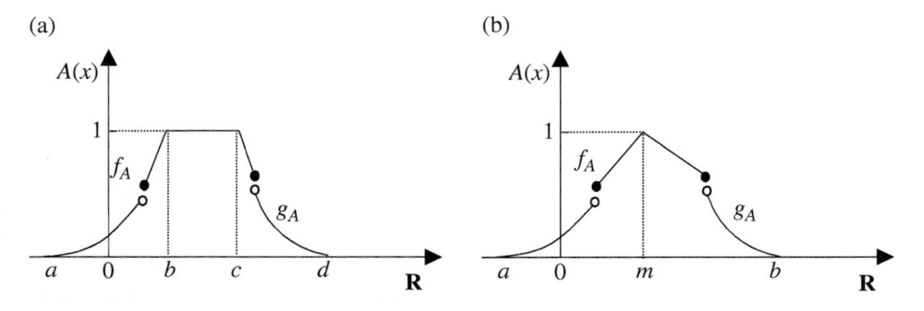

Figure 3.17 Canonical form of a fuzzy interval (a) and fuzzy number (b).

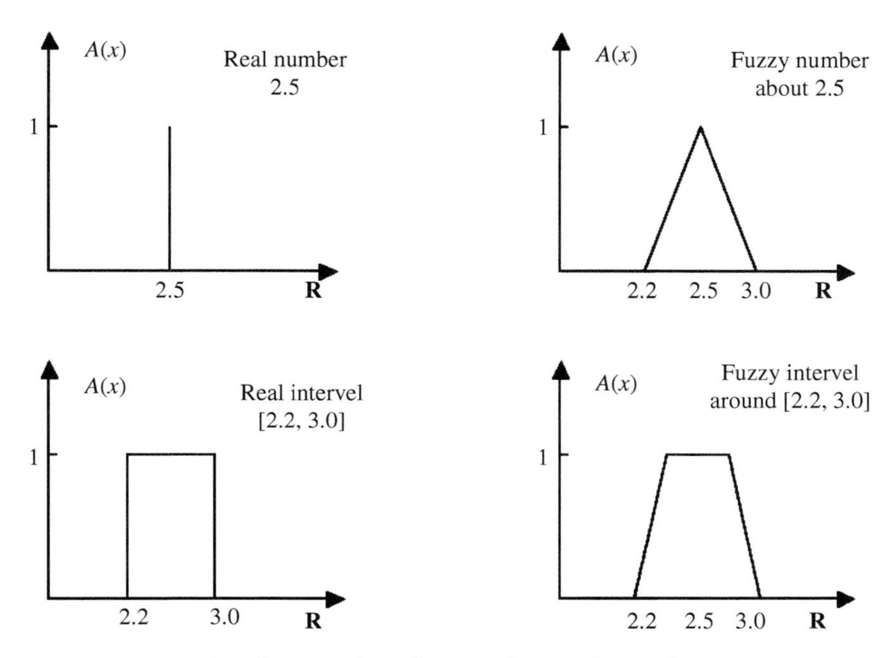

Figure 3.18 Examples of real numbers, fuzzy numbers, and intervals.

properties such as normality, unimodality, continuity, and boundness of support. As Figure 3.18 suggests, fuzzy intervals and numbers model our intuitive notion of approximate intervals and approximate numbers.

Before we move on to the discussion on operations on fuzzy numbers, let us introduce a few examples that motivate their use.

Consider that you traveled for two hours at speed of *about* 110 km/h. What was the distance you traveled? The speed is described in the form of some fuzzy set S whose membership function is given.

The next example is a more general version of this problem.

You traveled at speed of *about* 110 km/h for *about* 3 hours. What was the distance traveled? We assume that both the speed and time of travel are described by fuzzy sets.

In a certain manufacturing process, there are five operations completed in series. Given the nature of the manufacturing activities, the duration of each of them can be characterized by fuzzy sets T_1, T_2, …, and T_5. What is the time of realization of this process?

Basically, there exist two fundamental methods to carry out algebraic operations on fuzzy numbers. The first method is based on interval arithmetic and α-cuts while the second one employs the extension principle. The fundamentals of these two methods are discussed next.

3.7.3 Interval Arithmetic and α-Cuts

The first approach to compute with fuzzy numbers exhibits its roots in the framework of interval analysis, a branch of mathematics developed to deal with the calculus of tolerances. In this framework, the interest is in intervals of real numbers, $[a, b]$, a, $b\in\mathbf{R}$, such as [4, 6], [−1.5, 3.2] and so forth. The formulas developed to perform the basic arithmetic operations, namely, addition, subtraction, multiplication, and division are as follows (assuming that c, $d\neq 0$ for the division operation):

1. addition : $$[a,b] + [c,d] = [a + c, b + d] \tag{3.68}$$
2. subtraction : $$[a,b] - [c,d] = [a - d, b - c] \tag{3.69}$$
3. multiplication : $$[a,b].[c,d] = [\min(ac,ad,bc,bd), \max(ac,ad,bc,bd)] \tag{3.70}$$
4. division : $$[a,b]/[c,d] = \left[\min\left(\frac{a}{c},\frac{a}{d},\frac{b}{c},\frac{b}{d}\right), \max\left(\frac{a}{c},\frac{a}{d},\frac{b}{c},\frac{b}{d}\right)\right] \tag{3.71}$$

Now, let A and B be two fuzzy numbers and let $*$ be any of the four basic arithmetic operations. Thus, for any $\alpha\in(0, 1)$ the fuzzy set $A*B$ is computed via the α-cuts A_α and B_α of A and B, respectively:

$$(A*B)_\alpha = A_\alpha * B_\alpha \tag{3.72}$$

Recall that, by definition, the α-cuts A_α and B_α are closed intervals and therefore the formulas of interval operations can be applied for each value of α. When $*$ is/(division operation), we require that $0 \notin B_\alpha \ \forall\alpha\in(0, 1)$.

After the interval operation is performed for α-cuts, the use of the representation theorem leads us to the well-known relationship

$$A*B = \bigcup_{\alpha\in[0,1]} (A*B)_\alpha \tag{3.73}$$

In terms of the membership functions, we obtain

$$(A*B)(x) = \sup_{\alpha\in[0,1]} [\alpha(A*B)_\alpha(x)] = \sup_{\alpha\in[0,1]} \left[(A*B)^f_\alpha(x)\right] \tag{3.74}$$

where $(A*B)^f_\alpha(x) = \alpha\,(A*B)_\alpha(x)$.

Therefore, the interval arithmetic-α-cut method to perform fuzzy arithmetic is a generalization of interval arithmetic.

Example 3.6 If A and B are two triangular fuzzy number, denoted as $A(x, a, m, b)$ and $B(x, c, n, d)$, then their α-cuts are determined as

$$A_\alpha = [(m-a)\alpha + a, (m-b)\alpha + b]$$
$$B_\alpha = [(n-c)\alpha + c, (n-d)\alpha + d]$$

Now let $A = A(x, 1, 2, 3)$ and $B = B(x, 2, 3, 4)$. Then, the corresponding α-cuts are equal to

$$A_\alpha = [\alpha + 1, -\alpha + 3]$$
$$B_\alpha = [\alpha + 2, -2\alpha + 5]$$

Therefore

$$(A + B)_\alpha = [2\alpha + 3, -3\alpha + 8]$$
$$(A - B)_\alpha = [3\alpha - 4, -2\alpha + 1]$$
$$(\mathrm{AB})_\alpha = [(\alpha + 1)(\alpha + 2), (-\alpha + 3)(-2\alpha + 5)]$$
$$(A/B)_\alpha = [(\alpha + 1)/(-2\alpha + 5), (-\alpha + 3)(\alpha + 2)]$$
$$(A/B)_\alpha = [(-\alpha + 3)(\alpha + 2), [(\alpha + 1)/(-2\alpha + 5)]$$

Figure 3.19 shows the resulting fuzzy numbers $A + B$, $A - B$, AB and B/A, respectively.

The extension of the interval arithmetic and the use of α-cuts and the representation of fuzzy sets each fuzzy number can be regarded as a family of nested α-cuts. Subsequently, these α-cuts are used to reconstruct the resulting fuzzy number. In essence, the use of α-cuts is a sort of a brute-force method of computing with fuzzy numbers. However, α-cuts are becoming important to develop parametric representation of fuzzy numbers to control their shapes and associated approximation error (Stefanini et al. 2006).

3.7.4 Fuzzy Arithmetic and the Extension Principle

The second method of computing with fuzzy numbers dwells on the extension principle to extend standard operations on real numbers to fuzzy numbers (Chen and Chen 2009). Here, the fuzzy set $A{*}B$ expressed on $\mathbf{R}$ is defined using the extension principle

$$(A{*}B)(z) = \sup_{z = x*y} \min[A(x), B(y)],\ \forall z \in \mathbf{R} \tag{3.75}$$

In general, if t is a t-norm and $*: \mathbf{R}^2 \rightarrow \mathbf{R}$ is an operation on the real line, then operations on fuzzy numbers become

$$(A{*}B)(z) = \sup_{z = x*y} [A(x)\, t\, B(y)],\ \forall z \in \mathbf{R} \tag{3.76}$$

Figure 3.20 illustrates the addition $(A + B)$ of triangular fuzzy numbers A and B when using the minimum t-norm t_m and the drastic product t_d t-norm, respectively. Clearly, different choices of t-norms produce different results. In general, if $t_1 \leq t_2$ in the sense that $a\ t_1\ b \leq a\ t_2\ b$, $\forall a, b \in [0, 1]$, then

$$\sup_{z = x*y} [A(x)\, t_d\, B(y)] \leq \sup_{z = x*y} [A(x)\, t\, B(y)] \leq \sup_{z = x*y} [A(x)\, t_m\, B(y)]\ \forall z \in \mathbf{R} \tag{3.77}$$

(a)

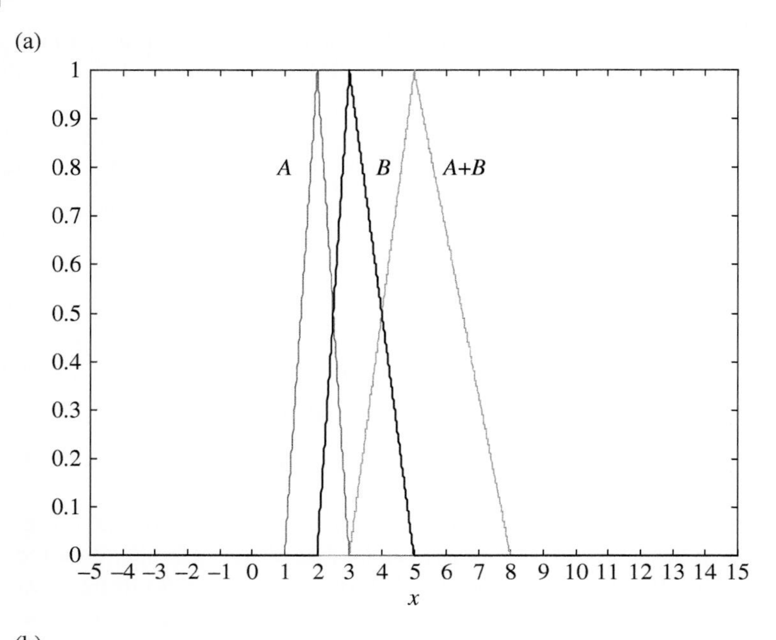

(b)

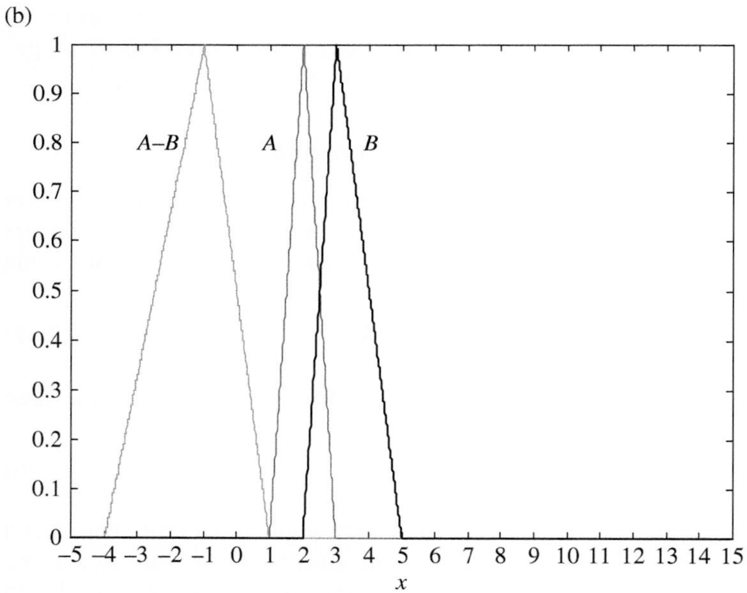

Figure 3.19 Algebraic operations on triangular fuzzy numbers.

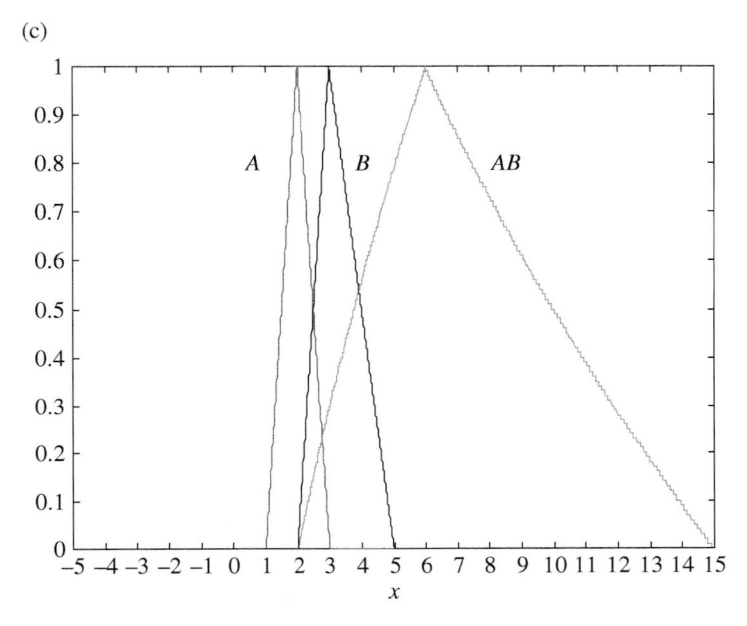

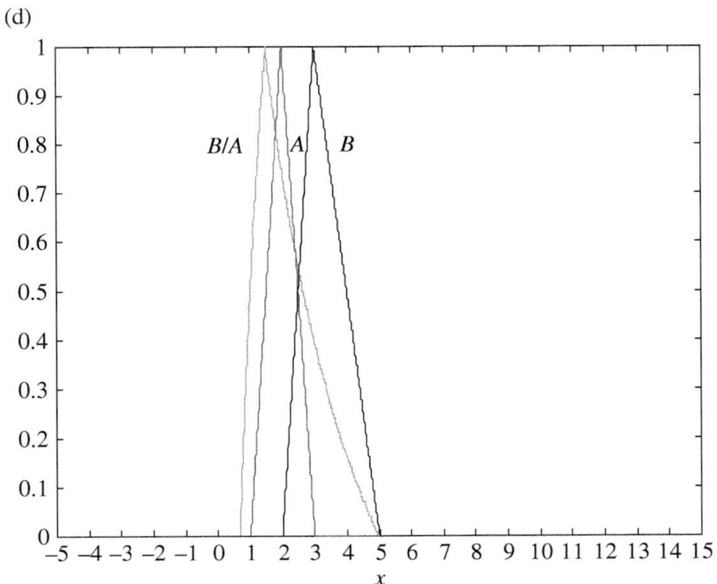

Figure 3.19 (Continued)

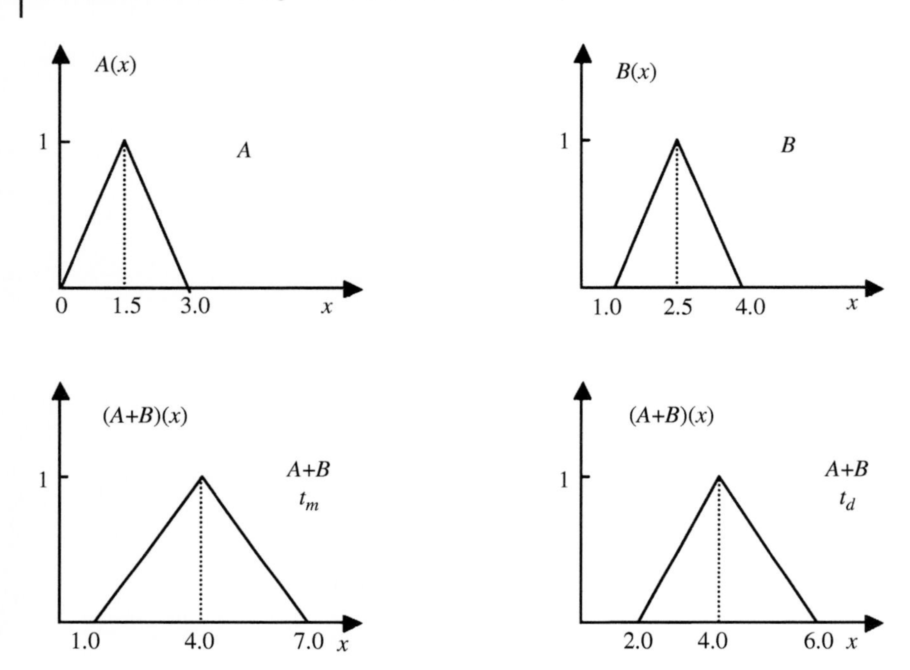

Figure 3.20 Algebraic operations: the use of the extension principle with different triangular norms.

therefore

$$^{t_d}(A*B), \leq {}^{t}(A*B)(z) \leq {}^{t_m}(A*B), \ \ \forall z \in \mathbf{R} \tag{3.78}$$

In the special case of the largest t-norm, which is minimum, t_m, the one we will concentrate on the remainder of this section, property 6 of Section 3.7 suggests a fundamental result as a basis to compute with fuzzy numbers under the framework of the extension principle.

Proposition For any fuzzy numbers A and B and a continuous monotone binary operation $*$ on **R**, the following equality holds for all α-cuts with $\alpha \in [0, 1]$:

$$(A*B)_\alpha = A_\alpha * B_\alpha \tag{3.79}$$

The proof of this proposition is given in Nguyen and Walker (1999). There are important consequences of the proposition:

1) Since A_α and B_α are closed and bounded for all α, $(A*B)_\alpha$ also is closed and bounded;
2) Because A and B are fuzzy numbers, they are normal and therefore $A*B$ is also normal.

These two observations clearly demonstrate that the extension principle produces a transformation that is a fuzzy number and therefore is a sound mechanism to perform algebraic operations with fuzzy numbers.

3) Computation of $A*B$ can be done by combining the increasing and decreasing parts of the membership functions of A and B.

Figure 3.21 offers a graphical visualization of this statement.

These results can be generalized to broader classes and choices of t-norms and operations with fuzzy quantities (Mares 1997; Klement et al. 2000; Stefanini 2009). Moreover, approximation schemes developed in the framework of interpolation of a fuzzy function (Perfilieva 2004). In what follows, we detail the basic operations with triangular fuzzy numbers because they are, by far, the most commonly used in practice. Moreover, the analysis focusing this class of fuzzy numbers reveals the most visible properties of fuzzy arithmetic.

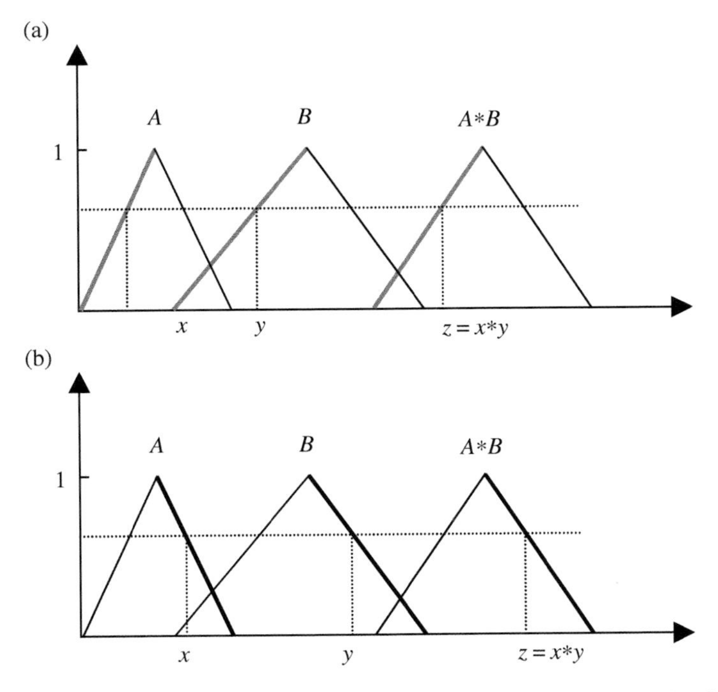

Figure 3.21 Combining increasing and decreasing parts of the membership functions of the fuzzy numbers *A* and *B*.

3.7.5 Computing with Triangular Fuzzy Numbers

Consider two triangular fuzzy numbers $A(x, a, m, b)$ and $B(x, c, n, d)$. More specifically, A and B are described by the following piecewise membership functions:

$$A(x) = \begin{cases} \frac{x-a}{m-a} & \text{if } x \in [a,m) \\ \frac{b-x}{b-m} & \text{if } x \in [m,b] \\ 0 & \text{otherwise} \end{cases} \qquad B(x) = \begin{cases} \frac{x-c}{n-c} & \text{if } x \in [c,n) \\ \frac{d-x}{d-n} & \text{if } x \in [n,d] \\ 0 & \text{otherwise} \end{cases} \tag{3.80}$$

Let us recall that the modal values m and n identify a dominant, typical value, while the lower and upper bounds, a or c and b or d, reflect the spread of the number. To simplify computing, for the time being we consider fuzzy numbers with positive lower bounds a, $c > 0$.

The plots of examples of triangular fuzzy numbers are given in Figure 3.22. These will be helpful in the clarification of detailed formulas.

3.7.6 Addition

The extension principle Eq. (3.49) applied to A and B to compute $C = A + B$ yields

$$C(z) = \sup_{z=x+y} \min[A(x), B(y)], \ \forall z \in R \tag{3.81}$$

The resulting fuzzy number is normal, that is, $C(z) = 1$ for $z = m + n$. Computations of the spreads of C can be done, according to statement 3 in Section 3.7.4, by treating the increasing and decreasing parts of the membership functions of A and B separately.

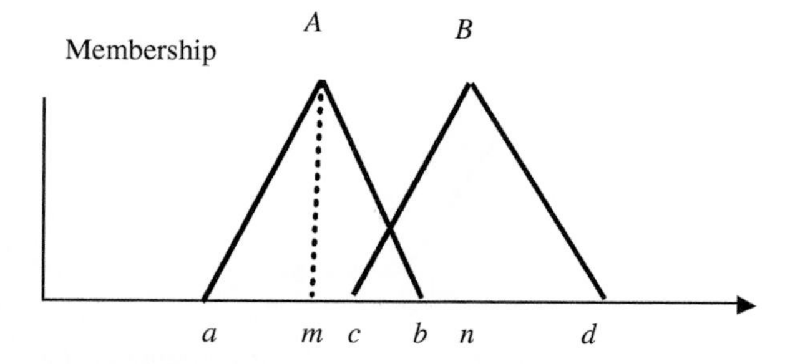

Figure 3.22 Examples of triangular fuzzy numbers A and B.

Consider first that $z < m + n$. In this situation, the calculation involves the increasing parts of the membership function of A and B. Note that there exist values x and y such that $x < m$ and $y < n$ for which we have

$$A(x) = B(y) = a, a \in [0, 1]. \tag{3.82}$$

Based on this relationship we derive

$$\frac{x - a}{m - a} = \alpha \tag{3.83}$$

along with

$$\frac{y - c}{n - c} = \alpha \tag{3.84}$$

for $x \in [a,m]$ and $y \in [c,n]$. Expressing x and y as functions of α we obtain

$$x = (m - a)\alpha + a \tag{3.85}$$

$$y = (n - c)\alpha + c \tag{3.86}$$

which are the same as the lower intervals we get using interval analysis, as it should.

Replacing the values of x and y in $z = x + y$ we have

$$z = x + y = (m - a)\alpha + a + (n - c)\alpha + c \tag{3.87}$$

that is,

$$\alpha = \frac{z - (a + c)}{(m + n) - (a + c)} \tag{3.88}$$

Notice that z has, as expected, the same lower limit value as the corresponding interval associated with the α-cut we use with interval analysis.

Proceeding similarly for the decreasing portions of the membership functions we obtain

$$\frac{b - x}{b - m} = \alpha \tag{3.89}$$

along with

$$\frac{d - y}{d - n} = \alpha \tag{3.90}$$

for $x \in [m,b]$ and $y \in [n,d]$. Again, expressing x and y as functions of α we get

$$x = (m - b)a + b \tag{3.91}$$

$$y = (n - d)a + d \tag{3.92}$$

Furthermore, replacing the values of x and y in $z = x + y$ we have

$$z = x + y = (m - b)a + b + (n - d)a + d \tag{3.93}$$

that is

$$\alpha = \frac{(b + d) - z}{(b + d) - (m + n)} \tag{3.94}$$

As expected, z has the same upper limit value as the corresponding interval associated with the α-cut we use with interval analysis.

Finally, from Eqs. (3.62) and (3.67) we obtain the membership function of $C = A + B$:

$$C(x) = \begin{cases} \dfrac{z - (a + c)}{(m + n) - (a + c)} & \text{if } z < m + n \\ 1 & \text{if } z = m + n \\ \dfrac{(b + d) - z}{(b + d) - (m + n)} & \text{if } z > m + n \end{cases} \tag{3.95}$$

Interestingly, C is also a triangular fuzzy number. To emphasize this fact, we use a concise notation

$$C = C(x, a + c, m + n, b + d). \tag{3.96}$$

Whenever several triangular fuzzy numbers are added, the result also is a triangular fuzzy number. In general, however, shape preserving does not hold for any shape fuzzy number and t-norms in the extension principle.

3.7.7 Multiplication

As with the addition, we look first at the increasing parts of the membership functions from which we get

$$x = (m - a)\alpha + a \tag{3.97}$$

$$y = (n - c)a + c \tag{3.98}$$

The product z of x and y becomes

$$z = xy = [(m - a)\alpha + a][(n - c)\alpha + c] \tag{3.99}$$

$$z = (m - a)(n - c)\alpha^2 + (m - a)\alpha c + a(n - c)\alpha + ac = f_1(\alpha) \tag{3.100}$$

If $ac \leq z \leq mn$, then the membership function of the fuzzy number $D = AB$ is an inverse of the function $f_1(\alpha)$, namely

$$D(z) = f_1^{-1}(z) \tag{3.101}$$

Similarly, consider the decreasing parts of the fuzzy numbers A and B:

$$x = (m - b)\alpha + b \tag{3.102}$$

$$y = (n - d)\alpha + d \tag{3.103}$$

$$z = xy = [(m-b)\alpha + b][(n-d)\alpha + d] \tag{3.104}$$

$$z = (m-b)(n-d)\alpha^2 + (m-b)\alpha\, d + b(n-d)\alpha + bd = f_2(\alpha) \tag{3.105}$$

As before, for any $mn \le z \le bd$ we have

$$D(z) = f_2^{-1}(z) \tag{3.106}$$

Notice that in this case the fuzzy number D is does not have a triangular membership function, which means that multiplication of triangular fuzzy numbers does not preserve the original shape. Instead, multiplication of piecewise linear membership functions produces a quadratic form of the resulting fuzzy number.

3.7.8 Division

Like multiplication, for the increasing parts of the membership functions

$$x = (m-a)\alpha + a \tag{3.107}$$

$$y = (n-c)\alpha + c \tag{3.108}$$

we compute the division $z = x/y$, which after replacing x and y

$$z = \frac{x}{y} = \frac{(m-a)\alpha + a}{(n-c)\alpha + c} = g_1(\alpha) \tag{3.109}$$

so that, for $a/c \le z \le m/n$, the fuzzy number $E = A/B$ has the following membership function:

$$E(z) = g_1^{-1}(\alpha) \tag{3.110}$$

Analogously, for the decreasing parts of the membership functions

$$x = (m-b)\alpha + b \tag{3.111}$$

$$y = (n-d)\alpha + d \tag{3.112}$$

and we obtain

$$z = \frac{x}{y} = \frac{(m-b)\alpha + b}{(n-d)\alpha + d} = g_2(\alpha) \tag{3.113}$$

Thus, for $m/n \le z \le b/d$, the membership function of $E = A/B$ is

$$E(z) = g_2^{-1}(\alpha) \tag{3.114}$$

Clearly, the membership function of E is a rational function. Hence, division, like multiplication, does not preserve shape of the triangular membership functions.

3.8 Conclusions

We have discussed various approaches and algorithmic aspects of the design of fuzzy sets. The estimation of membership functions is a multifaceted problem and the choice of a suitable method relies on the choice of the available experimental data and domain knowledge. For user-driven approaches, it is essential to evaluate and flag the consistency of the results. While some of the methods (the pairwise comparison) come with this essential feature, the results produced by the others have to be carefully inspected.

Transformation of fuzzy sets in the form of the extension principle and composition generalizes similar transformations in sets. They play an important role in providing further transformations through fuzzy relational equations, associative memories, and algebraic operations with fuzzy numbers.

Fuzzy numbers are convex and normal fuzzy sets are on the set of real numbers. Operations with fuzzy numbers can be developed with the help of the extension principle. In particular, standard fuzzy arithmetic can be approached choosing the min t-norm. Several other choices are possible, but practice has shown that standard fuzzy arithmetic is still the one with the highest applicability.

Exercises

3.1 In the horizontal mode of construction of a fuzzy set of *safe* speed on a highway, the yes-no evaluations provided by the panel of nine experts are the following:

x	20	50	70	80	90	100	110	120	130	140	150	160
No of yes responses	0	1	1	2	6	8	8	5	5	4	3	2

Determine the membership function and assess its quality by computing the corresponding confidence intervals. Interpret the results and identify the points of the universe of discourse that may require more attention.

3.2 In the vertical mode of membership function estimation, we are provided with the following experimental data:

α	0.3	0.4	0.5	0.6	0.7	0.8	0.9	1.0
range of **X**	[−2,13]	[−1,12]	[0, 11]	[1, 10]	[2,9]	[3,8]	[4, 7]	[5, 6]

Plot the estimated membership function and suggest its analytical expression.

3.3 In the calculations of the distance between a point and a certain geometric figure, we assumed that the boundaries of the figure are well-defined. How could you proceed with a more general case when the boundaries are not clearly defined, namely the figure itself is defined by some membership function, Figure 3.23. In other words, the figure is fully characterized by some membership function $R(\mathbf{x})$ where $\mathbf{x}$ is a vector of coordinates of $\mathbf{x}$. If $R(\mathbf{x}) = 1$, the point fully belongs to the figure while lower values of $R(\mathbf{x})$ indicate that $\mathbf{x}$ is closer to the boundary of R.

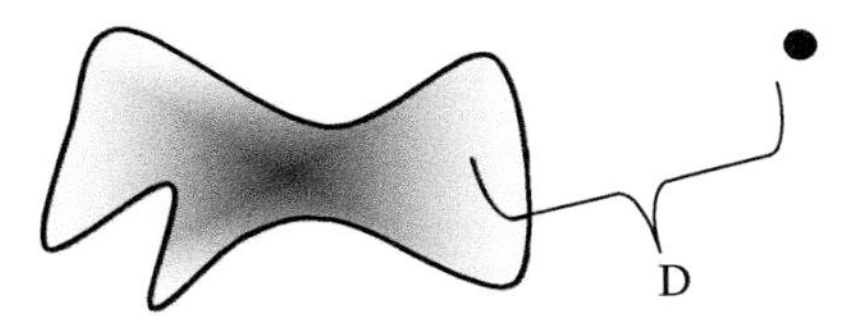

Figure 3.23 Forming a fuzzy set of distance between a geometric figure with fuzzy boundaries and a point.

3.4 Construct a fuzzy set describing a distance between the point of (5, 5) from a circle $x^2 + y^2 = 4$.

3.5 We maximize a function $f(x) = (x-6)^4$ in the range of [3, 10]. Suggest a membership function describing a degree of membership of the optimal solution that minimizes $f(x)$. What conclusion could you come up with based on the obtained form of the membership function?

3.6 The results of pairwise comparisons of four objects being realized in the scale of 1–5 are given in the following matrix form

$$\begin{bmatrix} 1 & 5 & 2 & 4 \\ 1/5 & 1 & 3 & 1/3 \\ 1/2 & 1/3 & 1 & 1/5 \\ 1/4 & 3 & 5 & 1 \end{bmatrix}$$

What is the consistency of the findings? Evaluate the effect of the lack of transitivity. Determine the membership function of the corresponding fuzzy set.

3.7 In the method of pairwise comparisons, we use different scales involving various levels of evaluation, typically ranging from five to nine. What impact could the number of these levels have on the produced consistency of the results? Could you offer any guidelines as how to achieve high consistency? What would be an associated tradeoff one should take into consideration here?

3.8 Construct a fuzzy set of *large* numbers for the universe of discourse of integer numbers ranging from 1 to 10. It has been found that the experimental results of the pairwise comparison could be described in the form

$$r(x,y) = \begin{cases} x - y & \text{if } x > y \\ 1 & \text{if } x = y \end{cases}$$

(for $x < y$ we consider the reciprocal version of this expression; that is, $1/(x - y)$).

3.9 In the FCM algorithm, the shape of the resulting membership function depends upon the value of the fuzzification coefficient (m). How does the mean value of the membership function relate with the values of "m." Run the FCM on the one-dimensional data set

$$\{1.3\ 1.9\ 2.0\ 5.5\ 4.9\ 5.3\ 4.5 - 1.3\ 0.0\ 0.3\ 0.8\ 5.1\ 2.5\ 2.4\ 2.1\ \ 1.7\}$$

considering that we have $c = 3$ clusters. Next, plot the relationship between the average of all membership grades and the associated fuzzification coefficient. For which values of "m" do the average of membership grades differ from 0.33 for less than δ? Consider several values of δ, say 0.2, 0.1, and 0.05. What could you tell about the impact of "m" on the resulting average?

3.10 Consider a family of car makes, say C_1, C_2, …, C_n. We are interested in forming fuzzy sets of economy, comfort, and safety, say A_{economy}, A_{comfort}, and A_{safety}. Use a method of a pairwise comparison to build the corresponding fuzzy sets. Next, using the method of pairwise comparison, evaluate the car makes with respect to the overall quality (which involves economy, comfort, and safety). Given the already constructed fuzzy sets of the individual attributes and the overall quality A_{overall}, what relationship could you establish between them?

3.11 Consider a fuzzy set of a *safe* speed on an average highway, Figure 3.24. How could this membership be affected when re-defining this concept in the following settings of (i) an autobahn (note that on these German highways there is no speed limit), and (ii) a snowy country road. Elaborate on the impact of various weather conditions on the corresponding

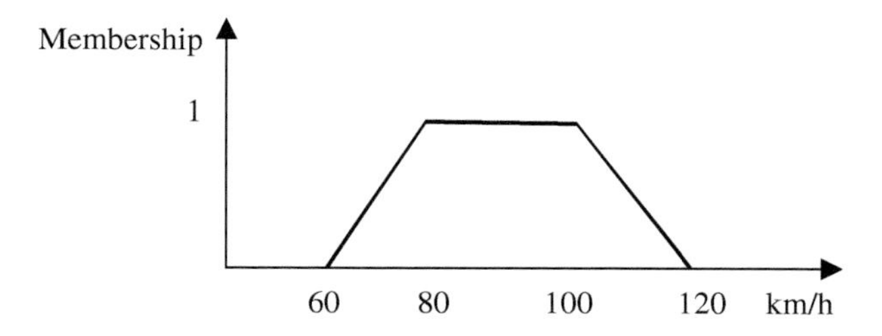

Figure 3.24 A fuzzy set of a *safe* speed on an average highway.

membership function. From the standpoint of the elicitation of the membership function, how could you transform the original membership function to address the needs of the specific context in which it is planned to be used?

$$G = \begin{bmatrix} 0.5 & 1.0 & 0.7 & 0.9 \\ 0.4 & 1.0 & 0.2 & 0.1 \\ 0.6 & 0.9 & 1.0 & 0.4 \end{bmatrix} \quad W = \begin{bmatrix} 0.9 & 0.3 & 0.1 & 0.7 & 0.6 & 1.0 \\ 0.1 & 0.1 & 0.9 & 1.0 & 1.0 & 0.4 \\ 0.0 & 0.3 & 0.6 & 0.9 & 1.0 & 0.0 \\ 1.0 & 0.0 & 0.0 & 0.0 & 1.0 & 1.0 \end{bmatrix}$$

3.12 Consider $\mathbf{X} = \{1, 2, 3, 4\}$ and the fuzzy set $A = \{0.1/1, 0.2/2, 0.7/3, 1.0/4\}$ defined in this space. Also, let $\mathbf{Y} = \{1, 2, 3, 4, 5, 6\}$. Given is a function $f: \mathbf{X} \rightarrow \mathbf{Y}$ such that $y = f(x) = x + 2$. Show that $B = f(A) = \{0.1/3, 0.2/4, 0.7/5, 1.0/6\}$.

3.13 Determine the α-cuts of the fuzzy set A whose membership function is equal to

$$A(x) = \begin{cases} 2x - x^2 & \text{if } 0 \le x \le 1 \\ 0 & \text{otherwise} \end{cases}$$

Let $f(x) = 2x - x^2$. Compute the image of the α-cuts of the fuzzy set A under "f." Sketch the transformations of the α-cuts graphically.

3.14 Develop, analytically, the membership function of the fuzzy number F that is the subtraction of fuzzy numbers A and B; namely, $F = A - B$.

3.15 Consider fuzzy numbers A and B whose membership functions are given in the form

$$A(x) = \begin{cases} e^{-(x-m)^2/k} & a \le x \le b \\ 0 & \text{otherwise} \end{cases}$$

$$B(x) = \begin{cases} 0 & \text{if } x \le a \\ \dfrac{x-a}{m-a} & \text{if } x \in [a,m] \\ \dfrac{b-x}{b-m} & \text{if } x \in [m,b] \\ 0 & \text{if } x \ge b \end{cases}$$

Show that their α-cuts are, given in the form

$$A_\alpha = \begin{cases} \left[m - \sqrt{\ln\left(\frac{1}{\alpha^k}\right)}, m - \sqrt{\ln\left(\frac{1}{\alpha^k}\right)}\right] & \text{if } \alpha \ge e^{-\left(\frac{-(a-m)^2}{k}\right)} \\ [a,b] & \text{iff } \alpha < e^{-\left(\frac{-(a-m)^2}{k}\right)} \end{cases}$$

and

$B_\alpha = [(m-a)\alpha + a, (m-b)\alpha + b], \forall \alpha \in [0,1].$

Sketch the membership functions of fuzzy sets of the addition, subtraction, multiplication, and division of A and B.

3.16 Are the parabolic fuzzy numbers $A, B, C \ldots$ whose membership functions come in the form

$$P(x,m,a) = \begin{cases} 1 - \left(\frac{x\text{-}m}{a}\right)^2 & \text{if } x \in [m\text{-}a, m+a] \\ 0 & \text{otherwise} \end{cases}$$

closed under addition operation? Justify your answer.

References

Arbelaitz, O., Gurrutxaga, I., Muguerza, J. et al. (2013). An extensive comparative study of cluster validity indices. *Pattern Recognition* 46 (1): 243–256.

Bezdek, J. (1981). *Pattern Recognition with Fuzzy Objective Function Algorithms.* New York: Plenum Press.

Chen, S.M. and Chen, J.H. (2009). Fuzzy risk analysis based on similarity measures between interval-valued fuzzy numbers and interval-valued fuzzy number arithmetic operators. *Expert Systems with Applications* 36 (3): 6309–6317.

Chen, M. and Wang, S. (1999). Fuzzy clustering analysis for optimizing fuzzy membership functions. *Fuzzy Sets and Systems* 103 (2): 239–254.

Civanlar, M. and Trussell, H. (1986). Constructing membership functions using statistical data. *Fuzzy Sets and Systems* 18 (1): 1–13.

Davies, D.L. and Bouldin, D.W. (1979). A clustering separation measure. *IEEE Transactions on Pattern Analysis and Machine Intelligence* 1 (2): 224–227.

Dishkant, H. (1981). About membership functions estimation. *Fuzzy Sets and Systems* 5 (2): 141–147.

Dombi, J. (1990). Membership function as an evaluation. *Fuzzy Sets and Systems* 35 (1): 1–21.

Dubois, D. and Prade, H. (1980). *Fuzzy Sets and Systems: Theory and Applications*. New York: Academic Press.

Dubois, D. and Prade, H. (1985). A review of fuzzy set aggregation connectives. *Information Systems* 36 (1–2): 85–121.

Duda, R.O., Hart, P.E., and Stork, D.G. (2001). *Pattern Classification*, 2e. Wiley.

Dyckhoff, H. and Pedrycz, W. (1984). Generalized means as a model of compensative connectives. *Fuzzy Sets and Systems* 14 (1): 143–154.

Gacek, A. and Pedrycz, W. (2015). Clustering granular data and their characterization with information granules of higher type. *IEEE Transactions on Fuzzy Systems* 23 (4): 850–860.

Hong, T. and Lee, C. (1996). Induction of fuzzy rules and membership functions from training examples. *Fuzzy Sets and Systems* 84 (1): 389–404.

Klement, P., Mesiar, R., and Pap, E. (2000). *Triangular Norms*. Dordrecht: Kluwer Academic Publishers.

Klir, G. and Yuan, B. (1995). *Fuzzy Sets and Fuzzy Logic: Theory and Applications*. Upper Saddle River: Prentice-Hall.

Krzanowski, W.J. and Lai, Y.T. (1988). A criterion for determining the number of groups in a data set using sum of squares clustering. *Biometrics* 44 (1): 23–34.

Kulak, O. and Kahraman, C. (2005). Fuzzy multi-attribute selection among transportation companies using axiomatic design and analytic hierarchy process. *Information Sciences* 170 (2–4): 191–210.

van Laarhoven, P. and Pedrycz, W. (1983). A fuzzy extension of Saaty's priority theory. *Fuzzy Sets and Systems* 11 (1–3): 199–227.

Mares, M. (1997). Weak arithmetics of fuzzy numbers. *Fuzzy Sets and Systems* 91 (2): 143–153.

Masson, M. and Denoeux, T. (2006). Inferring a possibility distribution from empirical data. *Fuzzy Sets and Systems* 157 (3): 319–340.

Medaglia, A., Fang, S., Nuttle, H., and Wilson, J. (2002). An efficient and flexible mechanism for constructing membership functions. *European Journal of Operational Research* 139 (1): 84–95.

Miller, G.A. (1956). The magical number seven plus or minus two: some limits of our capacity for processing information. *Psychological Review* 63 (1): 81–97.

Milligan, G.W. and Cooper, M. (1985). An examination of procedures for determining the number of clusters in a dataset. *Psychometrika* 50 (2): 159–179.

Nguyen, H. and Walker, E. (1999). *A First Course in Fuzzy Logic*. Boca Raton: CRC Press.

Pedrycz, W. (2005). *Knowledge-Based Clustering: From Data to Information Granules*. Hoboken, NJ: Wiley.

Pedrycz, W. (2013). *Granular Computing*. Boca Raton, Fl: CRC Press.

Pedrycz, W. and Gomide, F. (1998). *An Introduction to Fuzzy Sets: Analysis and Design*. Cambridge: MIT Press.

Pedrycz, A., Dong, F., and Hirota, K. (2009). Finite α cut-based approximation of fuzzy sets and its evolutionary optimization. *Fuzzy Sets and Systems* 160 (24): 3550–3564.

Rousseeuw, P. (1987). Silhouettes: a graphical aid to the interpretation and validation of cluster analysis. *Journal of Computational and Applied Mathematics* 20: 53–65.

Saaty, T. (1980). *The Analytic Hierarchy Process*. New York: McGraw Hill.

Saaty, T. (1986a). Scaling the membership functions. *European Journal of Operational Research* 25 (3): 320–329.

Saaty, T.L. (1986b). Axiomatic foundation of the analytic hierarchy process. *Management Science* 32 (7): 841–855.

Stefanini, L. (June 2009). A generalization of Hukuhara difference and division for interval and fuzzy arithmetic. *Fuzzy Sets and Systems* 161 (11): 1564–1584.

Stefanini, L., Sorin, L., and Guerra, M. (2006). Parametric representations of fuzzy numbers and application to fuzzy calculus. *Fuzzy Sets and Systems* 157 (18): 2423–2455.

Turksen, I. (1991). Measurement of membership functions and their acquisition. *Fuzzy Sets and Systems* 40 (1): 5–38.

Vendramin, L., Campello, R.J., and Hruschka, E.R. (2010). Relative clustering validity criteria: a comparative overview. *Statistical Analysis Data Mining* 3 (4): 209–235.

4

<*X, F*> Models of Multicriteria Decision-Making and Their Analysis

In this chapter, we concentrate on the construction, analysis, and application of models of multiobjective decision-making (in the form of <*X, F*> models). The basic concepts related to multicriteria decision-making are presented. The commonly utilized approaches to multiobjective decision-making are discussed. A great deal of attention is paid to the Bellman–Zadeh approach to decision-making in a fuzzy environment and its applications to multicriteria problems. Its application helps one to develop harmonious solutions, providing a rigorous as well as a computationally effective method of analyzing multiobjective models. The last circumstance opens a way for solving problems of short-term planning, operation, and control on a multicriteria basis. The questions of applying the ordered weighted average (OWA) operator are discussed. Its use is crucial to the implementation of several concepts of optimality, taking into account the level of optimism or pessimism of a DM. An important class of problems of multiobjective allocation of resources or their shortages is considered separately. The corresponding <*X, F*> models are constructed with the use of an original approach to the homogeneous and expert-acceptable formulation of specific objectives. The use of the presented results is illustrated by solving problems coming from power engineering area as well as problems of strategic planning.

4.1 Models of Multiobjective Decision-Making

When solving problems of multiobjective decision-making (problems of analyzing <*X, M*> models), a set of objective functions $F(x) = \{F_1(x), F_2(x), \ldots, F_q(x)\}$ is considered while the problem itself calls for simultaneous optimization of all objective functions, that is,

$$F_p(x) \rightarrow \underset{x \in L}{\text{extr}}, \quad p = 1, 2, \ldots, q \tag{4.1}$$

Multicriteria Decision-Making under Conditions of Uncertainty: A Fuzzy Set Perspective, First Edition. Petr Ekel, Witold Pedrycz, and Joel Pereira, Jr.

where $q \geq 2$ and L is a set of feasible solutions in $\mathbf{R}^n$. Depending on the nature of the problem under consideration, the term "extr" denotes the minimum or maximum.

As was indicated in Section 1.2, in solving multicriteria problems, we encounter uncertainty of goals, which is difficult to overcome and handle because "we simply do not know what we want". This will be pondered upon in the following considerations (Pedrycz et al. 2011).

From the point of view of the traditional (monocriteria) optimization, the problem formulated within the framework of the model Eq. (4.1), most likely, cannot be considered as being posed correctly. In particular, the notion of a complete optimal solution is considered in (Sakawa 1993). To be more specific, let us consider that all $F_p(x), p = 1, 2, \ldots, q$ are to be minimized. We say a point x^* is a complete optimal solution, if and only if there exists $x^* \in L$ such that $F_p(x^*) \leq F_p(x), p = 1, 2, \ldots, q$ for all $x \in L$. Also, the terms *ideal* solution or *utopia* point are equivalent because, in general, a complete optimal solution that simultaneously minimizes (or maximizes) all objective functions does not exist if these objective functions conflict each other. Thus, in reality the scalar concept of an "optimum" cannot be applied directly and has to be re-defined for multicriteria decision-making.

In terms of the analysis of the <*X*, *F*> models, the definition of optimality is not simple. The main difficulty is associated with the presence of conflicting objective functions, where the improvement in the sense of one objective function may lead to the deterioration in other objective function(s). For example, the maximization of reliability of power supply can be reached by a satiation of electrical networks with switching devices. However, this leads to an increase in network costs, working against the objective to minimize costs. Trade-offs exist between such conflicting objective functions, and the ultimate task is to find solutions that help one to effectively balance these trade-offs (Pedrycz et al. 2011). Such a balance is achieved when a solution cannot improve any objective function without degrading one or more objective functions. These solutions are referred as non-dominated solutions, efficient solutions, or Pareto optimal solutions, which have been briefly mentioned in Section 1.2 and are considered in more detail in the next section.

4.2 Pareto Optimal Solutions

Let us provide some important definitions (Hwang and Masud 1979; Zeleny 1982; Sakawa 1993; Ehrgott 2005; Pedrycz et al. 2011) that are helpful when talking about the analysis of <*X*, *F*> models. To make our considerations more focused, we consider that the problem at hand requires the minimization of the objective functions.

Domination: A solution x^1 dominates a solution x^2 if and only if

- x^1 is not worse than x^2 in all objective functions, that is $F_p(x^1) \leq F_p(x^2)$, $\forall p = 1, 2, \ldots, q$;
- x^1 is strictly better than x^2 in at least one objective functions, that is, $\exists p = 1, 2, \ldots, q$: $F_p(x^1) < F_p(x^2)$.

Similarly, for the objective space, a solution $F^1(x)$ dominates another solution $F^2(x)$, if $F^1(x)$ is not worse than $F^2(x)$ in all values of objective functions, and $F^1(x)$ is better than $F^2(x)$ in at least one of the values of objective functions.

It is evident that that x^1 is better than x^2 (i.e. x^1 dominates x^2), which happens when $F(x^1)$ dominates $F(x^2)$.

Weak domination: A solution x^1 weakly dominates a solution x^2 if and only if

- x^1 is not worse than x^2 in all objective functions, that is, $F_p(x^1) \leq F_p(x^2)$, $\forall p = 1, 2, \ldots, q$.

Pareto optimal solution: A point x^* is a Pareto optimal solution if a solution $x \neq x^* \in L$ that dominates it does not exist.

The Pareto optimal solutions, indicated previously, are also named non-dominated solutions and efficient solutions.

For the objective space, $F^*(x)$ is Pareto optimal, if x is a Pareto optimal solution.

All Pareto optimal solutions form a Pareto optimal solution set Ω^P (having a property such that solutions $x \in \Omega^P$ cannot be simultaneously improved on all objective functions). The corresponding points in the objective space form a Pareto optimal front Ω_F^P.

Weak Pareto optimal solution: A point x^* is a weak Pareto optimal solution if there does not exist a solution $x \neq x^* \in X$ that weakly dominates it.

All weak Pareto optimal solutions form a set of weak Pareto optimal solutions Ω^{WP}. The corresponding points in the objective space form a Pareto optimal frontier Ω_F^P.

Let Ω^{CO} denote a complete optimal solution set. Then, from these definitions, we can construct the following relations (Pedrycz et al. 2011):

$$\Omega^{CO} \subseteq \Omega^P \subseteq \Omega^{WP} \tag{4.2}$$

In principle, the concept of the Pareto optimal solution set is fundamental because a solution of a multicriteria decision-making problem must belong to this set. However, in general, its construction is usually a complicated and computationally cumbersome task. Diverse methods of building of Pareto optimal solution sets have been discussed (e.g. in Bentley and Wakefield 1998; Das and Dennis 1998; Deb 2001; Coelho et al. 2002; Statnikov and Matusov 2002; Konak et al. 2006).

In reality, the solution of multiobjective decision problems consists of several stages (Coelho 2000). However, many researchers tend to concentrate on issues related to the construction of the Pareto optimal solution sets, considering them as solutions to the multiobjective decision-making problems (Pedrycz et al. 2011).

Let us note that the Pareto optimal solutions do not provide any insight into the process of decision-making itself (a DM still has to choose manually a final solution or a preferred solution), since they are really a useful generalization of a utility function under conditions of minimum information (i.e. all objective functions are considered as having equal importance; in other words, a DM does not provide any preference of the objectives). Thus, the issue is how to incorporate DM preferences into the decision-making process.

Taking this into account, it should be stressed that although the step of analyzing $<X, M>$ models associated with determining the Pareto optimal solution set is useful, it does not permit one to obtain unique solutions to real-world problems. It is necessary to choose a concrete Pareto optimal solution through DMs' involvement in further information processing.

4.3 Approaches to Incorporating Decision-Maker Information

The possible way to classify approaches that help one to incorporate information of a DM is based on the moment (within the decision-making process) at which this information becomes presented and applied. In accordance with this criterion, three approaches exist: a priori, a posteriori, and interactive (for instance, Horn 1997; Coelho 2002; Pedrycz et al. 2011).

If preferences are expressed a priori, a DM has to define them in advance (before actually executing the decision process). In the procedures of the a priori type, it is directly or indirectly assumed that all information, permitting us to obtain the most preferable solution, is incorporated into a formal model, that is, in the description of a set of alternatives and objective functions and, consequently, can be extracted from the model, applying some transformations used in a constructive manner.

Procedures that are of the a posteriori type are usually associated with the availability of some system of hypotheses or axioms, which are to be verified for each individual situation of decision-making. These hypotheses or axioms are considered as additional and are not included in the formal model. If the verification of the axioms leads to a positive result, it is possible to construct a convincing mode to choose the most appropriate alternative. This verification is associated with obtaining and applying additional information provided by a DM.

Although a priori and a posteriori decision-making approaches are common in the literature related to decision-making (Coelho 2000), an *interactive* approach (it also uses additional information of a DM, however, as a progressive articulation of preferences) has been well evaluated by researchers (Gardiner and Steuer 1994) for reasons discussed, for example, in (Monarchi et al. 1973; Pedrycz et al. 2011).

When applying the interactive approach, procedures of successive improvement of solution quality are executed as a transition from $x_\alpha^0 \in \Omega^P \subseteq L$ to $x_{\alpha+1}^0 \in \Omega^P \subseteq L$ by considering information I_α of a DM at the αth step of the decision-making process:

$$x_1^0, F\left(x_1^0\right) \xrightarrow{I_1} \ldots \xrightarrow{I_{\alpha-1}} x_\alpha^0, F\left(x_\alpha^0\right) \xrightarrow{I_\alpha} \ldots \xrightarrow{I_{\omega-1}} x_\omega^0, F\left(x_\omega^0\right) \tag{4.3}$$

It is possible to distinguish two types of adaptation (taking this into account, the interactive approach itself is also called adaptive) in the process Eq. (4.3): computer to preferences of a DM and a DM to the problem. The first type of adaptation is based on information being received from a DM. The second type of adaptation is realized as a result of the steps $x_\alpha^0, F\left(x_\alpha^0\right) \xrightarrow{I_\alpha} x_{\alpha+1}^0, F\left(x_{\alpha+1}^0\right)$, which allow a DM to understand the relationships between his/her needs and possibilities of their satisfaction by the model Eq. (4.1). This explains the possibility to construct sufficiently universal procedures of multiobjective decision-making. The types of such procedures are implied by the variety of existing forms of additional information representation. For instance, it is possible to distinguish the following forms (Pedrycz et al. 2011):

- DM identifies significance of objective functions, that is, indicates proper assessments of weights of the criteria;
- DM identifies some desired levels of objective functions (goal values of objective functions, lower and/or higher admissible values of objective functions, admissible deviations from goal values of objective functions);
- DM compares sets of presented alternatives;
- information provided by DM includes different combinations of the first three types of reports.

The taxonomy of the approaches related to incorporating information of a DM in the decision-making process is not complete and is relatively conditional (Pedrycz et al. 2011). For instance, although many works in the field of multicriteria decision-making associate methods based on utility theory (Keeney and Raiffa 1976) with a posteriori approach, the authors of (Hwang and Masud 1979; Lai and Hwang 1996) link these methods with the a priori approach. Besides, there exist some decision-making procedures that cannot be uniquely related to one or another approach. Here, we can refer to the mixed procedures (a priori and a posteriori, or a posteriori and interactive). Finally, the same method of the groups of multiobjective decision-making methods discussed in the next section can be realized within the framework of different approaches to incorporating information of a DM in the decision-making process. However, the construction of certain taxonomy helps us adequately represent capabilities of diverse types of multicriteria decision-making methods as well as carefully identify their advantages and disadvantages.

4.4 Methods of Multiobjective Decision-Making

In the processes of the statement and solution of multiobjective decision-making problems, it is necessary to develop answers to some specific questions. Among these questions, it is important to identify the following:

4.4.1 Normalization of Objective Functions

In multiobjective decision-making problems, different objective functions may have different physical meaning and, consequently, are expressed in different units; in turn, their scales are not commensurable. Taking this into account, the comparison of quality of obtained solutions from the point of view of different objective functions is impossible. The operation of unifying scales of objective functions to a unique scale is called normalization.

4.4.2 Choice of the Principle of Optimality

In analyzing $\langle X, F \rangle$ models, the principle of optimality defines the properties of the desired solution and answers in which sense the desired solution excels all other possible solutions as well as offers guidelines as to the search for desired solutions. The principle of optimality is fundamental to multiobjective decision-making.

4.4.3 Consideration of Priorities of Objective Functions

Usually, considering the specificity of the problem, it becomes apparent that different objective functions have different importance, that is, one objective function has higher priority relative to another one. It is natural to take this information into consideration in the choice of the principle of optimality, assigning priority to more important objective functions. With this regard, the following question arises: define the formal description of the priority and the degree of its influence on the solution to the multiobjective problem.

The answers to the questions posed previously, and, subsequently, the development of methods of multiobjective decision-making, have been developed in different ways (for instance Hwang and Masud 1979; Zeleny 1982; Dubov et al.

1986; Lai and Hwang 1996; Rao 1996; Ehrgott 2005). We identify those in common use (Pedrycz et al. 2011):

- scalarization-based methods;
- methods based on imposing constraints on levels of objective functions, including a lexicographic method;
- methods of goal programming and of global criterion.

The first group of methods solving multiobjective decision-making problems is based on their reduction to scalar (monocriteria) problems by constructing some types of convolution (aggregation) (Kuhn and Tucker 1951; Zadeh 1963).

The simplest convolution is the following:

$$\Phi(x) = \sum_{p=1}^{q} f_p(x) \tag{4.4}$$

where $f_p(x) = F_p(x)$, $p = 1, 2, \ldots, q$ if all objective functions $F_p(x)$ are of a homogeneous character or exhibit the same semantics. However, if the objective functions have different meanings, it becomes impossible, as indicated before, to compare the quality of obtained solutions with respect to different objective functions.

It is possible to identify several requirements related to the normalization of objective functions. However, the most important of them (Pedrycz et al. 2011) is the necessity to assign equal values to $f_p\left(x_p^0\right)$ or $f_p\left(x_p^{00}\right)$ where $x_p^0 = \arg\min_{x\in L} F_p(x)$ for minimized objective functions and $x_p^{00} = \arg\max_{x\in L} F_p(x)$ for maximized objective functions. The inclusion of this requirement is helpful to compare objective functions on the basis of their numerical values.

The indicated requirement is satisfied by using the following normalization:

$$f_p(x) = \frac{\max\limits_{x\in L} F_p(x) \; - \; F_p(x)}{\max\limits_{x\in L} F_p(x) - \min\limits_{x\in L} F_p(x)} \tag{4.5}$$

if the objective function $F_p(x)$ is to be minimized. If the objective function $F_p(x)$ is to be maximized, then it is possible to utilize

$$f_p(x) = \frac{F_p(x) \; - \; \min\limits_{x\in L} F_p(x)}{\min\limits_{x\in L} F_p(x) - \min\limits_{x\in L} F_p(x)} \tag{4.6}$$

To construct Eqs. (4.5) or (4.6) for any objective function, it is necessary to solve the following monocriteria problems:

$$F_p(x) \rightarrow \min_{x \in L} \tag{4.7}$$

and

$$F_p(x) \rightarrow \max_{x \in L} \tag{4.8}$$

Considering this, it is possible to apply another type of normalization

$$f_p(x) = \frac{\min_{x \in L} F_p(x)}{F_p(x)} \tag{4.9}$$

if the corresponding $F_p(x)$ is to be minimized or use the normalization

$$f_p(x) = \frac{F_p(x)}{\max_{x \in L} F_p(x)} \tag{4.10}$$

if the corresponding $F_p(x)$ is to be maximized.

The construction of Eq. (4.9) requires determining only the solution to the problem Eq. (4.7) while the construction of Eq. (4.10) requires obtaining the solution to the problem Eq. (4.8). However, although the construction of Eq. (4.9) and Eq. (4.10) is more rational from the computational point of view, the quality of normalized functions Eqs. (4.9) and (4.10) may not be always acceptable. This is illustrated by Example 4.1 (Pedrycz et al. 2011).

Example 4.1 Let us construct the normalization functions Eqs. (4.5), (4.6), (4.9), and (4.10), when $\min_{x \in L} F_p(x) = 10$ and $\max_{x \in L} F_p(x) = 20$. The levels of $F_p(x)$ as well as the values of different $f_p(x)$ are shown in Table 4.1.

The results presented in Table 4.1 demonstrate that the use of Eqs. (4.5) and (4.6) provides $0 \leq f_p(x) \leq 1$ and does not modify a character of $F_p(x)$. At the same time, the application of Eqs. (4.9) and (4.10) leads to $0.5 \leq f_p(x) \leq 1$. Besides, the use of Eq. (4.9) does not permit one to keep a character of $F_p(x)$.

Table 4.1 Normalization of the objective function.

$F_p(x)$	10	11	12	13	14	15	16	17	18	19	20
$f_p(x)$ Eq. (4.4)	1	0.9	0.8	0.7	0.6	0.5	0.4	0.3	0.2	0.1	0
$f_p(x)$ Eq. (4.5)	0	0.1	0.2	0.3	0.4	0.5	0.6	0.7	0.8	0.9	1
$f_p(x)$ Eq. (4.8)	1	0.91	0.83	0.77	0.71	0.67	0.63	0.59	0.56	0.53	0.50
$f_p(x)$ Eq. (4.9)	0.50	0.55	0.60	0.65	0.70	0.75	0.80	0.85	0.90	0.95	1

The use of Boldur's method (Roy 1972) provides another way of the normalization. In particular, we can assign "values" or "utilities" v'_p and v''_p to $F_p\left(x_p^0\right)$ and $F_p\left(x_p^{00}\right)$, respectively, if $F_p(x)$ is to be minimized or to $F_p\left(x_p^{00}\right)$ and $F_p\left(x_p^0\right)$, respectively, if $F_p(x)$ is to be maximized. Then, accepting linear interpolation, it is possible to construct the following aggregation:

$$\Phi(x) = \sum_{p=1}^{q} \left[\alpha_p F_p(x) + \beta_p\right] \tag{4.11}$$

where the coefficients α_p and β_p can be obtained by solving the following system of equations:

$$\begin{cases} \alpha_p F_p\left(x_p^0\right) + \beta_p = v'_p \\ \alpha_p F_p\left(x_p^{00}\right) + \beta_p = v''_p \end{cases} \tag{4.12}$$

if $F_p(x)$ is to be minimized or by solving the following system of equations:

$$\begin{cases} \alpha_p F_p\left(x_p^{00}\right) + \beta_p = v'_p \\ \alpha_p F_p\left(x_p^0\right) + \beta_p = v''_p \end{cases} \tag{4.13}$$

if $F_p(x)$ is to be maximized.

It is not difficult to verify that if the "values" or "utilities" $v'_p = 1$ and $v''_p = 0$, the convolution Eq. (4.11) is reduced to Eq. (4.4).

If we have to differentiate the importance of different objective functions, one transforms Eq. (4.4) into

$$\Phi(x) = \sum_{p=1}^{q} \lambda_p f_p(x) \tag{4.14}$$

where $\lambda_p, p = 1, 2, \ldots, q$ are weighting factors or importance factors that reflect a relative importance of objective functions, which are to satisfy the following conditions:

$$\lambda_p \geq 0, \quad p = 1, 2, \ldots, q \tag{4.15}$$

and

$$\sum_{p=1}^{q} \lambda_p = 1 \tag{4.16}$$

In (Lu et al. 2008), the maximization of the convolution Eq. (4.14) is called the weighting function method.

The application of the convolutions Eqs. (4.4) and (4.14), which have found wide practical applications, corresponds to the principle of uniform optimality (Lyapunov 1972). At the same time, the application of convolutions of the type

$$\Phi(x) = \prod_{p=1}^{q} f_p(x) \tag{4.17}$$

as well as

$$\Phi(x) = \prod_{p=1}^{q} \lambda_p f_p(x) \tag{4.18}$$

or

$$\Phi(x) = \prod_{p=1}^{q} \left[f_p(x)\right]^{\lambda_p} \tag{4.19}$$

which have found some applications, corresponds to the principle of just compromise (Lyapunov 1972).

The methods based on imposing constraints on levels of objective functions are associated with specifying the desired levels for objective functions defined by the requirements of a DM (Benayoun et al. 1971; Pedrycz et al. 2011) and, then, by maximizing the convolutions, for example, those of the normalized objective functions Eq. (4.14).

The lexicographic method (Rao 1996; Ehrgott 2005) or a method of successive concessions (Podinovsky and Gavrilov 1975) can be related to the group that imposes constraints on the levels of objective functions.

When applying the lexicographic method, the objective functions are to be ranked and numbered in order of importance by a DM. Taking this into account, assume that the most important objective function is $F_1(x)$ while $F_q(x)$ is the least important objective function. Besides, for clarity of presentation, assume that all objective functions are to be minimized. Then, the original multiobjective problem can be replaced by a set of monocriteria problems. The first of them takes on the following form:

$$F_1(x) \to \min_{x \in L} \tag{4.20}$$

If x_1^* is the solution to the problem Eq. (4.20), then the second problem

$$F_2(x) \to \min_{x \in L} \tag{4.21}$$

is to be solved by taking into account the following additional constraint:

$$F_1\left(x_1^*\right) + \Delta F_1 \geq F_1(x) \tag{4.22}$$

where ΔF_1 is the concession on $F_1(x)$ to minimize $F_2(x)$.

If x_2^* is the solution to the problem Eqs. (4.21) and (4.22), then the concession ΔF_2 is defined to minimize $F_3(x)$ and in this way obtain x_3^*. The process of setting the concessions is continued to obtain $x_4^*,\ldots,x_q^*$. Thus, the point $x_p^*, p = 1, 2, \ldots, q$ is considered as the solution of the original multiobjective problem.

The idea of goal programming was first presented by Charnes et al. (1955), although the actual name first appear in Charnes and Cooper (1961). It has been further developed by Lee (1972), Ignizio (1976), and Romero (1991), among others. An annotated bibliography of goal programming of the period 1990–2000 is presented in Jones and Tamiz (2002).

The use of the goal programming method requires a DM to define goals for the objectives that he or she wishes to attain. A preferred solution is then defined as the one that minimizes deviations from the set of goals. Thus, if some goals $b_p, p = 1, 2, \ldots, q$ (goals can be $F_p\left(x_p^0\right)$ for minimized objective functions or $F_p\left(x_p^{00}\right)$ for maximized objective functions) are defined, then the problem of goal programming can be formulated as follows:

$$\text{minimize} \left[\sum_{p=1}^{q} \left(d_p^- + d_p^+\right)^t\right]^{\frac{1}{t}}, \quad t \geq 1 \tag{4.23}$$

subject to

$$x \in L \tag{4.24}$$

$$F_p(x) + d_p^- - d_p^+ = b_p, \quad p = 1, 2, \ldots, q \tag{4.25}$$

$$d_p^- d_p^+ = 0, \quad p = 1, 2, \ldots, q \tag{4.26}$$

$$d_p^-, d_p^+ \geq 0, \quad p = 1, 2, \ldots, q \tag{4.27}$$

where d_p^- and d_p^+ are the underachievement and overachievement of the pth goal, respectively. The value of t in Eq. (4.23) is based on the utility function chosen by a DM.

There exists a modification of goal programming, formulated within the framework of the model Eqs. (4.23)–(4.27), called priority, pre-emptive, or lexicographic goal programming, when a DM, in addition to defining goals for objectives, is able to give an ordinal ranking of these objectives. It is clear that this modification should be used when there exists a convincing priority ordering among the goals to be achieved. If a DM is more interested in direct comparisons of the objectives then weighted goal programming can be used by introducing weights in Eq. (4.23).

In reality, it is possible to state that goal programming measures the distance to the goals by using the sum (the weighted sum) of absolute distances from given goals. Taking this into account, it is possible to mention the so-called global criteria method (Lai and Hwang 1996), which differs from goal programming by the measurement of this distance by using the Minkowski metric.

Without a comprehensive discussion of the strengths and weaknesses of the indicated groups of methods (they are studied in detail in Hwang and Masud 1979; Zeleny 1982; Mashunin 1986; Lai and Hwang 1996; Ehrgott 2005, for example), it is necessary to indicate two fundamental weaknesses shared by all of them (Pedrycz et al. 2011).

The first weakness is associated with the ability of methods based on imposing constraints on levels of objective functions and methods of goal programming to produce solutions that are not Pareto optimal. This violates the basic concept of multicriteria decision-making. The second weakness is associated with the following considerations.

An important question in multicriteria decision-making regards the quality of the solution itself. An answer to this question is associated with the concept of harmonious solutions. In this concept, the solutions quality is considered high if the levels of satisfying objectives are equal or close to each other when the importance levels of the objective functions are equal (Ekel 2001; Ekel and Galperin 2003). It is not difficult to extend this concept for the case when the importance levels of the objective functions are different: the solutions are to be harmonious by taking into account the corresponding importance factors (Pedrycz et al. 2011; Ekel et al. 2016). From this point of view, the validity and advisability of the direction related to the principle of guaranteed results should be recorded (Lyapunov 1972). Other directions in multicriteria decision-making, in particular, those indicated previously, may lead to solutions with high levels of satisfaction of some criteria that are reached when assuring low levels of satisfaction of some other criteria. This situation could be completely unacceptable (for example, Ekel and Galperin 2003; Canha et al. 2007; Ekel et al. 2007).

The lack of clarity of the concept of "optimal solution" arises from the basic methodological complexity when solving multicriteria problems. When applying the Bellman–Zadeh approach to decision-making in a fuzzy environment (Bellman and Zadeh 1970) to solve multicriteria problems, this concept is defined with reasonable validity: the maximum degree of implementing goals serves as a criterion of optimality (Ekel 2002). This conforms to the principle of guaranteed result. The Bellman–Zadeh approach permits one to realize an effective (from the computational standpoint) as well as rigorous (from the standpoint of obtaining solutions $x \in \Omega^P \subseteq L$, at least, for convex Ω^P) method of analyzing multiobjective models. Finally, its use allows one to preserve a natural measure of uncertainty in decision-making and take into account indices, criteria, and constraints of qualitative character.

4.5 Bellman–Zadeh Approach to Decision-Making in a Fuzzy Environment and Its Application to Multicriteria Decision-Making

When applying the Bellman–Zadeh approach to decision-making in a fuzzy environment (Bellman and Zadeh 1970) for solving multicriteria problems, each objective function $F_p(x)$, $x \in L$, $p = 1, 2, \ldots, q$ is replaced by a fuzzy objective function or a fuzzy set $A_p = \left\{x, \mu_{A_p}(x)\right\}, x \in L, p = 1, 2, \ldots, q$. A fuzzy solution D with the setting up of the fuzzy sets A_p is obtained as a result of the intersection $D = \cap_{p=1}^{q} A_p$ with a membership function

$$\mu_D(x) = \bigwedge_{p=1}^{q} \mu_{A_p}(x) = \min_{p=1,2,\ldots,q} \mu_{A_p}(x), \quad x \in L \tag{4.28}$$

The use of Eq. (4.28) produced a solution x^0 providing the maximum degree

$$\max \mu_D(x) = \max_{x \in L} \min_{p=1,2,\ldots,q} \mu_{A_p}(x) \tag{4.29}$$

of belongingness to the fuzzy solution D and reduces the problem Eq. (4.1) to a search for

$$x^0 = \arg\max_{x \in L} \min_{p=1,2,\ldots,q} \mu_{A_p}(x) \tag{4.30}$$

To illustrate the application of Eq. (4.28)–(4.30), let us consider the simple Example 4.2.

Example 4.2 The membership functions of fuzzy objective functions A_1, A_2, and A_3 are presented in Table 4.2.

Applying Eq. (4.28), one constructs a fuzzy solution D, presented in Table 4.2, which in accordance with Eq. (4.30) gives rise to the solution $x^0 = 4$.

If we analyze multiobjective problems, to obtain Eq. (4.30), it is necessary to construct the membership functions $\mu_{A_p}(x), p = 1, 2, \ldots, q$ $p = 1, 2, \ldots, q$

Table 4.2 Membership functions of fuzzy objective functions and a fuzzy solution.

x	1	2	3	4	5	6	7	8	9	10
$A_1(x)$	0.1	0.2	0.8	1.0	0.8	0.7	0.5	0.3	0.2	0.1
$A_2(x)$	0.1	0.2	0.6	0.9	0.7	0.6	0.5	0.4	0.3	0.2
$A_3(x)$	0.4	0.6	1.0	0.8	0.7	0.6	0.5	0.4	0.3	0.2
$D(x)$	0.1	0.2	0.6	**0.8**	0.7	0.6	0.5	0.3	0.2	0.1

reflecting a degree of achieving their "own" optima by $F_p(x), x \in L, p = 1, 2, \ldots, q$. Taking into account the relationships Eqs. (4.5) and (4.6), we construct membership functions

$$\mu_{A_p}(x) = \left[\frac{\max_{x \in L} F_p(x) - F_p(x)}{\max_{x \in L} F_p(x) - \min_{x \in L} F_p(x)}\right]^{\lambda_p} \tag{4.31}$$

for the minimized objective functions or membership functions

$$\mu_{A_p}(x) = \left[\frac{F_p(x) - \min_{x \in L} F_p(x)}{\max_{x \in L} F_p(x) - \min_{x \in L} F_p(x)}\right]^{\lambda_p} \tag{4.32}$$

for the maximized objective functions.

Thus, the solution to problem Eq. (4.1) requires an analysis of $2q + 1$ monocriteria problems Eqs. (4.7), (4.8), and (4.29), respectively.

Since the solution x^0 belongs to $\Omega^P \subseteq L$, it is necessary to construct

$$\bar{\mu}_D(x) = \bigwedge_{p=1}^{q} \mu_{A_p}(x) \wedge \pi(x) = \min\left\{\min_{p=1,2,\ldots,q} \mu_{A_p}(x), \pi(x)\right\} \tag{4.33}$$

where $\pi(x) = \begin{cases} 1 \text{ if } x \in \Omega^P \\ 0 \text{ if } x \notin \Omega^P \end{cases}$

The procedures for solving the problem Eq. (4.29), discussed next, provide a way of obtaining $x^0 \in \Omega^P \subseteq L$ according to Eq. (4.33). Thus, this can be said about the equivalence of $\bar{\mu}_D(x)$ and $\mu_D(x)$. This circumstance permits one to give up the need to implement a cumbersome procedure for building $\Omega^P \subseteq L$.

The existence of additional conditions (indices, criteria, and/or constraints) of qualitative character, defined by linguistic variables reduces Eq. (4.30) to

$$x^0 = \arg \max_{X \in L} \min_{p=1,2,\ldots,q+s} \mu_{A_p}(x) \tag{4.34}$$

where $\mu_{A_p}(x), x \in L, p = q + 1, \ldots, s$ are membership functions of fuzzy values of linguistic variables, which reflect the nature of these additional conditions.

There are some theoretical justifications behind the validity of applying *min* operator in Eqs. (4.28)–(4.30), for example, refer to Bellman and Giertz 1974). Taking this into consideration, it is necessary to note that there exist many families of aggregation operators that may be used in place of the *min* operator. Thereby, it is possible to generalize Eq. (4.28) as follows:

$$\mu_D(x) = \text{agg}\left(\mu_{A_1}(x), \mu_{A_2}(x), \ldots, \mu_{A_q}(x)\right), \quad x \in L \tag{4.35}$$

Despite that some properties of these aggregation operators are well established, there is no clear and intuitively appealing interpretation of these properties. Likewise, there is a lack of a unifying interpretation of the operators themselves (Beliakov and Warren 2001). An important question emerges: among many types of aggregation operator, how do we select the one that is adequate for a particular problem to be solved? Although some selection criteria have been suggested in Zimmermann (1996), the majority of investigations focus on choosing the operators on the basis of some available experimental evidence. Thus, it is possible to assert that the selection of the operators, to a significant extent, is experience-based. Taking this into account, in Section 4.8 we discuss experiments not only demonstrating the use of the min operator, but also involving the product operator (which can be considered as one of the family of t-norm operators, Yager 1988, and which has found a quite visible position in analyzing decision-making problems, Pedrycz et al. 2011).

The application of the product operator reduces Eq. (4.28) to

$$\mu_D(x) = \prod_{p=1}^{q} \mu_{A_p}(x) \tag{4.36}$$

The use of Eq. (4.36) generates the following problem:

$$\max \mu_D(x) = \max_{x \in L} \min_{p=1,2,..,q} \mu_{A_p}(x) \tag{4.37}$$

whose solution comes in the form

$$x^0 = \arg \max_{x \in L} \prod_{p=1}^{q} \mu_{A_p}(x) \tag{4.38}$$

As was stressed previously, the important question in multicriteria decision-making is the quality of obtained solutions. It is considered high if the levels of satisfying objectives are equal or close to each other. However, in real-world applications, there are cases where the orientation to the principle of guaranteed results, indicated before, is to be considered too pessimistic: it is not reasonable to assume that a good alternative simultaneously satisfy all criteria (Pedrycz et al. 2011). A DM may consider such a requirement as being too hard (restrictive) and may prefer a softer one such as: a good alternative must satisfy "most" criteria, or "at least half" the criteria or a "few" criteria, for example. In reality, the indicated considerations reflect criteria of optimality. Thus, it is desirable to have an approach that would allow deviations from the observation of the principle of guaranteed results (Lyapunov 1972) (permitting mutual compensation between levels of satisfying objectives), however, would regulate these deviations on the basis of the level of optimism or pessimism of a DM. Taking this into account, it is necessary to indicate that the application of an OWA operator, originally proposed in (Yager 1988), permits one to implement several concepts of optimality using the same aggregation operator with different configurations.

4.6 OWA Operator Applied to Multiobjective Decision-Making

All groups of methods of multicriteria decision-making discussed previously allow a DM to prioritize objectives by the assignment of their importance factors. However, in a number of cases, it is not desirable or impossible to fully order the priorities of the objectives. In these cases, the ability of a DM to influence the decision-making processes is reduced and the solution is chosen according to the optimality concept realized within the applied decision-making method. This situation demands from a DM a deeper understanding of the optimality concepts implemented by each approach.

Taking this into consideration, it is necessary to use broader concepts to evaluate the quality of multicriteria solutions. It can be implemented on the basis of applying the OWA operator (Yager 1988) and its modifications. In particular, it is possible to use the OWA operator to implement several concepts of optimality, using the same approach, on the basis of the level of optimism or pessimism of a DM. The utility of the application of the OWA operator and its modifications is also associated with the need for risk modeling in multicriteria decision-making problems (Yager 1993; Malczewski 2006; Chen 2011).

The OWA operator of dimension q, which aggregates a set of q normalized values $a_1, a_2, \ldots, a_q$ is presented in the following form (Yager 1988):

$$\mathrm{OWA}\left(a_1, a_2, \ldots, a_q\right) = \sum_{i=1}^{q} w_i b_i \tag{4.39}$$

where b_i is the ith largest value among $a_1, a_2, \ldots, a_q$ and the set of weights in Eq. (4.39) satisfies the following conditions: $w_i \in [0, 1]$ and $\sum_{i=1}^{q} w_i = 1$.

Example 4.3 The aggregation of four normalized values $a_1 = 0.14$, $a_2 = 0.44$, $a_3 = 0.51$, and $a_4 = 0.25$, using Eq. (4.39) with $w_1 = 0.36$, $w_2 = 0.41$, $w_3 = 0.18$, and $w_4 = 0.05$, generates the following:

$$\begin{aligned} \mathrm{OWA}(0.14, 0.44, 0.51, 0.25) &= 0.51 \times 0.36 + 0.44 \times 0.41 + 0.25 \times 0.18 \\ &\quad + 0.14 \times 0.05 = 0.416 \end{aligned} \tag{4.40}$$

It should be emphasized that each weight w_i is associated with the ith ordered position rather than a particular element. Taking this into account, the fundamental aspect in applying the OWA operator is associated with the possibility to represent an optimality concept based on the corresponding weights. The OWA operator allows one to implement diverse aggregation operators. The important particular cases are shown in Table 4.3.

Table 4.3 Equivalence between OWA and other aggregation operators.

Aggregation operator	OWA weights
min	$w_1 = 1,\ w_i = 0,\ i = 2,3,\ldots,\ q$
max	$w_q = 1,\ w_i = 0,\ i = 1,\ 2,\ \ldots,\ q-1$
arithmetic mean	$w_i = 1/q,\ i = 1,\ 2,\ \ldots,\ q$

The aggregation operators presented in Table 4.3 are capable of reproducing rationally possible attitudes of a DM (Yager 1988). In particular, the max operator corresponds to an optimistic approach that allows a maximum mutual compensation between levels of satisfying objectives: it is enough to have any good evaluated criterion to compensate low levels of satisfying other criteria. This can be interpreted as a "pure" OR aggregation (Pedrycz et al. 2011). The min operator reproduces a pessimistic approach that does not deliver any compensation between criteria, always taking the worst evaluation associated with a certain criterion. This can be interpreted as a "pure" AND aggregation that corresponds to the principle of guaranteed results (Lyapunov 1972). The arithmetic mean operator allows a degree of intermediate compensation between the min and max operators.

The application of the OWA operator permits one to consider the notion of optimism or pessimism reflected by weights of the operator. The corresponding quantitative measure is considered in (Yager 1988). In particular, the evaluation of the level of optimism of a DM (also known as an orness degree) can be performed on the basis of the following expression:

$$\theta = \frac{1}{q-1}\sum_{i=1}^{q}(q-i)w_i \tag{4.41}$$

For example, if $w_1 = 1$ and $w_i = 0$, $i = 2,3,\ldots,q$ (min operator), then $\theta = 0$, while if $w_q = 1$ and $w_i = 0$, $i = 1, 2, \ldots, q-1$ (max operator), then $\theta = 1$. Figure 4.1 (Zarghami and Szidarovszky 2009) reflects the relation between the value θ and the optimism level of a DM.

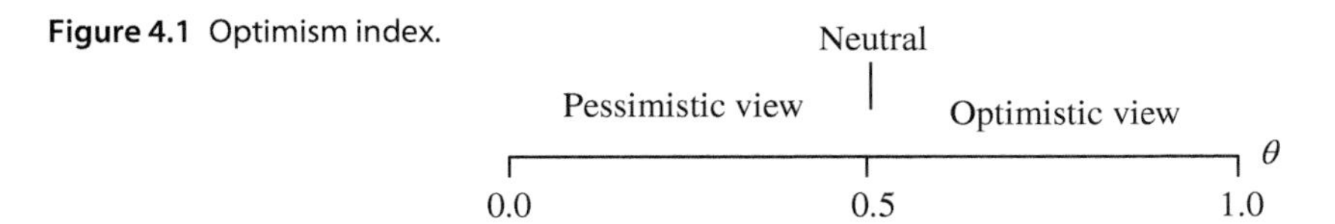

Figure 4.1 Optimism index.

There are several families of the OWA operators proposed in the literature (Yager 1993). Generally, these families differ in methods for determining the weights. Without the discussion of these methods (for instance, they are studied in Xu, 2005), we would like to distinguish the Slide OWA (*S*-OWA) operator (Yager and Filev 1994).

The *S*-OWA operator allows one to work with different levels of optimism and pessimism, which are based on its weights defined as follows:

$$w_i = \begin{cases} \frac{1}{q}(1-(\sigma+\beta))+\sigma, & i=1 \\ \frac{1}{q}(1-(\sigma+\beta)), & i=2,\dots,q-1 \\ \frac{1}{q}(1-(\sigma+\beta))+\beta, & i=q \end{cases} \tag{4.42}$$

where $\sigma, \beta \in [0, 1]$ with observing $\sigma + \beta \leq 1$. The parameters σ and β indicate the degree of optimism and pessimism, respectively, of a DM. When $\sigma = 1$ the S-OWA operator corresponds to the max operator and if β=1, the S-OWA operator corresponds to the min operator. Other values of σ and β, with $\sigma \neq \beta$ and $\sigma + \beta \leq 1$, allow different degrees of affinity with the max or min operators. The arithmetic average operator can be obtained when $\sigma = \beta = 0$, while the condition $\sigma = \beta \neq 0$ with $\sigma + \beta = 1$ indicates that a DM wants to simultaneously meet the minimum and maximum compensation, discarding the intermediate values.

One of the main attractions of applying the OWA operator in the context of multicriteria decision-making is associated with the fact that it permits a DM to indirectly specify the weights by applying linguistic quantifiers (Zadeh 1983; Yager 1988).

A fuzzy quantifier corresponds to a fuzzy set $Q(r)$, which reflects the level at which the portion of criteria $r \in [0, 1]$ satisfies the concept represented by Q. In Yager (1995), fuzzy quantifiers are used to define the weights of the OWA operators. In particular, in the decision-making process, a solution is to be chosen if there are at least r criteria that have good evaluations in relation to other solutions, according to the concept of a DM, which is reflected by Q, which is to satisfy the following conditions:

- $Q(0) = 0$
- $Q(1) = 1$
- if $r_1 > r_2$, then $Q(r_1) > Q(r_2)$

After choosing the appropriate quantifier, the weights of the OWA operator are obtained as follows:

$$w_i = Q\left(\frac{i}{q}\right) - Q\left(\frac{i-1}{q}\right), i = 1, 2, \dots, q \tag{4.43}$$

The fuzzy quantifiers Q can be represented in a natural language by the use of linguistic terms. In particular, Table 4.4 includes examples of some linguistic terms and the corresponding quantifiers (Liu and Han 2008).

In the case of applying the Bellman–Zadeh approach to decision-making in a fuzzy environment to analyzing multiobjective models, taking into account the results described in this section, the expression Eq. (4.30) is rewritten in the following form:

$$x^0 = \arg \max_{x \in L} \underset{p=1,2,\dots,q}{\text{OWA}} \mu_{A_p}(x) \tag{4.44}$$

The expression Eq. (4.33) is transformed to

$$\bar{\mu}_D(x) = \min \left\{ \underset{p=1,2,\dots,q}{\text{OWA}} \mu_{A_p}(x), \pi(x) \right\} \tag{4.45}$$

To illustrate the use of Eq. (4.44), let us consider the application of aggregation operators given in Table 4.5 to membership functions of three fuzzy objective functions of the example.

Table 4.4 Linguistic terms and linguistic quantifiers.

Linguistic term	Linguistic quantifier	θ
There exists	$Q(r) = \begin{cases} 0 & \text{if } r = 0 \\ 1 & \text{if } r \neq 0 \end{cases}$	1
All	$Q(r) = \begin{cases} 0 & \text{if } r \neq 1 \\ 1 & \text{if } r = 1 \end{cases}$	0
Most	$Q(r) = \begin{cases} 0 & \text{if } 0 \le r \le 0.3 \\ 2(r\text{-}0,3) & \text{if } 0.3 < r \le 0.8 \\ 1 & \text{if } 0.8 < r \le 1 \end{cases}$	0.45
At least half	$Q(r) = \begin{cases} 2r & \text{if } 0 \le r \le 0.5 \\ 1 & \text{if } 0.5 < r \le 1 \end{cases}$	0.75
As many as possible	$Q(r) = \begin{cases} 0 & \text{if } 0 \le r \le 0.5 \\ 2(r-0,5) & \text{if } 0.5 < r \le 1 \end{cases}$	0.25
Average	$Q(r) = r(0 \le r \le 1)$	0.5
More than j $\alpha = \frac{j}{q}$	$Q(r) = \begin{cases} 0 & \text{if } 0 \le r \le \alpha \\ \frac{r}{1-\alpha} & \text{if } \alpha < r \le 1 \end{cases}$	$\frac{1-\alpha}{2}$
At least j $\alpha = \frac{j}{q}$	$Q(r) = \begin{cases} \frac{r}{\alpha} & \text{if } 0 \le r \le \alpha \\ 1 & \text{if } \alpha < r \le 1 \end{cases}$	$1 - \frac{\alpha}{2}$

Table 4.5 OWA operators.

Operator	Linguistic term	w	θ
S_1	Average	[0, 33 0, 33 0, 33]	0.50
S_2	S-OWA (σ = 0, 5 and β = 0, 5)	[0, 50 0 0, 50]	0.50
S_3	As many as possible	[0 0, 33 0, 67]	0.25
S_4	All (min operator)	[0 0 1]	0

Example 4.5 The membership functions of fuzzy objective functions A_1, A_2, and A_3 are presented in Table 4.6. Applying Eq. (4.44), we construct fuzzy solutions D_{S_1}, D_{S_2}, D_{S_3}, and D_{S_4} presented in Table 4.6 as well, using the aggregation operators given in Table 4.5. The use of the operator S_1 provides $x^0 = 3$ and 4. The application of S_2 generates $x^0 = 6$; $x^0 = 5$ is a result of the use of S_3 and S_4.

Table 4.6 Membership functions of fuzzy objective functions and fuzzy solutions.

x	1	2	3	4	5	6	7	8	9	10
$A_1(x)$	0.1	0.2	1.0	1.0	0.7	0.6	0.5	0.7	0.2	0.1
$A_2(x)$	0.1	0.2	0.4	0.5	0.7	1.0	0.6	0.9	0.1	0.9
$A_3(x)$	0.4	0.6	1.0	0.9	0.7	0.6	0.5	0.6	0.3	0.2
$D_{S_1}(x)$	0.20	0.33	**0.80**	**0.80**	0.70	0.73	0.53	0.73	0.20	0.40
$D_{S_2}(x)$	0.25	0.40	0.70	0.75	0.70	**0.80**	0.55	0.75	0.20	0.50
$D_{S_3}(x)$	0.10	0.20	0.60	0.63	**0.70**	0.60	0.50	0.63	0.13	0.13
$D_{S_4}(x)$	0.10	0.20	0.40	0.50	**0.70**	0.60	0.50	0.60	0.10	0.10

4.7 Multiobjective Allocation of Resources and Their Shortages

Among various questions ("Where are we?", "Where do we want to go?", "What do we have in the front?", "What to do?", "How to do it?", "How are we doing?") related to different types of planning activity (strategic, innovation, new business, research and development, expansion, operational, maintenance, planning, etc.), two fundamental questions always arise: "What to do?" and "How to do it?"

To elaborate on answers to the first fundamental question ("What to do?") for some organizations, boards of directors, departments, and so on, it usually becomes necessary to:

- define objectives and establish goals for the corresponding levels of the planning hierarchy;
- evaluate, compare, choose, prioritize, and/or order solutions or alternatives (strategic actions, innovation projects, new business projects, research and development projects, expansion plans, operational strategies, maintenance actions, etc.).

The answers to the first fundamental question are to be based on the processing of information related to different perspectives or criteria, such as "investment attractiveness", "innovation level", "political effect", "originality level", "expansion efficiency", "operational complexity", "maintenance flexibility", and so on. Taking this into account, the evaluation, comparison, choice, prioritization, and/or ordering of solutions or alternatives are to be executed, considering quantitative information (based on measurable quantities with different levels of uncertainty) as well as qualitative information (based on knowledge, experience, and intuition of involved experts). The answers to the first fundamental question can be elaborated on by constructing and analyzing <*X*, *R*> models that are studied in Chapter 5.

The second fundamental question ("How to do it?"), which, in many cases, can also be independent of the first fundamental question, is associated with the rational allocation of various types of resource (financial, human, logistics, etc.) or their shortages between solutions or alternatives (strategic actions, innovation projects, new business projects, research and development projects, expansion plans, operational steps, maintenance actions, etc.; for simplicity, consumers) to achieve the stated objectives and goals in the maximal degree. Although this allocation, regarded from different points of view, is to be based on applying quantitative information as well qualitative information, in practice, it is carried out by taking into account only quantitative information, often completely ignoring its uncertainty. Considering this, in Stewart (2005), it is emphasized that models of multicriteria decision-making are essentially based on deterministic evaluations of the consequences of each action expressed in terms of each criterion, possibly subjecting final results and recommendations to a degree of sensitivity analysis. This approach may be justified when the primary source of complexity in decision-making is associated with the multicriteria nature of the problem rather than with the uncertain nature of individual consequences. Nevertheless, situations arise, for example, in strategic planning problems, when risks and uncertainties are as critical as the issue of conflicting management goals. In such situations, more formal modeling of the uncertainties becomes necessary (Stewart 2005).

Taking these observations into consideration, it should be noted that the construction and analysis of <X, F> models, combined with the results of (Ekel et al. 2008; Pereira Jr. et al. 2015; Ekel et al. 2016), can provide the full-edged answers to the second fundamental question.

In particular, the necessity was stressed (Roy 2010) for producing robust solutions in the analysis of multicriteria models. The results in (Ekel et al. 2008; Pereira Jr. et al. 2015; Ekel et al. 2016), associated with the fuzzy set-based generalization of the classic approach (Luce and Raiffa 1957; Raiffa 1968; Belyaev 1977; Webster 2003) to deal with the uncertainty of information in monocriteria decision-making to multicriteria problems, can serve for this purpose. These results are associated with combining two branches of mathematics of uncertainty: elements of game theory and fuzzy set theory. They allow one to carry out multicriteria analysis under conditions of uncertainty, based on the possibilistic approach (considering representative combinations of initial data, states of nature or scenarios), in accordance with the general scheme (Pereira Jr. et al. 2015). Without considering questions of producing robust solution, taking into account uncertainty of quantitative information as well as qualitative information (these questions are studied in detail in Chapters 6 and 7), next, we discuss ways of solving problems of multiobjective allocation of resources or their shortages under deterministic information (or for a concrete representative combination of initial data, state of nature, or scenario).

Taking this into account, it is necessary to indicate that methods leading to its solution, based on fundamental allocation principles (proportional allocation, optimal allocation, and the principle of inverse priorities, see Burkov and Kondrat'ev 1981) exhibit significant drawbacks (Pedrycz et al. 2011). They could be eliminated by engaging a multicriteria approach, which permits one to consider and maximize diverse positive consequences or minimize diverse negative consequences of allocating resources or their shortages. Our first results in this area are related to solving the set of problems of allocating power and energy shortages for different levels of territorial, temporal, and situational hierarchies of load management, generated by the impacts of the disaster at Chernobyl (Sklyarov et al. 1987). These results were implemented within a computing system used in energy utilities for allocating power and energy shortages, minimizing technological, economic, environmental, and social consequences associated with the limitation of consumers.

We can talk about the equivalence of the problems of allocating resources and allocating their shortages from different points of view (Pedrycz et al. 2011; Ekel et al. 2016). However, as shown next, the allocation of resource shortages permits one to realize a more flexible and practically more significant problem formulation and solution.

One of the fundamental questions in applying a multiobjective approach is the construction of objective functions on the basis of the corresponding objectives defined by a DM or, for instance, on the basis of specific strategic objectives, defined by strategic maps in the case of using the Balanced Scorecard methodology for strategic planning (Kaplan and Norton 1996; Kaplan and Norton 2004). Taking this into account, next, we discuss an approach to the homogeneous, clear, transparent, and acceptable for any expert, involved in the process of decision-making, formulation of specific objectives, which serves for constructing objective functions for the multiobjective models.

As it was indicated previously, we consider two classes of interrelated problems:

- multiobjective allocation of resources;
- multiobjective allocation of shortages of resources.

For the first class of problems, the homogeneous, clear, transparent, and acceptable for experts, participating in the process of decision-making, formulation of specific objectives, for example, for strategic planning, may be the following:

- Prevailing financial provision of projects generating a higher level of product supply abroad.
- Prevailing financial provision of projects generating a higher level of product for each invested U$ 1 000 000.

This form of formulating the specific objectives should be stressed given their universal character. For instance, the following objectives can be formulated for very much different area of maintenance planning in electrical distribution systems:

- Prevailing provision of repair teams for regions with greater lengths of distribution lines.
- Prevailing provision of repair teams for regions with a large number of consumers per 1 km^2.

In the case of multiobjective allocation of resource shortages, the objectives for strategic planning may be presented in the following form:

- Prevailing financial constraint of projects generating a lower level of product supply abroad.
- Prevailing financial constraint of projects generating a lower level of product for each invested U$ 1 000 000.00.

At the same time, in the case of allocation of shortages of resources for maintenance planning in electrical distribution systems, the following objectives can be formulated:

- Prevailing constraint of repair teams for regions with a shorter length of distribution lines.
- Prevailing constraint of repair teams for regions with a lower number of consumers per 1 km^2.

It is easy to note that the satisfaction of objectives related to multiobjective allocation of resources is associated with maximizing or minimizing the linear objective functions

$$F_p(X) = \sum_{i=1}^{n} c_{pi}x_i, \quad p = 1,2,\ldots,q \tag{4.46}$$

where x_i, p = 1, 2, …, q are variables, which correspond to sought resource volumes intended for the ith consumer; c_{pi}, p = 1, 2, …, q are specific indicators, which correspond to the pth specific objective, for the ith consumer.

At the same time, the satisfaction of objectives in the case of multiobjective allocation of shortages of resources is associated with maximizing or minimizing the linear objective functions

$$F_p(X) = \sum_{i=1}^{n} c_{pi}\Delta x_i, \quad p = 1,2,\ldots,q \tag{4.47}$$

The objective functions Eqs. (4.46) and (4.47) are not the only kind of objective functions, which can be used in multiobjective allocation of resources or their shortages. In particular, in Sklyarov et al. (1987), other types of objective function (linear, fractional, quadratic, etc.) are considered; sometimes they can better reflect the essence of specific objectives.

Three models for allocating resources or their shortages are discussed next.

4.7.1 Model 1: Allocation of Available Resources

We are given demands D_i, i = 1,2, …,n of consumers and a total available resource $R < \sum_{i=1}^{n} D_i$. Then, the problem exhibits the following structure:

$$F_p(x) \Rightarrow \max_{x \in L} \text{ or } \min_{x \in L}, \quad p = 1,2,\ldots,q \tag{4.48}$$

subject to the constraints

$$x_i \le D_i, \quad i = 1,2,\ldots,n \tag{4.49}$$

and

$$\sum_{i=1}^{n} x_i = R \tag{4.50}$$

4.7.2 Model 2: Allocation of Resource Shortages with Unlimited Cuts

Demands D_i, i = 1, 2, …, n of consumers are given and a total available resource is $R < \sum_{i=1}^{n} D_i$. Then, the resource shortage is $A = \sum_{i=1}^{n} D_i - R$ and the problem has the following structure:

$$F_p(\Delta x) \Rightarrow \max_{x \in L} \text{ or } \min_{x \in L}, \quad p = 1,2\ldots,q \tag{4.51}$$

subject to the constraints

$$\Delta x_i \geq 0, \quad i = 1, 2, \ldots, n \tag{4.52}$$

and

$$\sum_{i=1}^{n} \Delta x_i = A. \tag{4.53}$$

4.7.3 Model 3: Allocation of Resource Shortages with Limited Cuts

We are given demands D_i, $i = 1, 2, \ldots, n$ as well as minimally acceptable demands $D_i^m, i = 1, 2, \ldots, n$ of consumers. The problem is to achieve Eq. (4.51), by taking into account the constraints

$$0 \leq \Delta x_i \leq A_i = D_i - D_i^m, \quad i = 1, 2, \ldots, n \tag{4.54}$$

and those of Eq. (4.50).

All described models can be analyzed on the basis of applying the results described in Sections 4.5 and 4.6. Consider the specific features of this analysis.

The analysis of the Model 1 (allocation of available resources) is iterative. In particular, among consumers that did not receive the required resource volume ($x_i < D_i$), it is necessary to choose one with the least resource amount received, remove it from consideration, and solve the problem again. The procedure is to be repeated until the situation arises where only consumers remain satisfied with their demands.

The analysis of the Model 2 (allocation of resource shortages with unlimited cuts) is iterative as well. Among consumers that did not receive the required resource volume ($x_i < D_i$), it is necessary to choose one with the greatest resource cut, remove it from consideration, and solve the problem again. The procedure is to be repeated until the situation arises where consumers remain satisfied with their demands.

The Model 3 (allocation of resource shortages with limited cuts) is the most flexible: it allows diverse problem statements. Generally, its analysis is not iterative. However, in the case of the necessity of negotiating a final solution within the frameworks of $A_i = D_i - D_i^m, i = 1, 2, \ldots, n$ may transform the analysis into the iterative one.

Let us consider a general scheme of solving the problem formalized within Model 3, defined by the expressions Eqs. (4.51), (4.53), and (4.54).

To describe a general scheme, it is expedient to introduce a linguistic variable Q – *limitation for consumer* to provide a DM with the possibility to consider conditions that are difficult to formalize. Thus, the general scheme presumes the availability of a procedure for the construction of a term-set $T(Q)$ of the linguistic variable and membership functions for its fuzzy values. In addition, if the solution Δx_{α}^{0} with the values $\mu_{A_p}\left(\Delta x_{\alpha}^{0}\right), p = 1, 2, \ldots, q$ is not satisfactory, a DM has to have the possibility to correct it, passing $\Delta x_{\alpha+1}^{0}$ by modifying the importance of one or more objective functions. Considering this, we also assume the availability of the procedure for building and modifying a vector of importance factors $\Lambda = (\lambda_1, \lambda_2, \ldots, \lambda_q)$.

The general scheme of solving the problem described by Eqs. (4.51), (4.53), and (4.54), which has been used for implementing an adaptive interactive decision-making system (AIDMS1) (Pedrycz et al. 2011) is associated with the following sequence of steps:

1) Solution of the problems Eqs. (4.7) and (4.8) in order to obtain $\Delta x_{p}^{0}, p = 1, 2, \ldots, q$ and Δx_{p}^{00}, $p = 1, 2, \ldots, q$, respectively.
2) Construction of the membership functions defined by the expressions Eqs. (4.31) or (4.32).
3) Construction of an initial vector of the importance factors $\Lambda = (\lambda_1, \lambda_2 \ldots, \lambda_q)$.
4) In the case of using the OWA operator, the construction of the vector of the OWA operator weights $w = (w_1, w_2, \ldots, w_q)$.
5) Analysis of the availability of initial conditions defined by the linguistic variable. If these conditions are not available, then go to step 9; otherwise go to step 6.
6) Verification of compatibility of the initial conditions and, if necessary, their correction.
7) Solution of the problem Eq. (4.29) with the goal to obtain Δx_{α}^{0} defined by the expression Eq. (4.30). In the case of applying the OWA operator, the expressions of the type Eqs. (4.29) and (4.30) are to be modified accordingly.
8) Analysis of the current solution Δx_{α}^{0}. If a DM is satisfied with this solution, then go to step 11; otherwise go to step 9, taking $\alpha := \alpha + 1$.
9) Correction of the vector of the importance factors.
10) Insertion of additional conditions defined by the linguistic variables; then go to step 6.
11) Completion of computing – the solution Δx^{0} has been obtained.

The main functions of calculating kernel of this scheme are associated with obtaining $\Delta x_{p}^{0}, p = 1, 2, \ldots, q$ and $\Delta x_{p}^{00}, p = 1, 2, \ldots, q$ that are produced by solving the problems Eqs. (4.31) and (4.32) and by obtaining x^{0} according to the expression Eq. (4.30). The solution of the problems of the type Eqs. (4.7) and (4.8) is rather straightforward. The maximization of the expression of the type Eq. (4.28) can be based on a non-local search that comes as a modification of the Gelfand's and Tsetlin's "long valley" method (Raskin 1976).

Experimental evidence shows that variables in Eq. (4.28) can be divided into inessential and essential ones (Pedrycz et al. 2011). The change of inessential variables leads to essential variations of Eq. (4.28). The change of essential variables leads to inessential variations of Eq. (4.28). Thus, a structure of Eq. (4.28) may be considered as a multidimensional "long valley". If we apply direct search methods (Rao 1996), this circumstance requires the ascent from different initial points Δx_p^0 (Pareto points), if we minimize $F_p(\Delta x)$, or Δx_p^{00} (Pareto points), if we maximize $F_p(\Delta x)$, to find the most convincing solution Δx^0. This explains the application of a non-local search in the following procedure based on the results of Pedrycz et al. (2011):

1) The sequence $\{\Delta x^{(l)}\}$, $l = 1, 2, \ldots, q$ is constructed, starting from the points Δx_p^0, if we minimize $F_p(\Delta x)$, or Δx_p^{00}, if we maximize $F_p(\Delta x)$, obtained as a result of execution of Step 1 of the general scheme given before. This sequence satisfies the following property: $\min_{1 \le p \le q} \mu_{A_p}\left(\Delta x^{(l)}\right) \ge \min_{1 \le p \le q} \mu_{A_p}\left(\Delta x^{(l+1)}\right)$, $l = 1, \ldots, q-1$.
2) The local search for Δx^0 is executed from $\Delta x^{(1)}(l = 1)$. As a result of this search, we obtain a point $\Delta x^{(1)0}$ with the corresponding $\mu_{A_p}\left(\Delta x^{(1)0}\right), p = 1, 2, \ldots, q$.
3) The local search for Δx^0 is executed, starting from $\Delta x^{(l+1)}$. As a result of this search, we obtain a point $\Delta x^{(l+1)0}$ with the corresponding $\mu_{A_p}\left(\Delta x^{(l+1)0}\right), p = 1, 2, \ldots, q$.
4) The following verifications are carried out:
   a) if $\Delta x^{(1)0} \neq \Delta x^{(l+1)0}$, then go to Operation 5;
   b) if $\Delta x^{(1)0} = \Delta x^{(l+1)0}$ for $l \neq q-1$, then go to Operation 3, by incrementing $l := l + 1$;
   c) if $\Delta x^{(1)0} = \Delta x^{(l+1)0} = \Delta x^{(q)0}$, then go to Operation 8, by assuming $\Delta x^0 = \Delta x^{(1)0}$.
5) A line between points $\Delta x^{(t)0}$ and $\Delta x^{(t+1)0}$ is constructed to generate points $\Delta x_s^{(t,t+1)}, s = 1, 2, 3$ (see Figure 4.2). Among these points (if they are admissible from the position of observing the constraints Eqs. [4.53] and [4.54]), a point $\Delta x^{(t,t+1)0} = \arg\max_{t} \min_{1 \le p \le q} \mu_{A_p}\left(\Delta x_s^{(t,t+1)}\right)$ is selected to define a direction for future search.
6) The next local search for Δx^0 is executed, starting from $\Delta x^{(t,\,t+1)0}$. As a result of this search, we obtain a point $\Delta x^{(t+2)0}$ (see Figure 4.2).
7) We carrying out the analysis: if three "last" points $\Delta x^{(t)0}$, $\Delta x^{(t+1)0}$, and $\Delta x^{(t+2)0}$ differ on $\min_{1 \le p \le q} \mu_{A_p}\left(\Delta x^{(t)0}\right)$, $\min_{1 \le p \le q} \mu_{A_p}\left(\Delta x^{(t+1)0}\right)$, and $\min_{1 \le p \le q} \mu_{A_p}\left(\Delta x^{(t+2)0}\right)$ less than the desired level of accuracy, then go to

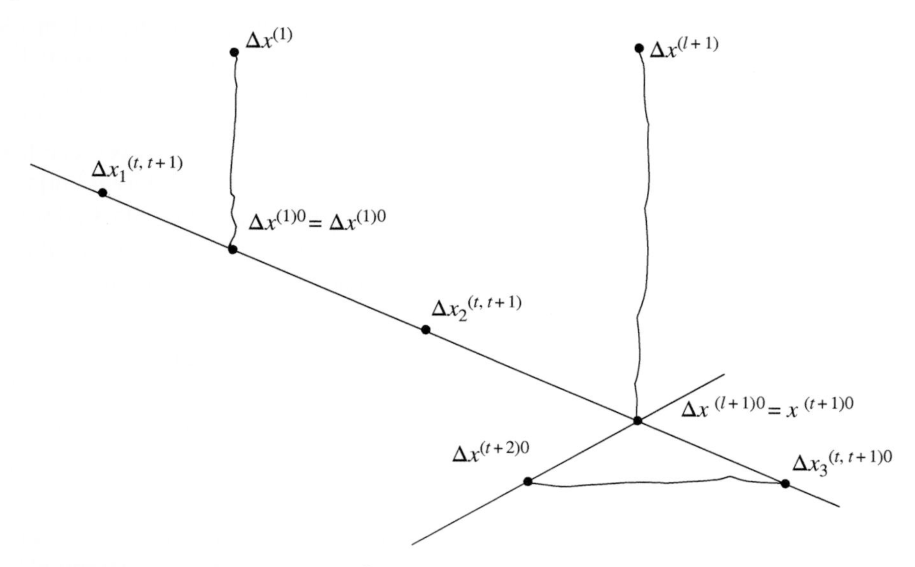

Figure 4.2 Non-local search for Δx^0.

Operation 8, taking $\Delta x^0 = \arg\max\left[\min_{1\le p\le q}\mu_{A_p}\left(\Delta x^{(t+2)}\right),\ \min_{1\le p\le q}\mu_{A_p}\left(\Delta x^{(t+2)}\right),\ \min_{1\le p\le q}\mu_{A_p}\left(\Delta x^{(t+2)}\right)\right]$; otherwise go to Operation 5, taking $\Delta x^{(t)0} := \Delta x^{(t,\ t+1)0}$ and $\Delta x^{(t+1)0} := \Delta x^{(t+2)0}$.

8) Calculations are completed – the solution $\Delta x^0 \in \Omega^P \subseteq L$ has been obtained.

The execution of Operations 2, 3, and 6 of the procedure is possible by making application of any search method (in particular, within the framework of the AIDMS1, modification of the univariate method has been used, see Rao 1996). If $\Delta x^{(k)}$ is a current point, the transition to $\Delta x^{(k+1)}$ is justified if

$$(\forall p = 1,2,\ldots,q) : \mu_{A_p}\left(x^{(k+1)}\right) \ge \min_{1\le p\le q}\mu_{A_p}\left(x^{(k)}\right) \tag{4.55}$$

In the case of

$$(\exists p = 1,2,\ldots,q) : \mu_{A_p}\left(\Delta x^{(k+1)}\right) < \min_{1\le p\le q}\mu_{A_p}\left(x^{(k)}\right) \tag{4.56}$$

the transition to $x^{(k+1)}$ is not expedient from the point of view of maximizing Eq. (4.29). This way of evaluating the expediency of the transition to the next point $x^{(m+1)}$ leads to the solution Eq. (4.30) that is Pareto if all inexpedient transitions are rejected.

The AIDMS1 (Pedrycz et al. 2011) includes a procedure for building and correcting the term-set $T(Q)$ and membership functions for fuzzy values of the linguistic variable Q – *limitation for consumer*. The initial term-set available for a DM is the following: $T(Q) = \langle$*near, approximately, slightly less, considerably less, slightly more,* and *considerably more*$\rangle$. The corresponding membership functions are defined as follows:

$$\mu_{A_p}(\Delta x_i) = e^{-k(R_i - x_i)^2} \tag{4.57}$$

$$\mu_{A_p}(\Delta x_i) = \begin{cases} 1 - e^{-k(R_i - x_i)^2} & , x_i \le R_i \\ 0 & , x_i > R_i \end{cases} \tag{4.58}$$

$$\mu_{A_p}(\Delta x_i) = \begin{cases} 1 - e^{-k(R_i - x_i)^2} & , x_i \ge R_i \\ 0 & , x_i < R_i \end{cases} \tag{4.59}$$

where k is a coefficient defined by a given solution accuracy and R_i is a "specific value" related to the corresponding condition (for example, *considerably less* than R_i) that is to be considered in the solution process.

The membership function described by Eq. (4.57) is related to the terms *near* and *approximately,* Eq. (4.58) – to *slightly less* and *considerably less,* and Eq. (4.59) – to *slightly more* and *considerably more.*

Besides, the AIDMS1 includes several procedures for constructing and correcting the vector $\Lambda = (\lambda_1, \lambda_2, \ldots, \lambda_q)$ of importance factors. One of them is based on the Saaty's pairwise comparison approach as discussed in Chapter 3.

Example 4.6 As an illustrative example, let us consider the problem of multiobjective allocation of a shortage of financial resources in strategic planning. The problem is to be resolved for the five projects within the framework of *Model 3,* discussed previously.

Taking into account the available resource R= 68 800.00 kU$ and the data in Table 4.7, it is not difficult to note that $A = \sum_{i=1}^{n} D_i - R$= 75 800.00 – 68 800.00 = 7000.00. Thereby, it is necessary to consider the following constraints:

$$\Delta x_1 + \Delta x_2 + \Delta x_3 + \Delta x_4 + \Delta x_5 = 7000.00 \tag{4.60}$$

$$0 \le \Delta x_1 \le 2500.00 \tag{4.61}$$

$$0 \le \Delta x_2 \le 1600.00 \tag{4.62}$$

$$0 \le \Delta x_3 \le 2100.00 \tag{4.63}$$

$$0 \le \Delta x_4 \le 1900.00 \tag{4.64}$$

and

$$0 \le \Delta x_5 \le 2800.00 \tag{4.65}$$

Table 4.7 Initial information for allocating a shortage of financial resources.

Project	D_i, kU$	D_i^m, kU$	A_i, kU$
1	15 500.00	13 000.00	2500.00
2	14 500.00	12 900.00	1600.00
3	15 000.00	12 900.00	2100.00
4	16 000.00	14 000.00	1900.00
5	14 800.00	12 000.00	2800.00

Table 4.8 Initial information for constructing objective functions.

Project	c_1, %	c_2, kU$	c_3, kU$	c_4
1	33.50	2350.00	15 500.00	1.90
2	22.00	2280.00	14 500.00	1.80
3	24.00	2050.00	15 500.00	2.05
4	12.50	2130.00	16 000.00	2.30
5	15.32	1980.00	14 800.00	2.55

The objectives are the following:

1) Prevailing financial constraint of projects generating a lower level of product supply abroad.
2) Prevailing financial constraint of projects generating a lower level of profit for each invested U$ 1 000 000.00.
3) Prevailing financial constraint of projects with a higher level of investment.
4) Prevailing financial constraint of projects with a higher level of relative risk coefficient.

The data in Table 4.8 help us to construct the following problem:

$$F_1(\Delta x) = 33.50\Delta x_1 + 22.00\Delta x_2 + 24.00\Delta x_3 + 12.50\Delta x_4 + 15.32\Delta x_5 \rightarrow \min \tag{4.66}$$

$$F_2(\Delta x) = 2,350.00\Delta x_1 + 2,280.00\Delta x_2 + 2,050.00\Delta x_3 + 2,130.00\Delta x_4 + 1,980.00\Delta x_5 \rightarrow \min \tag{4.67}$$

$$F_3(\Delta x) = 15,000.00\Delta x_1 + 14,500.00\Delta x_2 + 15,500.00\Delta x_3 + 16,000.00\Delta x_4 + 14,800.00\Delta x_5 \rightarrow \min \tag{4.68}$$

$$F_4(\Delta x) = 1.90\Delta x_1 + 1.80\Delta x_2 + 2.05\Delta x_3 + 2.30\Delta x_4 + 2.55\Delta x_5 \rightarrow \min \tag{4.69}$$

while observing the constraints in Eqs. (4.60)–(4.65).

To solve this problem, four different sets of weights of the OWA operator have been applied as shown in Table 4.9.

The results obtained on the basis of using each set of weights given in Table 4.9 are presented in Table 4.10. Analyzing the results given in Table 4.11, it is possible to observe that the levels of satisfying objectives reflect the essence of linguistic terms.

Table 4.9 Weights of the OWA operator.

Operator	Linguistic term	w	θ
S_1	Average	$[0.25 \quad 0.25 \quad 0.25 \quad 0.25]$	0.50
S_2	S-OWA (σ = 0, 5 e β = 0, 5)	$[0.50 \quad 0.00 \quad 0.00 \quad 0.50]$	0.50
S_3	As many as possible	$[0.00 \quad 0.00 \quad 0.50 \quad 0.50]$	0.25
S_4	All (*min* operator)	$[0.00 \quad 0.00 \quad 0.00 \quad 1.00]$	0

Table 4.10 Solutions determined with different sets of weights.

Operator	Δx_1	Δx_2	Δx_3	Δx_4	Δx_5
S_1	0.05	1600.00	2100.00	499.95	2800.00
S_2	854.05	1600.00	2100.00	0	2445.95
S_3	0	1600.00	2100.00	1052.50	2247.50
S_4	563.55	1599.98	2099.97	1497.60	1238.90

Table 4.11 Levels of satisfying objectives.

Operator	1	2	3	4
S_1	0.74	0.83	0.94	0.30
S_2	0.48	0.64	0.94	0.48
S_3	0.77	0.77	0.78	0.36
S_4	0.63	0.55	0.56	0.55

4.8 Practical Examples of Analyzing Multiobjective Problems

In this section, we concentrate on the utilization of the Bellman–Zadeh approach to decision-making in a fuzzy environment to solve the following power engineering problems in the multiobjective statement:

- power and energy shortage allocation as applied to load management;
- power system operation to realize dispatch on several objectives;
- optimization of network configuration in distribution systems;
- energetically effective voltage control in distribution systems.

Example 4.7 Different conceptions of load management (e.g. discussed in Talukdar and Gellings 1987; Prakhovnik et al. 1994) may be united by the following: the elaboration of control actions are performed on a two-stage basis. At the level of energy control centers, optimization of allocating power and energy shortages (natural, associated with inadequate installed power of generating sources and/or deficiency of primary energy, or with the economic advisability of load management) is carried out at different levels of territorial, temporal, and situational hierarchy of planning and operation. This allows one to draw up tasks for consumers. At the consumer level, control actions are realized in accordance with these tasks.

Thereby, the questions of power and energy shortage allocation are of fundamental importance in the family of load management problems. These questions are to be considered not only from the economic and technical points of view, but also from the social and ecological points of view. Besides, when resolving these questions, it is necessary to account for considerations related to forming incentives for consumers. The satisfaction of these requirements is possible on the basis of formulating and solving the problems within the framework of multiobjective models. This permits one to consider and minimize diverse implications of power and energy shortage allocation and to create incentives for consumers.

The application of the multiobjective approach to load management is also beneficial in providing a new look at problems of planning and operation of electricity markets (Stoft 2002). In particular, all market participants aspire to maximize their benefits (economical, technological, social, political, etc.). The goals of market participants, as a rule, come into conflict, which may be resolved by searching for a corresponding compromise. Its objective is the formation of mutually advantageous and harmonious relations between the market participants (Pedrycz et al. 2011).

The substantial analysis of problems of power and energy shortage allocation, systems of economic management, including taxation policy, as well as real, readily available reported and planned information, has led to the construction of a general set of objectives to solve these problems in a multiobjective statement. The complete list includes 17 types of objectives. Without listing all of them, it is possible to indicate the following selected objectives:

1. Prevailing constraint of consumers with a lower cost of production and/or given services on consumed 1 kWh of energy (achievement of a minimal drop in total production and/or given services).
4. Prevailing constraint of consumers with a lower level of payment in the state budget and/or a lower level of lease payment for basic production resources (funds) on consumed 1 kWh of energy.
12. Prevailing constraint of consumers with a higher level of energy possession coefficient of work on 1 kWh of energy consumed (achievement of a maximal drop in the number of workers, whose productivity and, consequently, salary is diminished because of limitations).
13. Prevailing constraint of consumers with a higher level of environmental pollution on consumed 1 kWh of energy.
15. Prevailing constraint of consumers with a lower value of the demand coefficient (primary limitation of consumers with greater possibilities of production out the peak time).
16. Prevailing constraint of consumers with a lower duration of using maximum load in 24 hours (primary limitation of consumers with greater possibilities in transferring maximum load in the daily interval).

The general set of goals is sufficiently complete because it is directed at decreasing diverse negative consequences for consumers and creating incentives for them. This set is universal because it can serve as the basis for constructing models at different levels of load management hierarchy by the aggregation of information and posterior decomposition of the problems in accordance with different indices. The concrete list of goals can be defined for every case by a DM, who can an individual or a group (for instance, the DM may be the leading organization of a country or state, a council of directors of enterprises, etc. whose decision regarding the list of goals can be considered as the legislative one at the corresponding level, see Pedrycz et al. 2011).

Consider the solution to the problems of power shortage allocation formalized within the framework of the model Eqs. (4.45), (4.50), and (4.51) for six consumers for A^1= 20 000 kW, A^2= 30 000 kW, A^3= 40 000 kW, A^4= 50 000 kW, and A^5= 60 000 kW considering the objectives 1, 12, 15, and 16 listed before. These objectives are described by the linear objective functions

$$F_p(x) = \sum_{i=1}^{6} c_{pi} \Delta x_i, \quad p = 1, 15, 16 \tag{4.70}$$

that are to be minimized and

$$F_{12}(x) = \sum_{i=1}^{6} c_{12,i}\Delta x_i \tag{4.71}$$

that is to be maximized.

Table 4.12 includes the initial information to solve the problems. The results obtained on the basis of using the min operator (Δx^0), product operator (Δx^{00}) as well as Boldur's method (Δx^{000}) for A^1= 20 000 kW and A^5= 60 000 kW are given in Table 4.13 and 4.14.

To reflect the quality of solutions obtained on the basis of different approaches Table 4.15 includes the mean magnitudes of absolute values $\delta(\Delta x)$ of deviations of membership function levels (satisfaction levels) $\mu_{A_p}(\Delta x^0)$ from their mean values $\hat{\mu}_{A_p}(\Delta x^0)$ calculated as follows:

Table 4.12 Initial information.

i	$c_{1,i}$, monetary units/kWh	$c_{12,i}$	$c_{15,i}$	$c_{16,i}$, hours	A_i, kW
1	1.50	5.40	0.63	15.30	14 000
2	4.10	6.20	0.33	17.20	6000
3	1.40	5.80	0.28	21.10	4000
4	2.20	5.30	0.21	18.50	7000
5	1.20	4.20	0.26	17.40	19 000
6	2.13	4.70	0.36	19.60	14 000

Table 4.13 Power shortage allocation.

i	$\Delta x^{1,0}$	$\Delta x^{1,00}$	$\Delta x^{1,000}$	$\Delta x^{5,0}$	$\Delta x^{5,00}$	$\Delta x^{5,000}$
1	5398	5804	0	13 020	14 000	14 000
2	2515	1104	0	5076	5731	6000
3	2399	870	0	3986	4000	4.000
4	950	6898	1000	6223	7000	7000
5	6738	5324	19 000	19 000	19 000	19 000
6	0	0	0	12 695	10 269	10 000

Table 4.14 Power shortage allocation.

p	1	2	3	4
$\mu_{A_p}(\Delta x^{1,0})$	0.604	0.605	0.605	0.606
$\mu_{A_p}(\Delta x^{1,00})$	0.615	0.590	0.633	0.630
$\mu_{A_p}(\Delta x^{1,000})$	0.974	0.020	0.951	0.596
$\mu_{A_p}(\Delta x^{5,0})$	0.428	0.431	0.428	0.428
$\mu_{A_p}(\Delta x^{5,00})$	0.366	0.700	0.353	0.714
$\mu_{A_p}(\Delta x^{5,000})$	0.321	0.750	0.357	0.741

Table 4.15 Mean deviations.

$\delta(\Delta x)$	A^1	A^2	A^3	A^4	A^5
$\delta(\Delta x^0)$	0	0.003	0.052	0.060	0.001
$\delta(\Delta x^{00})$	0.015	0.010	0.100	0.192	0.174
$\delta(\Delta x^{000})$	0.327	0.327	0.290	0.194	0.203

$$\delta(x) = \frac{1}{q}\sum_{p=1}^{q}\left|\mu_{A_p}(\Delta x^0) - \hat{\mu}_{A_p}(\Delta x^0)\right| \tag{4.72}$$

where

$$\hat{\mu}_{A_p}(\Delta x^0) = \frac{1}{q}\sum_{p=1}^{q}\mu_{A_p}(\Delta x^0) \tag{4.73}$$

In reality, the use of the expression Eq. (4.72) reflects the degree of deviation from the fully harmonious solution.

Table 4.15 covers the cases reflected in Table 4.13 and 4.14 as well as A^2= 30 000 kW, A^3= 40 000 kW, and A^4= 50 000 kW. The results given in Table 4.15 demonstrate that $\Delta x^0 \succ \Delta x^{00}$ and $\Delta x^0 \succ \Delta x^{000}$. The high quality of the solutions Δx^0 (from the point of view of their severity) is also confirmed by inequalities $\min_{1\le p\le q}\mu_{A_p}(\Delta x^0) > \min_{1\le p\le q}\mu_{A_p}(\Delta x^{00})$ and $\min_{1\le p\le q}\mu_{A_p}(\Delta x^0) > \min_{1\le p\le q}\mu_{A_p}(\Delta x^{000})$, which are observed for all cases. This confirms the validity of the utilization of the principle of guaranteed results based on applying min operator, which, as it was indicated before, is of the most noncompensatory behavior (Yager 1988).

The inequalities $\min_{1 \le p \le q} \mu_{A_p}(\Delta x^{00}) > \min_{1 \le p \le q} \mu_{A_p}(\Delta x^{000})$ also take place for all cases. This permits one to assert that decision-making in a fuzzy environment, even with a product operator, provides us with solutions more harmonious than on the basis of the traditional method.

To show the possibility of modifying solutions as a result of changing the importance of objective functions, assume, for example, for the case $A^1 = 20\,000$ kW, that the fourth objective function ($p = 16$) has the level "weak superiority" relative to other objective functions, which have the level "identical significance" relative to each other. The paired comparisons generate $\lambda_1 = \lambda_{12} = \lambda_{15} = 0.67$ and $\lambda_{16} = 0.33$. The solution corresponding to these important factors is the following: $x_1^{1,0} = 7,429$ kW, $x_2^{1,0} = 2,530$ kW, $x_3^{1,0} = 0$ kW, $x_4^{1,0} = 2,171$ kW, $x_5^{1,0} = 7,870$ kW, and $x_6^{1,0} = 0$ kW. This solution generates $\mu_{A_1}(\Delta x^{1,0}) = 0.493$, $\mu_{A_{12}}(\Delta x^{1,0}) = 0.386$, $\mu_{A_{15}}(\Delta x^{1,0}) = 0.386$, and $\mu_{A_{16}}(\Delta x^{1,0}) = 0.886$.

It is possible to increase to a higher degree the importance of the fourth objective function, using, for example, the level "strong superiority" relative to other objective functions.

In this case, we have $\lambda_1 = \lambda_{12} = \lambda_{15} = 0.5$, and $\lambda_{16} = 2.5$. The solution with these importance factors is the following: $x_1^{1,0} = 8.089$ kW, $x_2^{1,0} = 3,966$ kW, $x_3^{1,0} = 0$ kW, $x_4^{1,0} = 3$ kW, $x_5^{1,0} = 7,492$ kW, and $x_6^{1,0} = 0$ kW. This solution produces $\mu_{A_1}(\Delta x^{1,0}) = 0.306$, $\mu_{A_{12}}(\Delta x^{1,0}) = 0.329$, $\mu_{A_{15}}(\Delta x^{1,0}) = 0.168$, and $\mu_{A_{16}}(\Delta x^{1,0}) = 0.931$.

Example 4.8 The use of the results described previously permits one to apply the multicriteria approach to power system operation to realize dispatch on several objectives (for example, minimum of fuel cost, minimum of losses, maximum of security degree, minimum of environmental impact, etc.). This is illustrated by a case study (Berredo et al. 2011; Pedrycz et al. 2011) with the standard IEEE 30-bus system presented in Figure 4.3 (bus 1 is a slack bus) with considering objectives of minimizing losses $L(x)$, reducing sulfur oxide emissions $E_{SOx}(x)$, and reducing nitrogen oxide emissions $E_{NOx}(x)$.

The details of generator characteristics are given in Table 4.16. This also includes the coefficients (Nimura et al. 2001) for estimating levels of SOx and NOx emissions on the basis of the following expressions:

$$E_{SOx,i}(x_i) = a_{SOx,i}x_i^2 + b_{SOx,i}x_i + c_{SOx,i} \tag{4.74}$$

and

$$E_{NOx,i}(x_i) = a_{NOx,i}x_i^2 + b_{NOx,i}x + c_{NOx,i} \tag{4.75}$$

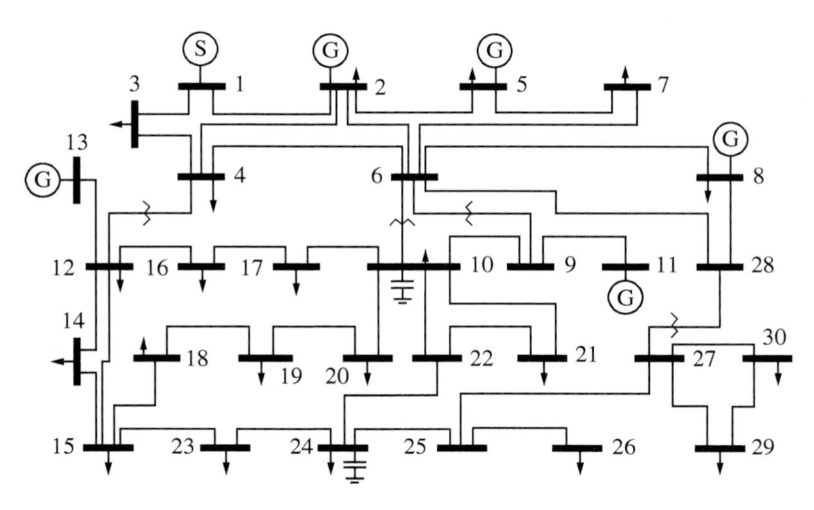

Figure 4.3 System diagram.

Table 4.16 Generator characteristics.

i	x_1	x_2	x_3	x_4	x_5
Bus	2	5	8	11	13
Fuel	hydro	gas	oil	coal	hydro
$a_{\text{SOx},\,i}$	0	0	0.010	0.015	0
$b_{\text{SOx},\,i}$	0	0	0.800	1.200	0
$a_{\text{NOx},\,i}$	0	0.010	0.015	0.030	0
$b_{\text{NOx},\,i}$	0	0.200	0.300	0.600	0
Limit	300	240	180	150	180

The initial loads of generators are the following: $x_1^{(0)} = 85.50$ MW, $x_2^{(0)} = 66.92$ MW, $x_3^{(0)} = 66.76$ MW, $x_4^{(0)} = 58.91$ MW, and $x_5^{(0)} = 48.67$ MW with $L(x^{(0)}) = 9.04$ MW, $E_{SOx}(x^{(0)}) = 220.72$ kg/h, and $E_{NOx}(x^{(0)}) = 284.52$ kg/h.

The consideration of Eqs. (4.74) and (4.75) creates no difficulties at all. At the same time, the presentation of the function $L(x)$ in an explicit form meets difficulties. The difficulties can be alleviated on the basis of applying procedures of sequential multiobjective optimization by using sensitivity models reflecting the loss change for each optimization step.

It is rational to construct the sensitivity models (Ekel et al. 2013) on the basis of experimental design techniques (Box et al. 1978; Jain 1991).

In particular, varying the loads of generators for the initial state on levels $x_i^{(0)} - \delta x_i^{(0)}$ and $x_i^{(0)} + \delta x_i^{(0)}$, $i = 1, 2, \ldots, 5$ and applying 2^{5-2} fractional design (Box et al. 1978; Jain 1991), it is possible to construct the following sensitivity model:

$$\bar{L}\left(X^{(0)}\right) = 0x_1 - 0.0814x_2 - 0.0494x_3 - 0.0475x_4 + 0x_5 \tag{4.76}$$

The linearized objective function Eq. (4.76) has been constructed for $\delta = 0.05$. Generally, the linearized objective function

$$\bar{L}\left(X^{(m)}\right) = c_1^{(m)}x_1 + c_2^{(m)}x_2 + c_3^{(m)}x_3 + c_4^{(m)}x_4 + c_5^{(m)}x_5 \tag{4.77}$$

together with the objective functions Eqs. (4.74) and (4.75) and the constraints for generators

$$\max\left\{0, x_i^{(m)} - \delta x_i^{(m)}\right\} \le x_i \le \min\left\{x_i^{(m)} + \delta x_i^{(m)}, x_i^{\max}\right\},\ i = 1, 2, \ldots, 5 \tag{4.78}$$

form the multiobjective problem for the mth step. In Eq. (4.78), $x_i^{\max}$ is a maximally permissible load of a generator.

Table 4.17 reflects the results obtained at successive steps of the sequential multiobjective optimization based on the application of the Bellman–Zadeh approach to analyzing multiobjective models.

Example 4.9 The most important functions of Distribution Management Systems are realized on the basis of solving problems of optimizing network configuration (network reconfiguration) and optimizing voltage control in distribution systems.

The problems of network reconfiguration are associated with altering distribution network topological structures by changing the state of network switches (in other words, by changing locations of network disconnections). These

Table 4.17 Results of successive steps of multiobjective optimization.

Step	$x_1^{(m)}$	$x_2^{(m)}$	$x_3^{(m)}$	$x_4^{(m)}$	$x_5^{(m)}$	$L(X^{(m)})$	$E_{SO_X}(X^{(m)})$	$E_{NO_X}(X^{(m)})$
0	85.50	66.92	66.76	58.91	48.67	9.04	220.72	284.52
1	85.86	70.26	67.25	56.67	46.71	8.91	215.23	281.81
2	84.34	73.78	66.36	53.91	48.38	8.78	205.41	274.69
3	82.44	77.47	65.20	51.31	50.35	8.60	195.73	268.60
4	83.56	81.35	64.94	48.74	48.18	8.55	188.25	265.70

problems are solved in long- and short-term planning and operation, and can be applied in design studies. An increased interest in solving these problems is associated with wide automation of distribution systems whose switches are remotely monitored and controlled. This makes it possible to solve the problems on-line in real time (Berredo et al. 2011).

Many works have been dedicated to their solution obtained on the basis of diverse approaches. For instance, we can cite Baran and Wu (1989); Broadwater et al. (1993); Carreno et al. (2008); Viswanadha Raju and Bijwe (2008); Zhu (2009); and Kumar and Jayabarathi (2012). Taking this into account, it is necessary to indicate that the majority of existing studies are focused on solving monocriteria problems (usually, power losses or energy losses are minimized). However, network reconfiguration problems are inherently multicriteria in nature because they have impact on reliability, service quality, and economical feasibility of power supply.

Taking this into account, the developed computing system named DNOS, designed to deal with multicriteria optimization of network configuration in distribution systems, permits one to consider and to minimize objective functions of power losses, energy losses, system average interruption frequency index (SAIFI) (IEEE 2004), system average interruption duration index (SAIDI) (IEEE 2004), undersupply energy, poor energy quality consumption (consumption of energy outside of permissible limits), and integrated overload of network elements in diverse combinations.

The solutions to monocriteria as well as multicriteria problems of optimizing network configuration are based on the use of the univariate method (Rao 1996) and its modifications, which are flexible and adaptable to different practical solution strategies as well as to a technology of representing information on network topology. It is worth noting that the solution to the multicriteria problems is associated with analyzing the conditions Eqs. (4.55) and (4.56) at each optimization step to make a decision on transition from the solution $x^{(k)}$ to the solution $x^{(k+1)}$.

Let us consider the results of solving a simple problem of optimizing configuration of a network 13.8 kV, which includes 77 busses and 87 branches (Berredo et al. 2011; Pedrycz et al. 2011). The following objective functions are considered and minimized: power losses ΔP, energy losses ΔW, and poor energy consumption ΔN.

The results of monocriteria optimization on ΔP, ΔW, and ΔN are presented in Table 4.18–4.20, respectively. Table 4.21 reflects the results of multicriteria optimization. To better observe the advantages of applying the multicriteria approach, Table 4.22 reflects comparative results based on monocriteria and multicriteria optimization. These results demonstrate that the application of the multicriteria approach leads to the harmonious solution with small deviations from locally optimal solutions obtained for each objective function.

Table 4.18 Levels of objective functions (monocriteria optimization on ΔP).

Objective function	Initial state	Optimal state	Objective function reduction, %
ΔP, kW	99.15	64.10	35.35
ΔW, kWh	1191.32	849.09	28.73
ΔN, kWh	502.42	1086.09	−116.17

Table 4.19 Levels of objective functions (monocriteria optimization on ΔW).

Objective function	Initial state	Optimal state	Objective function reduction, %
ΔP, kW	99.15	64.56	34.89
ΔW, kWh	1191.32	835.49	29.87
ΔN, kWh	502.42	349.97	30.34

Table 4.20 Levels of objective functions (monocriteria optimization on ΔN).

Objective function	Initial state	Optimal state	Objective function reduction, %
ΔP, kW	99.15	84.88	14.39
ΔW, kWh	1191.32	1146.49	3.76
ΔN, kWh	502.42	145.92	70.96

Table 4.21 Levels of objective functions (multicriteria optimization).

Objective function	Initial state	Optimal state	Objective function reduction, %
ΔP, kW	99.15	66.98	32.45
ΔW, kWh	1191.32	854.66	28.26
ΔN, kWh	502.42	188.28	52.53

Table 4.22 Levels of objective functions (monocriteria and multicriteria optimization).

Objective function	Optimization on ΔP	Optimization on ΔW	Optimization on ΔN	Multicriteria optimization
ΔP, kW	64.10	64.56	84.88	66.98
ΔW, kWh	849.09	835.48	1146.49	854.66
ΔN, kWh	1086.09	349.97	145.92	188.28

Example 4.10 As indicated previously, the optimization of voltage control in distribution systems is an important function of Distribution Management Systems.

The problem of optimizing modes of operation of basic means for voltage control in distribution systems is associated with choosing off-load taps for distribution transformers and conditions for operating tap changing under load transformers or voltage regulators of feeding substations (Berredo et al. 2011). The techniques for optimizing voltage control implemented within the framework of the developed computing system VCOS are directed at minimizing poor energy quality consumption on the basis of applying a synthesis of the integral criteria of voltage quality and calculations on permissible voltage levels (Berredo et al. 2011; Pedrycz et al. 2011).

In particular, the energy-weighted average voltage drops from buses of feeding substations to the centers of loads of low voltage networks of distribution transformers are used to choose their off-load taps.

The choice of conditions for operating tap changing under load transformers of feeding substations is considered as a stage that follows the choice of off-load taps for distribution transformer. However, if it is impossible to change off-load taps, this stage can be considered independent.

The optimal conditions for controlling a tap changing under load transformer of a feeding substation may be obtained (Berredo et al. 2011; Pedrycz et al. 2011) by providing the voltage addition at the bus of this substation that is equal in magnitude and opposite in sign to the power-weighted voltage levels in the centers of loads of low voltage networks of all distribution transformers for each step of loads curves. Thereby, it is possible to talk about the construction of a relationship $E_\psi = f_E(\psi)$ where $\psi = 1, 2, \ldots, \Psi$ is a step of load curves.

The regulation laws that are encountered in practice are the following (Berredo et al. 2011):

- voltage stabilization $E_S = a_S$;
- voltage control with current correction $E_I = a_I + b_I I$;
- voltage control with active power correction $E_P = a_P + b_P P$.

If we know, for example, the active load curve for a tap changing under load transformer of a feeding substation $P_\psi = f_P(\psi)$, $\psi = 1, 2, \ldots, \Psi$ and $E_\psi = f_E(\psi)$, $\psi = 1, 2, \ldots, \Psi$, and eliminating the parameter ψ, we can obtain a set of points in the system coordinates $P - E_\psi$. By applying the method of least squares to this set of points, the regulation law $E_P = a_P + b_P P$ can be obtained (Berredo et al. 2011; Pedrycz et al. 2011).

The described approach, as indicated previously, serves for minimizing poor energy quality consumption. However, according to a situational control hierarchy, the necessity of energetically effective voltage control can arise, considering load static characteristics. In particular, the results of De Steese et al.

(1990) demonstrate the possibility of significant reduction of peak load and energy consumption as a result of voltage reduction. This generates attempts to include this effect in the formulation of traditional distribution system problems, for instance (Milosevic and Begovic 2004). Thus, it is necessary to look at the second problem statement directed at minimizing poor energy quality consumption and reducing peak load and/or energy consumption (Berredo, et al. 2011; Pedrycz et al. 2011). The solution of this problem is one of functions of the VCOS system implemented on the basis of the results described previously.

Let us consider the results of solving the problem of optimizing voltage control in a 13.8 kV network that includes six feeders with 2629 distribution transformers (Berredo et al. 2011).

The voltage control law $E_P = a_P + b_P P$ and a reducing peak load consumption law $E_P^C = a_P^C + b_P^C P$, obtained with the use of the VCOS system, are visualized in Figure 4.4. In this case, where the objective functions have equal significance, the peak load is reduced by 6.68%; however, poor energy quality consumption is increased by 24.8%. At the same time, Figure 4.5 presents the solution, where the importance of the criterion of poor energy quality consumption has "absolute superiority" in relation to reducing peak load consumption. The law $E_P^C = a_P^C + b_P^C P$ of Figure 4.5 leads to reduction of peak load by 2.8% and an increase of poor energy quality consumption by 6.5%.

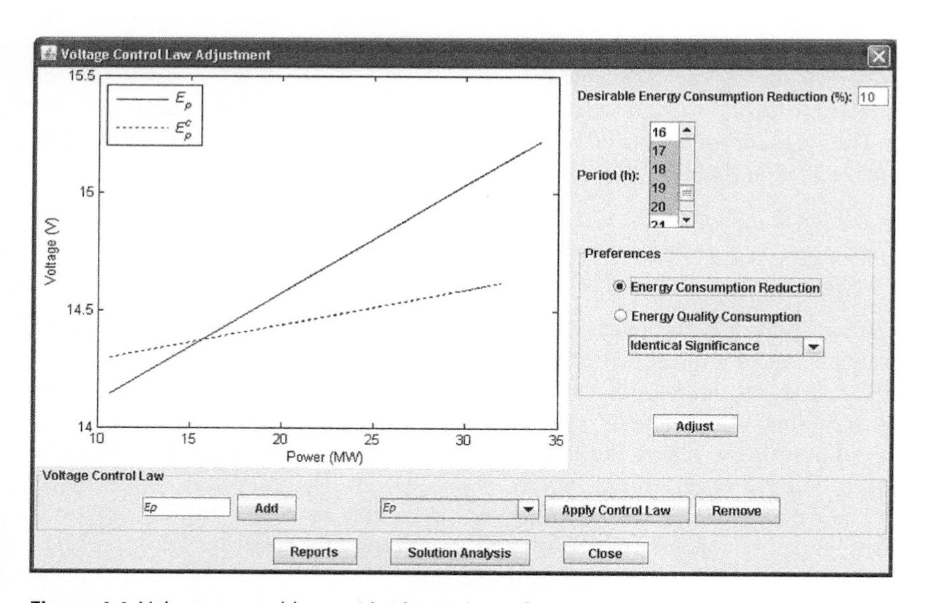

Figure 4.4 Voltage control laws with identical significance.

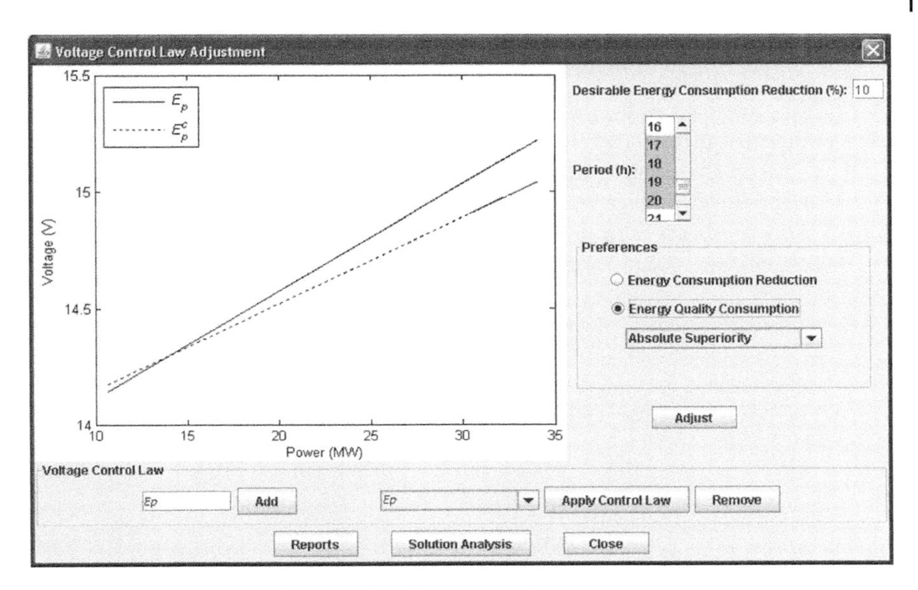

Figure 4.5 Voltage control laws with different significance.

4.9 Conclusions

We have concentrated on the analysis of models of multiobjective decision-making (<*X*, *F*> models). The basic ideas related to problems of multiobjective decision-making as well as the main classes of technique for their solution have been discussed. Much attention has been given to the Bellman–Zadeh approach to decision-making in a fuzzy environment to analyze multicriteria problems. Its application provides the concept of "optimal solution" with reasonable validity: here the maximum degree of satisfying all objectives serves as a criterion of optimality. This is consistent with the principle of guaranteed result, generating a constructive line in obtaining harmonious solutions on the basis of solving associated max-min problems. The questions of using the OWA operator have been discussed. Its application provides the possibility to implement several concepts of optimality in accordance with the level of optimism or pessimism of a DM. A special attention has been given to an important class of problem of multiobjective allocation of resources or their shortages. The corresponding <*X*, *F*> models are built with the use of an original approach to the homogeneous, clear, transparent, and expert-acceptable formulation of specific objectives. The use of the presented results has been illustrated by solving problems related to the power engineering area as well as problems of strategic planning.

Exercises

4.1 The characterization of eight alternatives for choosing a power supply scheme is given in Table 4.23. The comparison of alternatives is to be executed on the basis of the following objective functions: energy losses ΔW, undersupply energy ΔE, and poor energy consumption ΔN.

Try to reduce the number of alternatives on the basis of the notion of the optimal Pareto solutions.

Table 4.23 Levels of objective functions.

Alternative	ΔW	ΔE	ΔN
x_1	$10\,241.10^3$	312.10^3	1027.10^3
x_2	$11\,635.10^3$	300.10^3	1234.10^3
x_3	$10\,210.10^3$	314.10^3	1116.10^3
x_4	$10\,284.10^3$	316.10^3	1211.10^3
x_5	$11\,243.10^3$	324.10^3	1190.10^3
x_6	$10\,493.10^3$	331.10^3	1017.10^3
x_7	$10\,233.10^3$	318.10^3	1098.10^3
x_8	$10\,220.10^3$	320.10^3	1094.10^3

4.2 Solve the following problem of multiobjective optimization:

$$F_1(x) = 3x_1 + 11x_2 \rightarrow \min$$

$$F_2(x) = 9x_1 + 5x_2 \rightarrow \min$$

$$F_3(x) = 4x_1 + 5x_2 \rightarrow \min$$

subject to the constraints

$$0 \leq x_1 \leq 20$$

$$0 \leq x_2 \leq 10$$

$$x_1 + x_2 = 20$$

Apply the convolution Eq. (4.4) where the normalization of the objective function is realized on the basis of Eq. (4.5).

4.3 Solve the problem presented in Problem 4.2 by applying Boldur's method.

4.4 Solve the same problem (Problem 4.2) by applying the method based on imposing constraints on levels of objective functions to improve the quality of the solution (to provide a more harmonious solution).

4.5 Consider a problem (Borisov et al. 1990) of selecting a manager for an enterprise. There are five candidates for this position. They are evaluated on the basis of the following indicators: professional skills (C_1), organizational capabilities (C_2), work experience (C_3), authority (C_4), ability to work with people (C_5), and age (C_6). The results of evaluating the candidates are the following:

$$C_1 = \{0.9/x_1, 0.9/x_2, 0.6/x_3, 0.8/x_4, 0.5/x_5\}$$
$$C_2 = \{0.8/x_1, 0.9/x_2, 0.5/x_3, 0.7/x_4, 0.6/x_5\}$$
$$C_3 = \{0.7/x_1, 0.9/x_2, 0.8/x_3, 0.5/x_4, 0.3/x_5\}$$
$$C_4 = \{0.9/x_1, 0.8/x_2, 0.5/x_3, 0.6/x_4, 0.5/x_5\}$$
$$C_5 = \{0.9/x_1, 0.9/x_2, 0.4/x_3, 0.7/x_4, 0.6/x_5\}$$
$$C_6 = \{0.9/x_1, 0.4/x_2, 0.8/x_3, 0.7/x_4, 0.5/x_5\}$$

Try to select the best candidate assuming that all indicators have the same importance.

4.6 Consider a problem (Borisov et al. 1990) of selecting the location of a building for an enterprise. There are four alternatives. They are evaluated on the basis of the following indicators: proximity to a consumer (C_1), proximity to sources of raw materials (C_2), and availability of labor force (C_3). The results of evaluating the alternatives are the following:

$$C_1 = \{0.5/x_1, 0.7/x_2, 0.3/x_3, 0.6/x_4\}$$
$$C_2 = \{0.5/x_1, 0.4/x_2, 0.8/x_3, 0.4/x_4\}$$
$$C_3 = \{0.2/x_1, 0.1/x_2, 0.6/x_3, 0.9/x_4\}$$

Select the best alternative if the indicators have different importance reflected by the following matrix of pairwise comparisons:

$$R = \begin{bmatrix} 1 & 5 & 1/3 \\ 1/5 & 1 & 1/9 \\ 3 & 9 & 1 \end{bmatrix}$$

4.7 Solve the following problem of multiobjective optimization:

$$F_1(x) = 6x_1 + 9x_2 \to \min$$
$$F_2(x) = 8x_1 + 3x_2 \to \min$$

subject to the constraints

$$x_1 + x_2 = 10$$
$$x_1, x_2 \geq 0$$

Apply the Bellman–Zadeh approach and consider two cases:

a) the objective functions have identical importance,

b) the first objective function is three times more important than the second objective function.

4.8 Solve the following problem of multiobjective optimization:

$$F_1(x) = 3x_1 + 9x_2 \rightarrow \min$$

$$F_2(x) = 3x_1 + 5x_2 \rightarrow \max$$

subject to the constraints

$$x_1 + x_2 = 10$$

$$x_1, x_2 \geq 0$$

Apply the Bellman–Zadeh approach. Study two cases:

a) the objective functions have identical importance,

b) the first objective function is two times less important than the second objective function.

References

Baran, M. and Wu, F. (1989). Network reconfiguration in distribution systems for loss reduction and load balancing. *IEEE Transactions on Power Delivery* 4 (2): 1401–1407.

Beliakov, G. and Warren, J. (2001). Appropriate choice of aggregation operators in fuzzy decision support systems. *IEEE Transactions on Fuzzy Systems* 9 (6): 773–784.

Bellman, R. and Giertz, M. (1974). On the analytic formalism of the theory of fuzzy sets. *Information Sciences* 5 (2): 149–157.

Bellman, R.E. and Zadeh, L.A. (1970). Decision-making in a fuzzy environment. *Management Science* 17 (1): 141–164.

Belyaev, L.S. (1977). *A Practical Approach to Choosing Alternative Solutions to Complex Optimization Problems under Uncertainty*. IIASA, Laxenburg: IIASA.

Benayoun, R., de Montgolfier, J., Tergny, J., and Larichev, O.I. (1971). Linear programming with multiple objective functions: STEP method (STEM). *Mathematical Programming* 1 (3): 366–375.

Bentley, P.J. and Wakefield, J.P. (1998). Finding acceptable Pareto-optimal solutions using multiobjective genetic algorithms. In: *Soft Computing in Engineering Design and Manufacturing* (eds. P.K. Chawdhry, R. Roy and R.K. Pant), 231–240. Berlin/Heidelberg: Springer-Verlag.

Berredo, R.C., Ekel, P.Y., Martini, J.S.C. et al. (2011). Decision making in fuzzy environment and multicriteria power engineering problems. *International Journal of Electrical Power & Energy Systems* 33 (3): 623–632.

Borisov, A.N., Krumberg, O.A., and Fedorov, I.P. (1990) *Decision Making on the Basis of Fuzzy Models*, Riga (in Russian): Zinatne.

Box, G.E.P., Hunter, W.G., and Hunter, J.S. (1978). *Statistics for Experiments: An Introduction to Design, Data Analysis and Model Building.* New York: Wiley.

Broadwater, R.P., Khan, A.H., Shaalan, H.E., and Lee, R.E. (1993). Time varying load analysis to reduce distribution losses through reconfiguration. *IEEE Transactions on Power Delivery* 8 (1): 294–300.

Burkov, V. and Kondrat'ev, V. (1981). *Mechanisms of Functioning Organizational Systems.* Moscow (in Russian): Nauka.

Canha, L., Ekel, P., Queiroz, J., and Schuffner Neto, F. (2007). Models and methods of decision making in fuzzy environment and their applications to power engineering problems. *Numerical Linear Algebra with Applications* 14 (3): 369–390.

Carreno, E.M., Romero, R., and Padilha-Fertin, A. (2008). An efficient codification to solve distribution network reconfiguration for loss reduction problem. *IEEE Transactions on Power Systems* 23 (4): 1542–1551.

Charnes, A. and Cooper, W.W. (1961). *Management Models and Industrial Applications of Linear Programming.* New York: John Wiley & Sons.

Charnes, A., Cooper, W.W., and Ferguson, R. (1955). Optimal estimation of executive compensation by linear programming. *Management Science* 1 (2): 138–151.

Chen, T.Y. (2011). A multimeasure approach to optimism and pessimism in multiple criteria decision analysis based on Atanassov fuzzy sets. *Expert Systems with Applications* 38 (10): 12569–12584.

Coelho, C.A.C. (2000) Handling preferences in evolutionary multiobjective optimization: A survey. In Proceedings of the 2000 Congress on Evolutionary Computation, San Diego, pp. 30–37.

Coelho, C.A.C. (2002). Evolutionary multi-objective optimization: Critical Review. In: *Evolutionary Optimization* (eds. R. Sarker, M. Mohammadian and X. Yao), 117–146. Boston: Kluwer.

Coelho, C.A.C., Van Veldhuizen, D.A., and Lamont, G.B. (2002). *Evolutionary Algorithms for Solving Multi-Objective Problems.* New York: Kluwer.

Das, I. and Dennis, J.E. (1998). Normal-boundary intersection: a new method for generating the Pareto surface in nonlinear multicriteria optimization problems. *SIAM Journal of Optimization* 8 (4): 631–657.

De Steese, J.G., Merrick, S.B., and Kennedy, B.W. (1990). Estimating methodology for a large regional application of conservation voltage reduction. *IEEE Transactions on Power Systems* 5 (3): 862–870.

Deb, K. (2001). *Multi-Objective Optimization Using Evolutionary Algorithms.* Chichester: John Wiley & Sons.

Dubov, Y.A., Travkin, C.J., and Yakimets, V.N. (1986). *Multicriteria Models for Forming and Choosing System Alternatives.* Moscow (in Russian): Nauka.

Ehrgott, M. (2005). *Multicriteria Optimization.* Berlin: Springer-Verlag.

Ekel, P.Y. (2001). Methods of decision making in fuzzy environment and their applications. *Nonlinear Analysis: Theory Methods and Applications* 47 (5): 979–990.

Ekel, P.Y. (2002). Fuzzy sets and models of decision making. *Computers and Mathematics with Applications* 44 (7): 863–875.

Ekel, P.Y. and Galperin, E.A. (2003). Box-triangular multiobjective linear programs for resource allocation with application to load management and energy market problems. *Mathematical and Computer Modelling* 37 (1): 1–17.

Ekel, P.Y., Menezes, M., and Schuffner Neto, F. (2007). Decision making in fuzzy environment and its application to power engineering problems. *Nonlinear Analysis: Hybrid Systems* 1 (4): 527–536.

Ekel, P.Y., Martini, J.S.C., and Palhares, R.M. (2008). Multicriteria analysis in decision making under information uncertainty. *Applied Mathematics and Computation* 200 (2): 501–516.

Ekel, P., Junges, M., Kokshenev, I., and Parreiras, R. (2013). Sensitivity and functionally oriented models for power system planning, operation, and control. *International Journal of Electrical Power & Energy Systems* 45 (1): 489–500.

Ekel, P., Kokshenev, I., Parreiras, R. et al. (2016). Multiobjective and multiattribute decision making in a fuzzy environment and their power engineering applications. *Information Sciences* 360-361 (1): 100–119.

Gardiner, L.R. and Steuer, R.E. (1994). Unified interactive multiple objective programming: An open architecture for accommodating new procedures. *Journal of the Operational Research Society* 45 (12): 1456–1466.

Horn, J. (1997). Multicriteria decision making and evolutionary computation. In: *Handbook of Evolutionary Computation, Vol. 1* (eds. T. Bäck, D.B. Fogel and Z. Michalewicz), F1.9.1–F.1.9.15. Oxford: Oxford University Press.

Hwang, C.L. and Masud, A.S.M. (1979). *Multiple Objective Decision Making: Methods and Applications*. Heidelberg: Springer-Verlag.

IEEE (2004). *Guide for Electric Power Distribution Reliability Indices: IEEE Standard 1366TM-2003*. New York: IEEE.

Ignizio, J.P. (1976). *Goal Programming and Extensions*. Lexington: Lexington Books.

Jain, R. (1991). *The Art of Computer Systems Performance Analysis: Techniques for Experimental Design, Measurement, Simulation, and Modeling*. New York: John Wiley & Sons.

Jones, D.F. and Tamiz, M. (2002). Goal programming in the period 1990-2000. In: *Multiple Criteria Optimization: State of the Art Annotated Bibliographic Surveys* (eds. M. Ehrgott and X. Gandibleux), 129–170. Boston: Kluwer.

Kaplan, R.S. and Norton, D.P. (1996). *The Balanced Scorecard: Translating Strategy into Action*. Boston: Harvard Business Press.

Kaplan, R.S. and Norton, D.P. (2004). *Strategy Maps: Converting Intangible Assets into Tangible Outcomes*. Boston: Harvard Business Press.

Keeney, R.L. and Raiffa, H. (1976). *Decision with Multiple Objectives: Preferences and Value Tradeoffs*. New York: John Wiley & Sons.

Konak, A., Coit, D.W., and Smith, A.R. (2006). Multi-objective optimization using genetic algorithms: A tutorial. *Reliability Engineering & System Safety* 91 (9): 992–1007.

Kuhn, H.W. and Tucker, A.W. (1951) Nonlinear programming. In Proceedings of the Second Berkeley Symposium on Mathematical Statistics and Probability, Berkeley, CA, pp. 481–492.

Kumar, K.S. and Jayabarathi, T. (2012). Power system reconfiguration and loss minimization for an distribution systems using bacterial foraging optimization algorithm. *International Journal of Electrical Power & Energy Systems* 36 (1): 13–17.

Lai, Y.L. and Hwang, C.L. (1996). *Fuzzy Multiple Objective Decision Making: Methods and Applications.* Berlin: Springer-Verlag.

Lee, S.M. (1972). *Goal Programming for Decision Analysis.* Philadelphia: Auerback.

Liu, X. and Han, S. (2008). Orness and parameterized RIM quantifier aggregation with OWA operators: A summary. *International Journal of Approximate Reasoning* 48 (1): 77–97.

Lu, J., Zhang, G., Ruan, D., and Wu, F. (2008). *Multi-Objective Group Decision Making: Methods, Software and Applications with Fuzzy Set Techniques.* London: Imperial College Press.

Luce, R.D. and Raiffa, H. (1957). *Games and Decisions.* New York: John Wiley & Sons.

Lyapunov, A.A. (ed.) (1972). *Operational Research. Methodological Aspects.* Moscow (in Russian): Nauka.

Malczewski, J. (2006). Integrating multicriteria analysis and geographic information systems: the ordered weighted averaging (OWA) approach. *International Journal of Environmental Technology and Management* 6 (1/2): 7–19.

Mashunin, Y.K. (1986). *Methods and Models of Vector Optimization.* Moscow (in Russian): Nauka.

Milosevic, B. and Begovic, M. (2004). Capacitor placement for conservative voltage reduction on distribution feeders. *IEEE Transactions on Power Delivery* 19 (3): 1360–1367.

Monarchi, D.E., Kisiel, C.C., and Duckstein, L. (1973). Interactive multiobjective programming in water resources: A case study. *Water Resources Research* 9 (4): 837–850.

Nimura, T, Moreira, F.A, Nakashima, T., Cavati, C.R., and Yokoyama, R. (2001) Multiple attribute performance evaluation of independent power producers in a deregulated power system based on fuzzy sets, *In: Proceedings of the International Conference on Intelligent System Application to Power Systems,* Rio de Janeiro, pp. 122–126.

Pedrycz, W., Ekel, P., and Parreiras, R. (2011). *Fuzzy Multicriteria Decision-Making: Models, Methods, and Applications.* Chichester: John Wiley & Sons.

Pereira, J.G. Jr., Ekel, P.Y., Palhares, R.M., and Parreiras, R.O. (2015). On multicriteria decision making under conditions of uncertainty. *Information Sciences* 324 (1): 44–59.

Podinovsky, V.V. and Gavrilov, V.M. (1975). *Optimization on Sequentially Utilized Criteria*. Moscow (in Russian): Sovetskoe Radio.

Prakhovnik, A.V., Ekel, P.Y., and Bondarenko, A.F. (1994). *Models and Methods of Optimizing and Controlling Modes of Operation of Electric Power Supply Systems*. Kiev (In Ukrainian): ISDO.

Raiffa, H. (1968). *Decision Analysis*. Reading: Addison-Wesley.

Rao, S. (1996). *Engineering Optimization*. New York: John Wiley & Sons.

Raskin, L.G. (1976). *Analysis of Complex Systems and Elements of Optimal Control Theory*. Moscow (in Russian): Sovetskoe Radio.

Romero, C. (1991). *Handbook of Critical Issues in Goal Programming*. Oxford: Pergamon Press.

Roy, B. (1972). Décisions avec critères multiples: Problèmes et méthodes. *Merta International* 11 (2): 121–151.

Roy, B. (2010). To better respond to the robustness concern in decision aiding: four proposals based on a twofold observation. In: *Handbook of Multicriteria Analysis* (eds. C. Zopounidis and P.M. Pardalos), 3–24. Berlin, Heidelberg: Springer.

Sakawa, M. (1993). *Fuzzy Sets and Interactive Multiobjective Optimization*. New York: Plenum Press.

Sklyarov, V.F., Prakhovnik, A.V., and Ekel, P.Y. (1987). On the multicriteria power consumption control. *Electronic Modeling* 9 (1): 61–65. (in Russian).

Statnikov, R.B. and Matusov, J.B. (2002). *Multicriteria Analysis in Engineering*. Dordrecht: Kluwer.

Stewart, T. (2005). Dealing with uncertainties in MCDA. In: *Multiple Criteria Decision Analysis – State of the Art Annotated Surveys, International Series in Operations Research and Management Science* (eds. J. Figueira, S. Greco and M. Ehrgott), 445–470. New York: Springer.

Stoft, S. (2002). *Power System Economics: Designing Markets for Electricity*. New York: John Wiley & Sons.

Talukdar, S. and Gellings, C. (1987). *Load Management*. New York: IEEE Press.

Viswanadha Raju, G.K. and Bijwe, P.R. (2008). An efficient algorithm for minimum loss reconfiguration of distribution systems based on sensitivity and heuristics. *IEEE Transactions on Power Systems* 23 (3): 1280–1287.

Webster, T.J. (2003). *Managerial Economics: Theory and Practice*. London: Academic Press.

Xu, Z. (2005). An overview of methods for determining OWA weights. *International Journal of Intelligent Systems* 20 (8): 843–865.

Yager, R.R. (1988). On ordered weighted averaging aggregation operators in multicriteria decision making. *IEEE Transactions on Systems, Man, and Cybernetics* 18 (2): 183–190.

Yager, R. (1993). Families of OWA operators. *Fuzzy Set and Systems* 59 (2): 125–148.

Yager, R.R. (1995). Multicriteria decision making using fuzzy quantifiers. In: *Proceedings of 1995 Conference on Computational Intelligence for Financial Engineering*. New York, NY, pp. 42–46.

Yager, R.R. and Filev, D.P. (1994). Parameterized and-uke and or-like OWA operators. *International Journal of General Systems* 22 (3): 297–316.
Zadeh, L.A. (1963). Optimality and non-scalar-valued performance criteria. *IEEE Transactions on Automatic Control* 8 (1): 59–60.
Zadeh, L.A. (1983). A computational approach to fuzzy quantifiers in natural languages. *Computers and Mathematics with Applications* 9 (1): 149–184.
Zarghami, M. and Szidarovszky, F. (2009). Revising the OWA operator for multi criteria decision making problems under uncertainty. *European Journal of Operational Research* 198 (1): 259–265.
Zeleny, M. (1982). *Multiple Criteria Decision Making*. New York: McGraw-Hill.
Zhu, J. (2009). *Optimization of Power System Operation*. Hoboken: John Wiley & Sons.
Zimmermann, H.J. (1996). *Fuzzy Set Theory and Its Application*. Boston: Kluwer.

5

<X, R> Models of Multicriteria Decision-Making and Their Analysis

In this chapter, we present an introduction to preference modeling realized in terms of binary fuzzy relations. In dealing with $<X, R>$ models, which serve for multiattribute decision-making, a fundamental question arises of how can one construct fuzzy preference relations? In practice, a decision-maker (DM) can directly assess fuzzy preference relations. The corresponding techniques are considered. A natural and convincing approach to constructing fuzzy preference relations based on the ordering of fuzzy quantities is discussed as well. Since any involved expert or any included criterion may require different formats for representing preferences (five main types of preference format are considered in this chapter), the questions of their conversion into fuzzy preference relations on the basis of so-called transformation functions are considered. We discuss methods used to analyze problems of multicriteria evaluation, comparison, choice, prioritization, and/or ordering of alternatives. Two types of situation exist that give rise to these problems. The first one is related to the direct statement of multiattribute decision-making problems when the consequences associated with solutions to problems cannot be estimated with a single criterion. The second class is related to problems that may be solved on the basis of a single criterion or several criteria; however, if the uncertainty of information does not permit a unique solution to be obtained, it is possible to include additional criteria and thereby convert these problems into multiattribute tasks. Diverse techniques of multiattribute analysis in a fuzzy environment are discussed. Although these techniques are directly related to individual decision-making, they can be applied to procedures of group decision-making (diverse aspects of their application in group decision-making are discussed, for instance, in Ekel et al. 2009; Parreiras et al. 2010, 2012a, 2012b; Pedrycz et al. 2011). The use of the presented results is illustrated by solving practical problems coming from different areas.

Multicriteria Decision-Making under Conditions of Uncertainty: A Fuzzy Set Perspective, First Edition. Petr Ekel, Witold Pedrycz, and Joel Pereira, Jr.

5.1 Introduction to Preference Modeling with Binary Fuzzy Relations

As stated in Chapter 2, the binary fuzzy relation consists of a fuzzy set with a bidimensional membership function $R: X \times X \rightarrow [0, 1]$. In essence, such a relation associates each ordered pair of elements (X_k, X_l), where $X_k, X_l \in X$, with an entry $R(X_k, X_l)$ coming from the unit interval that reflects the degree to which elements X_k and X_l are in the relation R.

In the case of dealing with discrete (finite) sets of alternatives in preference modeling, the binary fuzzy relations can be represented (Pedrycz et al. 2011) in two ways:

- a square matrix R of dimension of elements of X, where each entry R_{kl} corresponds to $R(X_k, X_l)$;
- a weighted graph where each element from X corresponds to a node and the relations between the elements are represented as arcs, in such a way that $R(X_k, X_l)$ corresponds to an arc oriented from X_k toward X_l.

Example 5.1 Consider a set $X = \{X_1, X_2, X_3, X_4\}$, where all possible pairs of elements are interrelated as follows: $R(X_1, X_1) = R(X_2, X_2) = R(X_3, X_3) = R(X_4, X_4) = 1$, $R(X_1, X_2) = 0.6$, $R(X_1, X_3) = 0.4$, $R(X_1, X_4) = 0.4$, $R(X_2, X_1) = 0.8$, $R(X_2, X_3) = 1$, $R(X_2, X_4) = 0.7$, $R(X_3, X_1) = 0.7$, $R(X_3, X_2) = 0.5$, $R(X_3, X_4) = 0.5$, $R(X_4, X_1) = 0.2$, $R(X_4, X_2) = 0$, and $R(X_4, X_3) = 0.6$. These dependencies can be represented as the following fuzzy binary relation

$$R = \begin{bmatrix} 1 & 0.6 & 0.6 & 0.4 \\ 0.8 & 1 & 1 & 0.7 \\ 0.7 & 0.5 & 1 & 0.5 \\ 0.2 & 0 & 0.6 & 1 \end{bmatrix} \quad (5.1)$$

or as the weighted graph in Figure 5.1.

It is possible to put in correspondence to a generic binary fuzzy relation R, its inverse (or transpose) relation R^{-1}, its complementary relation R^c, and its dual relation R^d, which are defined, respectively, as follows (Fodor and Roubens 1994b; Pedrycz et al. 2011):

$$R^{-1}(X_k, X_l) = R(X_l, X_k) \quad (5.2)$$

$$R^c(X_k, X_l) = 1 - R(X_k, X_l) \quad (5.3)$$

$$R^d(X_k, X_l) = 1 - R(X_l, X_k) = \left(R^{-1}(X_k, X_l)\right)^c \quad (5.4)$$

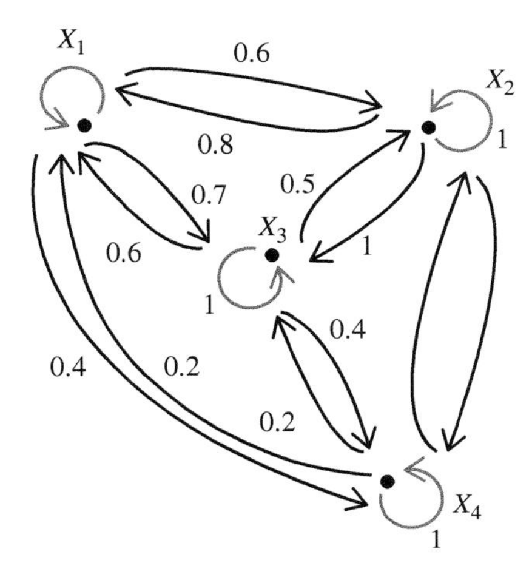

Figure 5.1 Relation R of Example 5.1 represented as a graph.

Example 5.2 The application of Eqs. (5.2)–(5.4) to matrix R of Example 5.1 generates the following matrices:

$$R^{-1} = \begin{bmatrix} 1 & 0.8 & 0.7 & 0.2 \\ 0.6 & 1 & 0.5 & 0 \\ 0.6 & 1 & 1 & 0.6 \\ 0.4 & 0.7 & 0.5 & 1 \end{bmatrix} \tag{5.5}$$

$$R^{c} = \begin{bmatrix} 0 & 0.4 & 0.6 & 0.6 \\ 0.2 & 0 & 0 & 0.3 \\ 0.3 & 0.5 & 0 & 0.5 \\ 0.8 & 1 & 0.4 & 0 \end{bmatrix} \tag{5.6}$$

$$R^{d} = \begin{bmatrix} 0 & 0.2 & 0.3 & 0.8 \\ 0.4 & 0 & 0.5 & 1 \\ 0.4 & 0 & 0 & 0.4 \\ 0.6 & 0.3 & 0.5 & 0 \end{bmatrix} \tag{5.7}$$

The operations of intersection, union, and complement of fuzzy relations were considered in Chapter 2.

The properties characterizing the binary fuzzy relations, which are of interest from the point of view of preference modeling, are considered in (Öztürk et al. 2005; Pedrycz et al. 2011) as well as in Chapter 2.

Among these properties, it is necessary to distinguish the properties related to the family of T-transitivities. In particular, as indicated in Chapter 2, the basic idea of transitivity is that the strength of the direct relationship between two elements should not be weaker than their indirect relationship involving other elements. In the case of preference modeling, transitivity can be adopted as a consistency condition related to the following consideration (Pedrycz et al. 2011): if someone says that X_k is better than X_j and that X_j is better than X_l, then it is expected that this person prefers X_k to X_l at least at a minimum strength and not that they prefer X_l to X_k.

From the practical point of view, depending on the selected transitivity property, the corresponding consistency condition may be more rigorous or more relaxed. The family of T-transitivities includes the following conditions (Pedrycz et al. 2011):

- the min-transitivity condition for fuzzy preference relations

$$R(X_k,X_l) \geq \min\left(R(X_k,X_j),R(X_j,X_l)\right) \tag{5.8}$$

- the product-transitivity condition

$$R(X_k,X_l) \geq R(X_k,X_j),R(X_j,X_l) \tag{5.9}$$

- the Lukasiewicz-transitivity condition

$$R(X_k,X_l) \geq \max\left(R(X_k,X_j) + R(X_j,X_l) - 1,0\right) \tag{5.10}$$

Example 5.3 Let us consider $R(X_1, X_3) = 0.4$ and $R(X_3, X_4) = 0.5$ from Example 5.1. Suppose that the element $R(X_1, X_4)$ is missing. It can be estimated, for instance, by the min-transitivity condition as follows: $R(X_1, X_4) \geq 0.4$.

Let us discuss questions of preference modeling on the basis of using binary fuzzy preference relations.

If a DM is asked to compare two alternatives $X_k, X_l \in X$ and determine which one of these alternatives he/she prefers, it is possible to expect one of the following answers:

- X_k and X_l are indifferent;
- X_k is strictly better than X_l;
- X_l is strictly better than X_k;
- X_k and X_l are incomparable (a DM is not able to compare the alternatives).

Taking this into account, three types of judgment can be distinguished: indifference, strict preference, and incomparability. They can be modeled with the use of binary fuzzy relations in such a way that the membership function of each binary fuzzy relation reflects the credibility (or intensity) of the judgment with quantifying into the interval [0, 1] (Pedrycz et al. 2011). The coherence between the model and the corresponding judgment is provided by some basic properties that each binary fuzzy relation has to have in accordance with the judgment that it is to reflect.

Let us consider the indifference, strict preference, and incomparability judgments as binary fuzzy relations (Pedrycz et al. 2011).

The judgment of indifference is used in situations where a DM believes that both alternatives satisfy equally his/her interests. Indifference can be modeled as a binary fuzzy relation I with the following properties:

- reflexivity: a DM is always indifferent to X_k and X_k;
- symmetry: a statement "X_k is indifferent to X_l" is equivalent to a statement "X_l is indifferent to X_k."

The judgment of strict preference is utilized when a DM can define which one is the better one of two alternatives. The strict preference can be modeled as a binary fuzzy relation P, which is to satisfy the following conditions:

- irreflexivity: a DM cannot strictly prefer X_k to X_k;
- asymmetry: a DM cannot strictly prefer X_k to X_l and X_l to X_k, at the same time.

Incomparability is used when a DM cannot express his/her opinion and in situations where a DM is asked about his/her preference and the answer is "I do not know," due to missing or uncertain information or as a consequence of the existence of conflicting information. The judgment of incomparability is reflected by means of a binary fuzzy relation J with the following properties:

- irreflexivity: a DM cannot say that X_k is incomparable to X_k;
- symmetry: a statement "X_k is incomparable to X_l" is equivalent to a statement "X_l is incomparable to X_k."

In the solution of decision-making problems based on fuzzy preference modeling, it is also important to consider a fuzzy nonstrict preference relation R (also called a fuzzy weak preference relation in literature). This relation is a starting point in analyzing models associated with binary fuzzy relations (Orlovsky 1978; Fodor and Roubens 1994a; Ekel 2002).

The relation $R(X_k, X_l)$ represents the degree to which X_k weakly dominates X_l (X_k is at least as good X_l or X_k is not worse than X_l). In a somewhat loose sense, $R(X_k, X_l)$ also represents the degree of truth of the statement "X_k is preferred over X_l" or the statement "X_k is at least as good as X_l" (Kulshreshtha and Shekar 2000).

The relation R is a reflective one (Pedrycz et al. 2011) and can be defined as the union of strict preference and indifference as follows:

$$R = P \cup I \tag{5.11}$$

Equivalently, in a less intuitive way, Eq. (5.11) (Pedrycz et al. 2011) can be stated in the following form:

$$R^d = P \cup J \tag{5.12}$$

With the use of the operations on fuzzy sets it is possible to define P, I, and J exclusively in terms of R. The corresponding results are given, for instance, in Fodor and Roubens (1993) and Pedrycz et al. (2011).

There are several ways for building the fuzzy nonstrict preference relations (some of them are discussed next). Considering this, it is necessary to indicate that regardless of the method applied to construct the fuzzy nonstrict preference relation, the fuzzy strict preference relation as well as the fuzzy indifference relation can be obtained from that relation on the basis of applying several ways. The most popular way that, probably, is one of the first introduced for this goal, is presented in Orlovsky (1978, 1981). In particular, the fuzzy strict preference relation can be presented in the following form:

$$P(X_k, X_l) = \max\left(R(X_k, X_l) - R(X_l, X_k), 0\right) \tag{5.13}$$

At the same time, the fuzzy indifference relation can be represented as follows:

$$I(X_k, X_l) = \min\left(R(X_k, X_l), R(X_l, X_k)\right) \tag{5.14}$$

Example 5.4 Let us consider the fuzzy nonstrict preference relation

$$R = \begin{bmatrix} 1 & 1 & 1 & 0.8 \\ 0.8 & 1 & 1 & 0.7 \\ 0.7 & 0.5 & 1 & 0.6 \\ 1 & 1 & 1 & 1 \end{bmatrix} \tag{5.15}$$

Applying Eq. (5.13) to Eq. (5.15), it is possible to obtain

$$P = \begin{bmatrix} 0 & 0.2 & 0.3 & 0 \\ 0 & 0 & 0.5 & 0 \\ 0 & 0 & 0 & 0 \\ 0.2 & 0.3 & 0.4 & 0 \end{bmatrix} \tag{5.16}$$

The use of Eq. (5.14) provides

$$I = \begin{bmatrix} 1 & 0.8 & 0.7 & 0.8 \\ 0.8 & 1 & 0.5 & 0.7 \\ 0.7 & 0.5 & 1 & 0.6 \\ 0.8 & 0.7 & 0.6 & 1 \end{bmatrix} \tag{5.17}$$

The way of constructing fuzzy strict preference relations and indifference relations on the basis of Eqs. (5.13) and (5.14), respectively, is not the only one. Several studies have proposed other methods. Among the most important results in this field, it is possible to refer to Ovchinnikov (1981), Roubens (1989), and Ovchinnikov and Roubens (1991), for instance. As an example, in Ovchinnikov (1981), it is proposed that

$$P(X_k, X_l) = \begin{cases} R(X_k, X_l) & \text{if } R(X_k, X_l) > R(X_l, X_k) \\ 0 & \text{if } R(X_k, X_l) \leq R(X_l, X_k) \end{cases} \tag{5.18}$$

and the indifference relation I the same as in Orlovsky (1978, 1981).

An overview of works indicated previously, as well as other similar studies, is presented in De Baets and Fodor (1997) with conclusions that these works correspond to independent efforts to define fuzzy preference relations. A more formal class of results can be found (for instance, in Alsina 1985; Ovchinnikov and Roubens 1992; Fodor and Roubens 1994b; Burfardi 1998; Llamazares 2003; Fodor and Rudas 2006; Fodor and Baets 2008). These results are based on applying axiomatic methods to extending the classical (Boolean) preference models to the fuzzy environment, in an attempt to derive the fuzzy strict preference, indifference, and incomparability relations from a fuzzy nonstrict preference relation, without losing the fuzzy counterparts of the Boolean preference structures (Pedrycz et al. 2011). One of the important practical conclusions of works indicated before and, in particular, of Fodor and Roubens (1994b), is the validity of the results of Orlovsky (1978, 1981) on constructing fuzzy strict and indifference relations on the basis of Eqs. (5.13) and (5.14). Finally, taking into account the results of Fodor and Roubens (1994a, 1994b), it is possible to define the fuzzy incomparability relation in the following form:

$$J(X_k, X_l) = \min\left(1 - R(X_k, X_l), 1 - R(X_l, X_k)\right) \tag{5.19}$$

5.2 Construction of Fuzzy Preference Relations

As was indicated before, a fuzzy nonstrict preference relation $R(X_k, X_l)$ reflects the degree to which the alternative X_k is at least as good as X_l by means of its membership function $\mu_R(X_k, X_l) : X \times X \to [0, 1]$.

When fuzzy preference relations are constructed by direct assessment, it is supposed that a DM is capable of indicating to what extent X_k is better (for instance, more or less, more intelligent, more attractive, etc.) than X_l by providing a subjective value from the unit interval. Different encoding schemes can be utilized to reflect the preference strength of one alternative over another. Following (Pedrycz et al. 2011), we discuss two schemes next. One of them is directed at the construction of a nonreciprocal fuzzy preference relation

NR (Orlovsky 1978; Fodor and Roubens 1994b; Ekel 2002), which has a correspondence with the notion of fuzzy nonstrict preference relation discussed in the previous section. Besides, a natural and convincing approach to constructing fuzzy preference relations based on the ordering of fuzzy quantities is discussed next, and also generates nonreciprocal fuzzy preference relations.

Another encoding scheme helps one to build an additive reciprocal fuzzy preference relation *RR*, which is a fuzzy preference relation satisfying (Tanino 1984; Chiclana et al. 1998; Chiclana et al. 2001) the following property of additive reciprocity:

$$RR(X_k, X_l) + RR(X_l, X_k) = 1 \quad \forall X_k, X_l \in X \tag{5.20}$$

The encoding scheme for constructing the additive reciprocal fuzzy preference relation can be presented (Pedrycz et al. 2011) in the following form:

- $RR(X_k, X_l) = 0.5$ means that X_k is indifferent to X_l;
- $0.5 < RR(X_k, X_l) \leq 1$ means that X_k is preferred to X_l;
- $0 \leq RR(X_k, X_l) < 0.5$ means that X_l is preferred to X_k;
- the entries of the main diagonal are filled with 0.5, since each element is equal to itself and, as a result, indifferent to itself.

It is natural that when a DM provides a value $RR(X_k, X_l)$, the value of $RR(X_l, X_k)$ can be obtained, taking into account Eq. (5.20), as $RR(X_l, X_k) = 1 - RR(X_k, X_l)$. The authors of (Pedrycz et al. 2011) claim that there are no preferred eliciting procedures to assist a DM to directly define additive reciprocal fuzzy preference relations. Thus, a DM has to articulate preferences based on his/her own intuition and capabilities of quantifying coherently preference strengths with the rules presented previously.

Besides, the additive reciprocal fuzzy preference relations accommodate intransitivity (Tanino 1984; Chiclana et al. 2004). However, it is desirable to collect consistent fuzzy preference relations, since, in general, the methods for the analysis of such relations also require them to be consistent in order to guarantee high-quality outcomes (Pedrycz et al. 2011). The additive transitivity property that is one of the most intuitively appealing conditions for attesting the consistency of additive reciprocal fuzzy preference relations is given in Tanino (1984) and Herrera-Viedma et al. (2004) in the following form:

$$\begin{gathered} RR(X_k, X_j) - 0.5 = (RR(X_k, X_l) - 0.5) + (RR(X_l, X_j) - 0.5) \\ \forall k, j, l \in \{1, 2, \ldots, n\} \end{gathered} \tag{5.21}$$

The elicitation process for constructing additive reciprocal fuzzy preference relations requires $n(n-1)/2$ pairwise comparisons. However, by enforcing additive transitivity, it is also possible to collect only $(n-1)$ pairwise comparisons and estimate the missing comparisons with the use of certain techniques such as the one proposed by in Herrera-Viedma et al. (2004). If a DM provides all pairwise comparisons and they do not satisfy Eq. (5.21), it is possible to identify which pairwise comparisons are to be reviewed by a DM. Another way involves using an automated method to repair (enhance) the provided judgments (without the need to ask a DM to review the corresponding judgments), by modifying them as slightly as possible, just to guarantee an acceptable level of consistency (Pedrycz et al. 2011).

Example 5.5 Four apartments (namely, X_1, X_2, X_3, and X_4) are considered for rental with taking into account their infrastructure convenience. A DM believes that X_1 and X_4 are the best alternatives, considering the transport accessibility, proximity of schools, grocery stores, restaurants, pharmacy stores, and cinemas. Besides, these alternatives are located in quiet streets. X_2 is characterized by the lack of a nearby school. Taking into account that the family has two children of 9 and 11 years old, this is a significant drawback. X_3 is characterized by the absence of nearby cinemas, which is much smaller flaw.

Taking these considerations into account, the DM provided the following comparisons between the pairs of alternatives, based on the infrastructure convenience:

- $RR(X_1, X_2) = 1$ and $RR(X_2, X_1) = 0$, since X_1 is extremely better than X_2;
- $RR(X_1, X_3) = 0.8$ and $RR(X_3, X_1) = 0.2$, since X_1 is strongly better than X_3;
- $RR(X_1, X_4) = 0.5$ and $RR(X_4, X_1) = 0.5$, since X_1 is as good as X_4;
- $RR(X_2, X_3) = 0.4$ and $RR(X_3, X_2) = 0.6$, since X_3 is to a moderate extent better than X_2;
- $RR(X_2, X_4) = 0$ and $RR(X_4, X_2) = 1$, since X_4 is extremely better than X_2;
- $RR(X_3, X_4) = 0.2$ and $RR(X_4, X_3) = 0.8$, since X_4 is strongly better than X_3.

The results of these comparisons are reflected by the following additive reciprocal fuzzy preference relation:

$$RR = \begin{bmatrix} 0.5 & 1 & 0.8 & 0.5 \\ 0 & 0.5 & 0.4 & 0 \\ 0.2 & 0.6 & 0.5 & 0.2 \\ 0.5 & 1 & 0.8 & 0.5 \end{bmatrix} \quad (5.22)$$

Although Eq. (5.22) does not fully satisfy the condition Eq. (5.21), its quality is good enough (the reader is invited to verify this).

The encoding scheme for constructing the nonreciprocal fuzzy preference relation is associated (Pedrycz et al. 2011) with the following conditions:

- if $NR(X_k, X_l) = 1$ and $NR(X_l, X_k) = 1$, then X_k is indifferent to X_l;
- if $NR(X_k, X_l) = 1$ and $NR(X_l, X_k) = 0$, then X_k is strictly preferred to X_l;
- if $NR(X_k, X_l) = 0$ and $NR(X_l, X_k) = 1$, then X_l is strictly preferred to X_k;
- if $NR(X_k, X_l) = 0$ and $NR(X_l, X_k) = 0$, then X_l is strictly preferred to X_k;
- the entries of the main diagonal are filled with 1, as each element is equal to itself and, as a result, indifferent to itself.

Intermediate judgments are allowed as well. Unlike (Pedrycz et al. 2011), they can be presented as follows:

- if $NR(X_k, X_l) = 1$ and $0 \leq NR(X_l, X_k) < 1$, then X_k is weakly preferred to X_l;
- if $0 \leq NR(X_k, X_l) < 1$ and $NR(X_l, X_k) = 1$, then X_l is weakly preferred to X_k.

Example 5.6 Taking into account the considerations related to the four apartments, which are the subject of Example 5.5, it is possible to provide (to construct the nonreciprocal fuzzy preference relation) the following comparisons between the pairs of alternatives:

- $NR(X_1, X_2) = 1$ and $NR(X_2, X_1) = 0$, since X_1 is extremely better than X_2;
- $NR(X_1, X_3) = 1$ and $NR(X_3, X_1) = 0.3$, since X_1 is strongly better than X_3;
- $NR(X_1, X_4) = 1$ and $NR(X_4, X_1) = 1$, since X_1 is as good as X_4;
- $NR(X_2, X_3) = 0.8$ and $NR(X_3, X_2) = 1$, since X_3 is to a moderate extent better than X_2;
- $NR(X_2, X_4) = 0$ and $NR(X_4, X_2) = 1$, since X_4 is extremely better than X_2;
- $NR(X_3, X_4) = 0.3$ and $NR(X_4, X_3) = 1$, since X_4 is strongly better than X_3.

The results of the comparisons given here are reflected by the following nonreciprocal fuzzy preference relation:

$$NR = \begin{bmatrix} 1 & 1 & 1 & 1 \\ 0 & 1 & 0.8 & 0 \\ 0.3 & 1 & 1 & 0.3 \\ 1 & 1 & 1 & 1 \end{bmatrix} \tag{5.23}$$

Example 5.7 The application of Eq. (5.13) to Eq. (5.23) leads to the construction of the following fuzzy strict preference relations:

$$P = \begin{bmatrix} 0 & 1 & 0.7 & 0 \\ 0 & 0 & 0 & 0 \\ 0 & 0.2 & 0 & 0 \\ 0 & 1 & 0.7 & 0 \end{bmatrix} \tag{5.24}$$

At the same time, applying Eq. (5.14) to Eq. (5.23), we can build the following fuzzy indifference preference relation:

$$I = \begin{bmatrix} 1 & 0 & 0.3 & 1 \\ 0 & 1 & 0.8 & 0 \\ 0.3 & 0.8 & 1 & 0.3 \\ 1 & 0 & 0.3 & 1 \end{bmatrix} \tag{5.25}$$

With respect to the transitivity of nonreciprocal fuzzy preference relations, it should be mentioned that the min-transitivity, estimated by Eq. (5.8), is one of the most common consistency conditions. Among the most utilized consistency conditions, it can also be indicated the weak-transitivity condition (Pedrycz et al. 2011) given as follows:

$$\begin{aligned}&\text{If } RR\left(X_k,X_j\right) \geq RR\left(X_j,X_k\right) \text{ and } RR\left(X_j,X_l\right) \geq RR\left(X_l,X_j\right) - 0.5),\\&\text{then } RR(X_k,X_l) \geq RR(X_l,X_k) \quad \forall X_k X_j, X_l \in X\end{aligned} \tag{5.26}$$

A natural and rational (from the fundamental as well as psychological points of view) approach for deriving fuzzy preference relations is the use of fuzzy estimates provided by a DM to evaluate each alternative. In essence, the use of this approach is associated with the need to compare or rank fuzzy numbers to choose the best (largest or smallest) or worst (smallest or largest) among them on the basis of applying the corresponding techniques (Pedrycz et al. 2011).

Various works can be indicated that are dedicated to the techniques for comparing or ranking fuzzy numbers (for instance, Jain 1976; Baas and Kwakernaak 1977; Baldwin and Guild 1979; Orlovsky 1981; Yager 1981; Dubois and Prade 1983; Lee and Li 1988; Tseng and Klein 1989; Chen and Hwang 1992; Fortemps and Roubens 1996; Cheng 1998; Horiuchi and Tamura 1998; Raj and Kumar 1999, Modarres and Sadi-Nezhad 2001; Chu and Tsao 2002; Facchinetti 2002; Liu and Han 2005; Abbasbandy and Asady 2006; Wang and Lee 2008; Abbasbandy and Hajjari 2009; Chen and Chen 2009; Wang and Luo 2009; Saadi-Nezhad and Shahnazari-Shahrezaei 2013; Destercke and Couso 2015).

The authors of Chen and Hwang (1992) have proposed a way for classifying the groups of techniques related to the ordering of fuzzy quantities. In particular, the following groups of techniques have been classified:

- preference relations;
- use of fuzzy mean and spread characteristics;
- fuzzy scoring techniques;
- linguistic methods.

Among these classes, the authors Horiuchi and Tamura (1998) consider the construction of fuzzy preference relations by means of pairwise comparisons as being the most practical and justified approach. Considering this, it is worth distinguishing the fuzzy number ranking index proposed by Orlovsky (1981). It is based on the concept of a membership function of a generalized preference relation.

In particular, if $F(X_k)$ and $F(X_l)$ are fuzzy sets reflecting evaluations of the objective function F or the attribute F for alternatives X_k and X_l, respectively, the quantity $\eta\{\mu[F(X_k)], \mu[F(X_l)]\}$ is the degree of preference

$\mu[F(X_k)] \succeq \mu[F(X_l)]$, while $\eta\{\mu[F(X_l)], \mu[F(X_k)]\}$ is the degree of preference $\mu[F(X_l)] \succeq \mu[F(X_k)]$. Then, the membership functions of the generalized preference relations $\eta\{\mu[F(X_k)], \mu[F(X_l)]\}$ and $\eta\{\mu[F(X_l)], \mu[F(X_k)]\}$ take the following forms:

$$\eta\{\mu[F(X_k)],\mu[F(X_l)]\} = \sup_{F(X_k),F(X_l)\in F} \min\{\mu[F(X_k)], \mu[F(X_l)],\mu_R[F(X_k),F(X_l)]\} \tag{5.27}$$

$$\eta\{\mu[F(X_l)],\mu[F(X_k)]\} = \sup_{F(X_k),F(X_l)\in F} \min\{\mu[F(X_l)], \mu[F(X_k)],\mu_R[F(X_l),F(X_k)]\} \tag{5.28}$$

respectively.

In Eqs. (5.27) and (5.28), $\mu_R[F(X_k), F(X_l)]$ and $\mu_R[F(X_l), F(X_k)]$ are the membership functions of the corresponding fuzzy preference relations that, respectively, reflect the essence of the preferences of X_k over X_l and X_l over X_k (for instance, "more attractive," "more flexible," etc.).

When F can be measured on a numerical scale and if the essence of preference behind relation R is coherent with the natural order ($\leq$) along the axis of measured values of F, then Eqs. (5.27) and (5.28), respectively, are reduced to the following expressions:

$$\eta\{\mu[F(X_k)], \mu[F(X_l)]\} = \sup_{\substack{F(X_k),\, F(X_l)\in F \\ F(X_k)\leq F(X_l)}} \min\{\mu[F(X_k)], \mu[F(X_l)]\} \tag{5.29}$$

$$\eta\{\mu[F(X_l)], \mu[F(X_k)]\} = \sup_{\substack{F(X_k),\, F(X_l)\in F \\ F(X_l)\leq F(X_k)}} \min\{\mu[F(X_k)], \mu[F(X_l)]\} \tag{5.30}$$

when F is a minimization criterion or attribute.

If F is a maximization criterion or attribute, then the relationships Eqs. (5.29) and (5.30) are to be modified. In particular, Eq. (5.29) is to be written for $F(X_k) \geq F(X_l)$ and Eq. (5.30) is to be written for $F(X_l) \geq F(X_k)$.

The expressions Eqs. (5.29) and (5.30) are consistent with the Baas–Kwakernaak (Baas and Kwakernaak 1977), Baldwin–Guild (Baldwin and Guild 1979), and one of the Dubois–Prade (Dubois and Prade 1983) fuzzy number ranking indices.

Example 5.8 Assume we are given the alternatives X_1 and X_2 with their respective fuzzy values $F(X_1)$ and $F(X_2)$ of the objective function, as illustrated in Figure 5.2. With the use of Eqs. (5.29) and (5.30), we evaluate the degrees of the preferences of X_1 over X_2 ($X_1 \succeq X_2$) and of X_2 over X_1 ($X_2 \succeq X_1$), in order to select the smallest value between $F(X_1)$ and $F(X_2)$. For illustrative purposes, Table 5.1 shows the corresponding membership functions, given just for some selected points distributed in the universe of discourse.

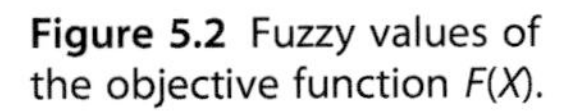

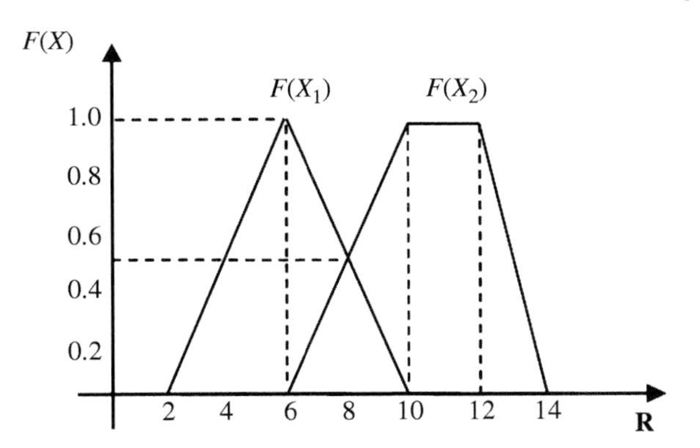

Figure 5.2 Fuzzy values of the objective function $F(X)$.

Table 5.1 Example 5.8: Membership functions of $F(X_1)$ and $F(X_2)$.

R	**2**	**4**	**6**	**8**	**10**	**12**	**14**
$\mu[F(X_1)]$	0	0.50	1	0.50	0	0	0
$\mu[F(X_2)]$	0	0	0	0.50	1	1	0

Table 5.2 Example 5.8: Cartesian product of $F(X_1)$ and $F(X_2)$.

$F(X_1)\rightarrow$ $F(X_2)$ $\downarrow$	**0/2**	**0.50/4**	**1/6**	**0.50/8**	**0/10**	**0/12**	**0/14**
0/2	**0**	0	0	0	0	0	0
0/4	0	**0**	0	0	0	0	0
0/6	0	0	**0**	0	0	0	0
0.50/8	0	0.50	0.50	**0.50**	0	0	0
1/10	0	0.50	1	0.50	**0**	0	0
1/12	0	0.50	1	0.50	0	**0**	0.35
0/14	0	0	0	0	0	0	**0**

The formal application of Eqs. (5.29) and (5.30) is associated with the construction of the Cartesian product of $F(X_1)$ and $F(X_2)$, as presented in Table 5.2.

The entries located on the main diagonal (marked in bold) and below are related to the area with $F(X_1) \leq F(X_2)$. Taking this into account, according to Eq. (5.29), it is possible to find $\eta\{\mu[F(X_1)], \mu[F(X_2)]\} = 1$. On the other hand, the entries located on the main diagonal and above are associated with $F(X_2) \leq F(X_1)$. By applying Eq. (5.30), we obtain $\eta\{\mu[F(X_2)], \mu[F(X_1)]\} = 0.50$. Therefore, we have that X_1 is equally good (small) or better (smaller) than X_2. In particular, it is possible to interpret the obtained results as follows: X_2 is equivalent to X_1 with the degree $\eta\{\mu[F(X_2)], \mu[F(X_1)]\} = 0.50$. At the same time, X_1 is strictly better than X_2 with the degree $\eta\{\mu[F(X_1)], \mu[F(X_2)]\}$-$\eta\{\mu[F(X_2)], \mu[F(X_1)]\} = 0.50$.

By the way, the fact that the level of the intersection of $F(X_1)$ and $F(X_2)$ corresponds to 0.50 (see Figure 5.2) permits one to define $\eta\{\mu[F(X_1)], \mu[F(X_2)]\} = 1$ and $\eta\{\mu[F(X_2)], \mu[F(X_1)]\} = 0.50$ without the need to construct the Cartesian product of $F(X_1)$ and $F(X_2)$. This way of determining $\eta\{\mu[F(X_k)], \mu[F(X_l)]\}$ and $\eta\{\mu[F(X_l)], \mu[F(X_k)]\}$ can be used in many practical situations.

Let us consider another example where membership functions of the compared fuzzy quantities have more than one intersection.

Example 5.9 We are given the alternatives X_1 and X_2 with fuzzy quantities of the objective function $F(X_1)$ and $F(X_2)$, which are presented in Figure 5.3. It is necessary to evaluate the degrees of the preferences X_1 over X_2 ($X_1 \succeq X_2$) and X_2 over X_1($X_2 \succeq X_1$) to select the largest value between $F(X_1)$ and $F(X_2)$. The corresponding membership functions are given in Table 5.3.

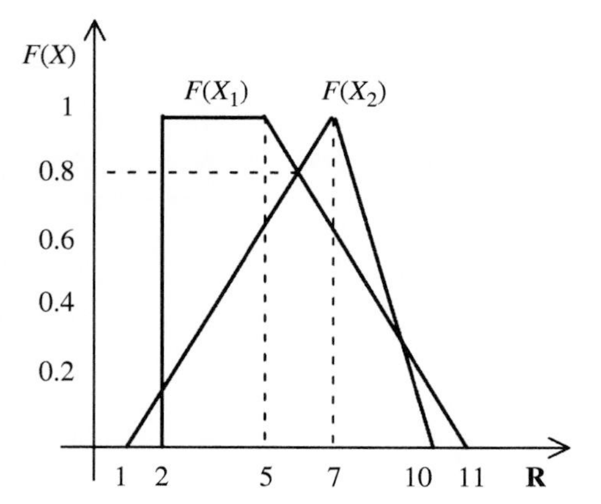

Figure 5.3 Comparison of alternatives with trapezoidal membership functions.

Table 5.3 Example 5.9: Membership functions of $F(X_1)$ and $F(X_2)$.

R	1	2	3	4	5	6	7	8	9	10	11
$F(X_1)$	0	1	1	1	1	0.83	0.67	0.50	0.34	0.17	0
$F(X_2)$	0	0.17	0.34	0.50	0.67	0.83	1	0.67	0.34	0	0

Table 5.4 Example 5.9: Cartesian product of $F(X_1)$ and $F(X_2)$.

$F(X_1)\rightarrow$ $F(X_2)$ $\downarrow$	0/1	1/2	1/3	1/4	1/5	0.83/6	0.67/7	0.50/8	0.34/9	0.17/10	0/11
0/1	**0**	0	0	0	0	0	0	0	0	0	0
0.17/2	0	**0.17**	0.17	0.17	0.17	0.17	0.17	0.17	0.17	0.17	0
0.34/3	0	0.34	**0.34**	0.34	0.34	0.34	0.34	0.34	0.34	0.17	0
0.50/4	0	0.50	0.50	**0.50**	0.50	0.50	0.50	0.50	0.34	0.17	0
0.67/5	0	0.67	0.67	0.67	**0.67**	0.67	0.67	0.50	0.34	0.17	0
0.83/6	0	0.83	0.83	0.83	0.83	**0.83**	0.67	0.50	0.34	0.17	0
1/7	0	1	1	1	1	0.83	**0.67**	0.50	0.34	0.17	0
0.67/8	0	0.67	0.67	0.67	0.67	0.67	0.67	**0.50**	0.34	0.17	0
0.34/9	0	0.34	0.34	0.34	0.34	0.34	0.34	0.34	**0.34**	0.17	0
0/10	0	0	0	0	0	0	0	0	0	**0**	0
0/11	0	0	0	0	0	0	0	0	0	0	**0**

The Cartesian product of $F(X_1)$ and $F(X_2)$, constructed on the basis of Eqs. (5.29) and (5.30) is presented in Table 5.4.

One can note that $\eta\{\mu[F(X_1)], \mu[F(X_2)]\} = 0.83$ and $\eta\{\mu[F(X_2)], \mu[F(X_1)]\} = 1$. Besides, as shown in Example 5.8, we can conclude that $F(X_1)$ is equivalent to $F(X_2)$ with the degree $\eta\{\mu[F(X_1)], \mu[F(X_2)]\} = 0.83$. At the same time, $F(X_2)$ is strictly larger (better) than $F(X_1)$ with the degree $\eta\{\mu[F(X_2)], \mu[F(X_1)]\}$-$\eta\{\mu[F(X_1)], \mu[F(X_2)]\} = 0.17$.

In such a manner, on the basis of the relations between Eqs. (5.29) and (5.30) (or, generally, between Eqs. (5.27) and (5.28)), it is possible to express the degree of preference of any of the alternatives compared and, therefore, to construct nonreciprocal fuzzy preference relations. Utilization of this approach is well founded. However, it is important to indicate that, in practice, there are situations where the fuzzy quantities $F(X_k)$ and $F(X_l)$ have trapezoidal membership functions that are located in such a way that it is not possible to distinguish X_k from X_l. For instance, we can observe that for the situation shown in Figure 5.4 the alternatives are indistinguishable since

$$\eta\{\mu[F(X_1)], \mu[F(X_2)]\} = \eta\{\mu[F(X_2)], \mu[F(X_1)]\} = \alpha \tag{5.31}$$

As will be further discussed in this chapter, in such situations, the algorithms applied to solve the problems with fuzzy coefficients do not allow one to obtain unique solutions because they "stop" when conditions like Eq. (5.31) arise. This should be considered to be a natural consequence of the existence of decision

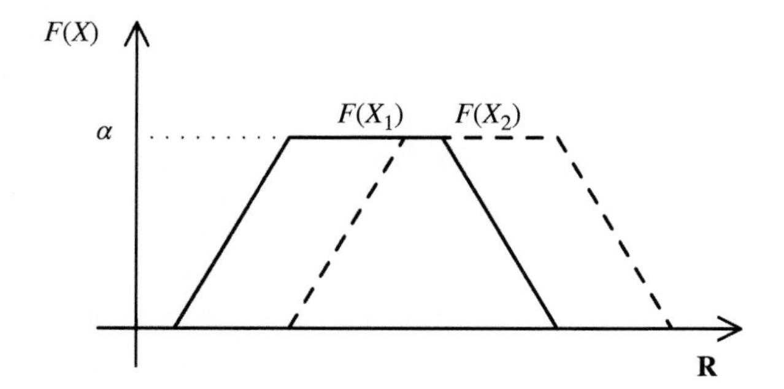

Figure 5.4 Alternatives with trapezoidal membership functions.

uncertainty regions, produced by a combination of the uncertainty and the relative stability of optimal solutions (Ekel and Popov 1985; Popov and Ekel 1987). In this connection, other indices may be used as additional means for the ranking of fuzzy numbers.

The reviews on techniques used for ranking of fuzzy numbers can be found in Dubois and Prade (1999) and Wang and Kerre (2001).

The authors Wang and Kerre (2001) enumerate more than 35 fuzzy number ranking indices and conclude the following: unlike the case of real numbers, fuzzy quantities have no natural order. The basic idea behind the methods for the ordering of fuzzy quantities consists of converting each fuzzy quantity into a real number and realizing the comparison of fuzzy quantities on the basis of the resulting real numbers. However, each approach for realizing such a conversion focuses on an intricate aspect inherent to fuzzy quantities. As a consequence, each approach suffers from some weakness associated with the loss of information inherent to the conversion of a fuzzy quantity to a single real number. The authors Wang and Kerre (2001) support this point of view by citing Freeling (1980): "by reducing the whole of our analysis to a single number, we are losing much of the information we have purposely been keeping throughout calculations." The authors Cheng (1998) and Lee-Kwang (1999) share this opinion too. Cheng (1998) also indicates that many of indices produce different rankings for the same problem. The authors Cheng (1998), Ekel et al. (1998), Lee-Kwang (1999), and Chen and Lu (2002) underline that fuzzy number ranking indices occasionally result in choices that appear inconsistent with intuition. Chen and Lu (2002) indicate that the majority of methods for the ranking of fuzzy numbers suppose that membership functions of fuzzy numbers are normalized. However, this limitation is not always adequate. Tseng and Klein (1989) indicate that the ranking methods may not reflect the preferences of interests of a DM. Further, many techniques of ordering help one only to observe an order among fuzzy quantitatives; however, they do not permit one to measure the degree of dominance among them, requiring a significant number of calculations (Chen and Klein 1997). Finally, it is necessary to mention that the majority of indices for the ranking of fuzzy quantities has been proposed with the aspiration to obligatory distinguish the alternatives, which is always questionable because the uncertainty of information creates inherent decision uncertainty regions. Taking this into account, the possibility of identifying situations where the compared alternatives cannot be distinguished should be considered a merit of the fuzzy number ranking index based on the concept of a membership function of a generalized preference relation.

Considering this, the fuzzy number ranking index based on the idea of a membership function of a generalized preference relation is used in the present book for ordering of fuzzy quantities and, on this basis, for the construction of fuzzy preference relations.

5.3 Preference Formats

Fuzzy preference relations are not a unique form of preference representation. In real-world applications, every professional involved in any decision process has their own perception of the problem, a different way of thinking, and usually has access to different information sources. As a consequence, it is natural to meet circumstances where every DM selects a different format to express their own preferences. Furthermore, several factors may lead a DM to select a different format for expressing their own preferences for each criterion. Among these factors, we can list the following (Pedrycz et al. 2011):

- Each criterion comes with its significance (a fundamental feature that provides significance to the difference between two degrees evaluated on this criterion). Depending on whether this significance has a quantitative or qualitative character, the use of certain preference formats can make the preference elicitation process easier and also more reliable.
- Each criterion is associated with information arising from different sources and with information having different levels of uncertainty.
- A DM may find that his/her preference strengths can be better reflected or quantified by a specific preference format.
- The fact that a DM may possess previous knowledge or experience in expressing a specific preference format can motivate him/her to choose it again.

Considering this, it should be noted that, for instance, the authors of Zhang et al. (2004, 2007) distinguish eight preference formats that can be used to establish preferences among analyzed alternatives. With the availability of different formats, a DM can select the one that makes him/her feel more comfortable for articulating his/her own preferences (Pedrycz et al. 2011). In this book, we deal with five preference formats that, in our opinion, cover most part of real situations in preparing information for decision-making. They include:

- ordering of the alternatives;
- utility values;
- fuzzy estimates;
- multiplicative preference relations;
- fuzzy preference relations.

The last preference format was considered before. Next, we briefly consider other formats.

5.3.1 Ordering of Alternatives

In certain cases, a DM is ready to provide a direct ranking of alternatives (from the point of view of some indicators, criteria, etc.) in accordance with his/her preferences. Sometimes this fact should be regarded as positive. In other cases, this capacity of a DM is to be considered as a necessary measure. In particular, when a DM has difficulties in assessing the strength of preferences quantitatively, it is advantageous to use information of purely ordinal character. By asking a DM to provide a complete ranking of the alternatives, he/she is released from having to quantify the difference in his/her preference strengths between any two alternatives. In this way, the chances of deriving recommendations based on incorrect information are reduced (Pedrycz et al. 2011).

The ordering of alternatives from best to worst can be represented as an array $O = \{O(X_1)\ \ O(X_2)\ldots O(X_n)\}$, with $O(X_k)$ being a permutation function that returns the position of alternative X_k among the integer values $\{1,2,\ \ldots,n\}$ (Chiclana et al. 1998).

Example 5.10 A DM is to provide the ordering of six alternatives $X = \{X_1, X_2, X_3, X_4\}$ from the best (first position) to the worst (last position) on a given criterion F. Table 5.5 shows DM judgments that permitted the construction of the following ordered array: $O = \{3\ 1\ 2\ 4\}$.

5.3.2 Utility Values

The terms "utility function" and "value function" are related to two types of preference models. Utility theory deals with preference models for risky decisions that is, decisions involving alternatives whose consequences are

Table 5.5 Ordering of alternatives.

Alternatives	X_1	X_2	X_3	X_4
Positions	Third	First	Second	Fourth

uncertain and (as a consequence) involve risks (Pedrycz et al. 2011). The value theory is considered as a simplification of utility theory for dealing with decision under certainty. In this book, we focus on preference models based on utility functions and value functions, along with their preference eliciting procedures. However, we consider only the preference elicitation process of value functions, indicating Keeney and Raiffa (1976) and Von Winterfeldt and Edwards (1986) for preference elicitation techniques associated with the construction of utility functions. The elicitation techniques for these two types of model are different enough. However, there is no need to make a distinction among them from the point of view of their conversion to fuzzy preference relations on the basis of applying transformation functions (Pedrycz et al. 2011). Taking this into account, for simplification, we use the term "utility" to refer to both types of model to provide a text coherent with some relevant references on the corresponding transformation functions (Chiclana et al. 1998, 2001; Zhang et al. 2007).

Let us consider the representation of DM preferences with the use of the preference function named the utility function $U(X)$ (by convention, the highest value of the utility function is equal to 1 and its lowest value is equal to 0). In the literature, it is possible distinguish two main types of utility function: ordinal and cardinal (Pedrycz et al. 2011). The ordinal utility function is related to the ordering of the alternatives rather than reflecting the preference strength of one alternative over another. In real-world applications, the ordinal utility function is usually modeled as a monotonically increasing (or decreasing) function, such as a maximizing profit function (or a minimizing cost function), defined over the significance axis of the considered criterion (Dyer 2005).

It is assumed that the ordinal utility function preserves the preference ordering of alternatives in such a way that

$$\text{if } U(X_k) > U(X_l), \text{then } X_k \text{ is preferred to } X_l \tag{5.32}$$

$$\text{if } U(X_k) = U(X_l), \text{then } X_k \text{ is indifferent to } X_l \tag{5.33}$$

Since in the use of ordinal utility functions the ranking of the numbers is all that matters, any monotonic transformation of this function is considered equivalent to it. Thus, the main weakness of ordinal utility functions is associated with the fact that different ordinal utility functions can be utilized for reflecting the same ordering of the alternatives (Pedrycz et al. 2011). In the aggregation across the criteria in the multicriteria analysis, each one of these admissible functions may lead to different outcomes. However, this ambiguity can be reduced by using the measurable or cardinal utility function for capturing the strength of preferences.

The most commonly utilized cardinal utility function (Farquhar and Keller 1989; Belton 1999; Dyer 2005) is based on differences in preference strengths in such a way that, given a measurable utility function, if we have that

$$U(X_k) - U(X_l) > U(X_j) - U(X_k) \tag{5.34}$$

then we can conclude that the difference in the preference between X_k and X_l is greater than the difference in preference between X_j and X_k.

It should be noted that the ratio between two preference degrees expressed as cardinal utilities based on interval scales is not meaningful, only the ratio between their differences is significant. The ratio between preference strengths only makes sense when utilities are measured on a ratio scale. This is the most rigorous type of preference measure, being admissible only when they are measured on an appropriate scale with an absolute zero (Pedrycz et al. 2011). Considering this, one can state the preference for any alternative X_k over another alternative X_l as a ratio $U(X_k)/U(X_l)$ between their respective utilities; that is, it is possible to determine how many times alternative X_k is better (or worse) than the other alternative X_l by means of this ratio.

The interval scale cardinal utility function can be determined by several ways (for instance, Winterfeldt and Edwards 1986; Belton 1999). Here, we consider one of them: the direct rating (Pedrycz et al. 2011). In its use, a DM is supposed to compare differences in preference strengths in order to determine the utility of each alternative.

The essence of the direct rating technique consists of obtaining utility values, taking as a reference only two anchor points in such a way that it is unnecessary to define a scale for characterizing other performances rather than of the alternatives being evaluated (Belton 1999). It is associated with the following sequence of steps.

- *Step 1.* A DM identifies two anchors, which correspond to the worst evaluated alternative and the best evaluated alternative from the point of view of the considered criterion. The values 0 and 1 are, respectively, assigned to them.
- *Step 2.* A DM rates the remaining alternatives in between the extreme points of the scale in such a way that the spacing between the alternatives reflects the strength of preferences of one alternative over another.
- *Step 3.* A DM should review the assessments and, whenever necessary, update them (the process stops only if a DM is in complete accordance with the elicited utility values). If the criterion under consideration can be captured by an attribute that can be measured on a numerical scale, it is possible to plot the assessed points and draw a smooth curve passing through them.

Example 5.11 A DM has modified the considerations related to the four apartments, which are the subject of Example 5.5. In particular, taking into account that X_1 is closer than X_4 to the municipal park, a DM, in Step 1 of applying the direct rating technique, identifies the apartments X_1 and X_2 as the anchors (X_1 is the best alternative and X_2 is the worst one).

In Step 2, a DM thinks that the difference in preference of X_1 over X_4 is much lower than the difference in preference of X_4 over X_2, which is characterized by

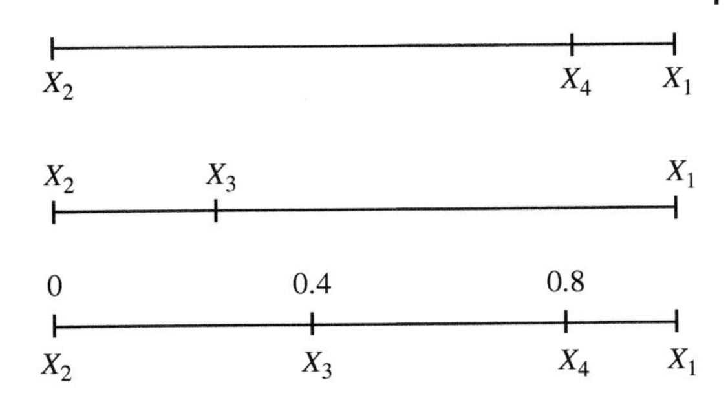

Figure 5.5 Utilities elicited by applying the direct rating technique.

the lack of nearby school (the family has two children, 9 and 11 years old). Besides, a DM thinks that the difference in preference of X_1 over X_3 is more or less equal to difference in preference of X_3 over X_2. Finally, a DM judges that X_4 is preferred to X_3. X_3 is characterized by the absence of nearby cinemas, which is more significant than the long distance to the municipal park. These considerations are reflected by Figure 5.5. In Step 3, the given assessments are confirmed with the preferences of a DM.

5.3.3 Fuzzy Estimates

The elements of X can be evaluated with the use of fuzzy estimates $L = \{l(X_1), l(X_2), \ldots, l(X_n)\}$, where $l(X_k)$ is the fuzzy estimate associated with alternative X_k completed from the point of view of a given criterion F. The fuzzy estimate $l(X_k)$ refers to a fuzzy number that can be directly specified by a DM or indirectly specified by means of linguistic terms from a set S such as, for instance, $S(F)$ ={*low velocity, average velocity, high velocity*}. The linguistic terms are to be converted into fuzzy estimates as required to perform the analysis of the problem (Pedrycz et al. 2011). Although it is possible to conclude that the use of linguistic terms makes the preference elicitation process more intuitive, it is important to indicate that the effectiveness of the elicitation can be diminished due to the existence of differences between numerical interpretation of the linguistic terms in the experts' minds and their numerical representation in the model being utilized (Pedrycz et al. 2011). In this context, the techniques discussed in Chapter 3 for constructing and equalizing fuzzy sets may be helpful for reducing this type of elicitation error.

Example 5.12 The DM utilized linguistic terms from a set $S(F)$ = {very poor, poor, average, good, very good} to evaluate the infrastructural convenience for each apartment from the previous example. The set of linguistic terms along with their respective representation through fuzzy sets are shown in Figure 5.6. The elicited preferences are given in Table 5.6.

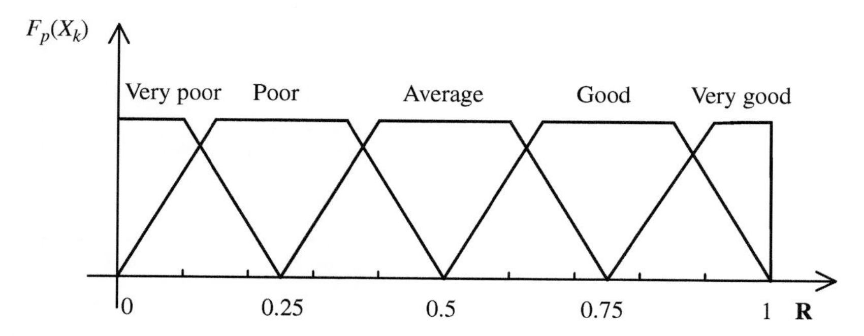

Figure 5.6 Set of linguistic terms.

Table 5.6 Evaluation of alternative by means of linguistic terms.

Apartment	Evaluation
X_1	Very good
X_2	Poor
X_3	Average
X_4	Good

5.3.4 Multiplicative Preference Relations

The multiplicative preference relation can be represented as a $n \times n$ positive reciprocal matrix MR reflecting the preference intensity ratio between the alternatives in accordance with the Analytic Hierarchy Process (AHP) approach (Saaty 1980) presented in Chapter 3 as Saaty's priority method. Each entry $MR(X_k, X_l)$ of this matrix represents a preference intensity ratio and can be interpreted as: "X_k is $MR(X_k, X_l)$ times more dominant than X_l" (Saaty 1980) or, following Chiclana et al. (2001), as "X_k is $MR(X_k, X_l)$ times as good as X_l."

The elicitation process within the AHP approach permits a DM to express preferences verbally, applying the corresponding linguistic terms or, numerically, utilizing different ratio scales (Saaty 1980). If a DM uses linguistic terms, the judgments are to be converted into numbers to realize the analysis of the decision-making problem. In this way, regardless of how the elicitation process is carried out, it is necessary to define an adequate ratio scale (Pedrycz et al. 2011). The selection of a proper ratio scale is to be done by considering the entire set of objects about which ratio comparisons are to be performed (Harker and Vargas 1987; Salo and Hämälainen 1997).

Under the situation of multiplicative reciprocity, once a DM provides $MR(X_k, X_l)$, the value of $MR(X_l, X_k)$ is automatically inferred as $MR(X_l, X_k) = 1/MR(X_k, X_l)$.

Although multiplicative preference relations accommodate intransitivity, it is desirable to collect judgments with as much consistency as possible, since the methods for the analysis of such relations usually require them to be transitive in order to guarantee results of high quality (Pedrycz et al. 2011). As was already indicated in Chapter 3, the perfect consistency of a multiplicative preference relation is reflected as the satisfaction of the multiplicative transitivity property, that is,

$$MR(X_k, X_j) = MR(X_k, X_l) \cdot MR(X_l, X_j) \quad \forall j, k, l \in \{1, 2, ..., n\} \tag{5.35}$$

In the elicitation process, it is necessary to collect $n(n-1)/2$ pairwise comparisons or else, by enforcing multiplicative transitivity, it is possible to collect only $(n-1)$ pairwise comparisons and estimate the missing ones with the use of the condition Eq. (5.35). As described in Chapter 3, when a DM provides all pairwise comparisons, if they do not satisfy Eq. (5.35) it is possible to identify the inconsistent pairwise comparisons so that a DM can review them. It is also possible to apply an automated method to improve the consistency of the constructed multiplicative preference relations (Zeshui and Cuiping 1999).

Example 5.13 Let us consider the preference elicitation process by means of a multiplicative preference relation, based on the 1–9 Saaty scale considered in Chapter 3, taking into account the same infrastructural convenience for the four apartments. The pairwise judgments provided by a DM, for instance, are the following:

- X_1 is two times better than X_2;
- X_1 is four times better than X_3;
- X_1 is eight times better than X_4;
- X_2 is two times better than X_3;
- X_2 is four times better than X_4;
- X_3 is eight times better than X_4.

These judgments are reflected in the form of the following matrix:

$$MR = \begin{bmatrix} 1 & 2 & 4 & 8 \\ 1/2 & 1 & 2 & 4 \\ 1/4 & 1/2 & 1 & 2 \\ 1/8 & 1/4 & 1/2 & 1 \end{bmatrix} \tag{5.36}$$

To verify the consistency of the obtained multiplicative preference relation, we can utilize the index of inconsistency proposed by Saaty and considered in Chapter 3. Since the maximal eigenvalue of the matrix Eq. (5.36) is 4, the index inconsistency calculated applying (3.10) is

$$\nu = \frac{(4-4)}{3} = 0 \tag{5.37}$$

In this way, we can speak about the absolute consistency of Eq. (5.36).

Example 5.14 Suppose that a DM changed his/her opinions. The new judgments are the following:

- X_1 is eight times better than X_2;
- X_1 is four times better than X_3;
- X_1 is two times better than X_4;
- X_2 is two times better than X_3;
- X_2 is four times better than X_4;
- X_3 is eight times better than X_4.

These judgments are reflected as follows:

$$MR = \begin{bmatrix} 1 & 8 & 4 & 2 \\ 1/8 & 1 & 2 & 4 \\ 1/4 & 1/2 & 1 & 2 \\ 1/2 & 1/4 & 1/2 & 1 \end{bmatrix} \tag{5.38}$$

The maximal eigenvalue of the matrix Eq. (5.38) is Eq. (5.35). The corresponding index of inconsistency is

$$\nu = \frac{(5,35-4)}{3} = 0.45 \tag{5.39}$$

Therefore, as ν is much higher than 0.1, the inconsistency level of Eq. (5.38) is not acceptable.

5.4 Transformation Functions and Their Application to Unifying Different Preference Formats

In procedures of decision-making, when different preference formats are utilized, the information is to be uniformed with the use of adequate transformation functions to be aggregated. These transformation functions serve for converting heterogeneous preference information, which may be quantitative or qualitative, two valued or fuzzy, ordered or nonordered, ordinal or cardinal, and even based on different types of scale, into fuzzy preference relations (Pedrycz et al. 2011).

The previous section showed how fuzzy preference relations can be obtained from the evaluation of the alternatives on the basis of fuzzy estimates. Next, we discuss questions of constructing fuzzy preference relations from the evaluation of the alternatives on the basis of other preference formats as well as quantitative information. First, we consider certain transformation functions for converting the preference information from the ordering of the alternatives, utility values, multiplicative preference relations, and nonreciprocal fuzzy preference relations into the additive reciprocal format. We will not consider all the existing transformation functions but just some of them. Then, we derive from those selected transformation functions other transformation functions, which can be used for converting the preference information from those different formats (including the additive reciprocal fuzzy preference relations) into the nonreciprocal format (Pedrycz et al. 2011).

5.4.1 Transformation of the Ordered Array into the Additive Reciprocal Fuzzy Preference Relation

The following transformation function, which helps one to convert the ordered array into the additive reciprocal fuzzy preference relation, is proposed in Chiclana et al. (1998):

$$RR(X_k, X_l) = \frac{1}{2}\left(1 + \frac{O(X_l) - O(X_k)}{n-1}\right) \tag{5.40}$$

Some properties of this transformation function are discussed in Chiclana et al. (1998) and Pedrycz et al. (2011). In particular, the use of Eq. (5.40) produces the additive reciprocal fuzzy preference relation that satisfies the additive transitivity.

Example 5.15 Let us consider the ordered array of Example 5.10, presented in Table 5.5. For instance, applying Eq. (5.40) to the pair X_2 and X_3, we can obtain the following results:

$$RR(X_2, X_3) = \frac{1}{2}\left(1 + \frac{2-1}{3}\right) = 0.67 \tag{5.41}$$

$$RR(X_3, X_2) = \frac{1}{2}\left(1 + \frac{1-2}{3}\right) = 0.33 \tag{5.42}$$

The application of Eq. (5.40) to other pairs of alternatives permits us to construct the following additive reciprocal fuzzy preference relation:

$$RR = \begin{bmatrix} 0.50 & 0.17 & 0.33 & 0.67 \\ 0.83 & 0.50 & 0.67 & 1 \\ 0.67 & 0.33 & 0.50 & 0.83 \\ 0.33 & 0 & 0.17 & 0.50 \end{bmatrix} \tag{5.43}$$

The reader can see that the additive reciprocal fuzzy preference relation (5.43) is coherent with the array provided by the DM.

5.4.2 Transformation of the Utility Values into the Additive Reciprocal Fuzzy Preference Relation

If the utility values are estimated on a ratio scale, the additive reciprocal fuzzy preference relation can be derived (Tanino 1984; Chiclana et al. 1998; Pedrycz et al. 2011) from the corresponding array by applying the following correlations:

$$RR(X_k, X_l) = \begin{cases} \dfrac{U(X_k)}{U(X_k) + U(X_l)} & \text{if } U(X_k) + U(X_l) \neq 0 \\ 0.5 & \text{if } U(X_k) = U(X_l) = 0 \end{cases} \tag{5.44}$$

$$RR(X_k, X_l) = \begin{cases} \dfrac{[U(X_k)]^2}{[U(X_k)]^2 + [U(X_l)]^2} & \text{if } U(X_k) + U(X_l) \neq 0 \\ 0.5 & \text{if } U(X_k) = U(X_l) = 0 \end{cases} \tag{5.45}$$

It is necessary to indicate that the transformation functions Eqs. (5.44) and (5.45) generate the additive reciprocal fuzzy preference relations that do not satisfy the additive transitivity, but satisfy the multiplicative transitivity (Tanino 1984; Pedrycz et al. 2011).

As the example from Pedrycz et al. (2011) presented next shows, the main difference between Eqs. (5.44) and (5.45) is associated with the fact that the strength of the additive reciprocal fuzzy preference relation calculated on the basis of Eq. (5.45) tends to be farther from the indifference judgment than the corresponding strength of the additive reciprocal fuzzy preference relation calculated with the use of Eq. (5.44).

Example 5.16 The application of Eqs. (5.44) and (5.45) to the utility array $U = \{0.2\ 0.4\ 0.1\ 0.3\}$ provides the following results:

$$RR = \begin{bmatrix} 0.50 & 0.33 & 0.67 & 0.40 \\ 0.67 & 0.50 & 0.8 & 0.57 \\ 0.33 & 0.2 & 0.50 & 0.25 \\ 0.6 & 0.43 & 0.75 & 0.50 \end{bmatrix} \tag{5.46}$$

$$RR = \begin{bmatrix} 0.50 & 0.20 & 0.80 & 0.31 \\ 0.80 & 0.50 & 0.94 & 0.64 \\ 0.50 & 0.06 & 0.50 & 0.10 \\ 0.69 & 0.36 & 0.90 & 0.50 \end{bmatrix} \tag{5.47}$$

As can be seen, the additive reciprocal fuzzy preference relations Eqs. (5.46) and (5.47) are coherent with the utility array. However, in the use of Eq. (5.47), more than in applying Eqs. (6.24), the strength of nonstrict preference of each alternative over another tends to the more distant from the judgment of indifference, being higher or lower than 0.50.

If we process a vector of cardinal utility values, defined on an interval scale normalized in [0, 1], the additive reciprocal fuzzy preference relation can be obtained by applying (Tanino 1984; Chiclana et al. 1998; Pedrycz et al. 2011) as follows:

$$RR(X_k, X_l) = \frac{1}{2}(1 + U(X_k) - U(X_l)) \tag{5.48}$$

Example 5.17 Let us recall the utility values from the Example 5.11, that is $U = \{1\ 0\ 0.5\ 0.8\}$. Applying Eq. (5.48), we obtain the following additive reciprocal fuzzy preference relation:

$$RR = \begin{bmatrix} 0.50 & 1 & 0.75 & 0.60 \\ 0 & 0.50 & 0.25 & 0.10 \\ 0.25 & 0.75 & 0.50 & 0.35 \\ 0.40 & 0.90 & 0.65 & 0.50 \end{bmatrix} \tag{5.49}$$

The reader can note that Eq. (5.49) is coherent with the vector assessed by the DM.

5.4.3 Transformation of the Multiplicative Preference Relation into the Additive Reciprocal Fuzzy Preference Relation

The authors of Chiclana et al. (2001) and Herrera-Viedma et al. (2004) have proposed the following transformation function to convert the multiplicative preference relation into the additive reciprocal fuzzy preference relation:

$$RR(X_k, X_l) = \frac{1}{2}\left(1 + \log_m MR(X_k, X_l)\right) \tag{5.50}$$

where m is an upper limit of the ratio scale.

Accepting the scale used in the AHP approach (Saaty 1980) with $m = 9$ it is possible to transform Eq. (5.50) as follows:

$$RR(X_k, X_l) = \frac{1}{2}\left(1 + \log_9 MR(X_k, X_l)\right) \tag{5.51}$$

It is necessary to indicate that if the multiplicative preference relation satisfies multiplicative transitivity, then Eq. (5.50) generates the additive reciprocal fuzzy preference relation that verify the property of additive transitivity (Herrera-Viedma et al. 2004).

Example 5.18 Let us transform the multiplicative preference relation Eq. (5.36) in Example 5.13 into the additive reciprocal fuzzy preference relation on the basis of applying of Eq. (5.51):

$$RR = \begin{bmatrix} 0.5 & 0.66 & 0.82 & 0.97 \\ 0.34 & 0.5 & 0.66 & 0.82 \\ 0.18 & 0.34 & 0.5 & 0.66 \\ 0.03 & 0.18 & 0.34 & 0.5 \end{bmatrix} \quad (5.52)$$

5.4.4 Transformation of the Nonreciprocal Fuzzy Preference Relation into the Additive Reciprocal Fuzzy Preference Relation

The following three transformation functions, which permit one to convert the nonreciprocal fuzzy preference relation into the additive reciprocal fuzzy preference relation, are considered and analyzed in Pedrycz et al. (2011):

$$RR(X_k, X_l) = \frac{1}{2}(1 + NR(X_k, X_l) - NR(X_l, X_k)) \quad (5.53)$$

$$RR(X_k, X_l) = \frac{NR(X_k, X_l)}{NR(X_k, X_l) + NR(X_l, X_k)} = \frac{1}{1 + \dfrac{NR(X_l, X_k)}{NR(X_k, X_l)}} \quad (5.54)$$

$$RR(X_k, X_l) = \frac{[NR(X_k, X_l)]^2}{[NR(X_k, X_l)]^2 + [NR(X_l, X_k)]^2} = \frac{1}{1 + \left(\dfrac{NR(X_l, X_k)}{NR(X_k, X_l)}\right)^2} \quad (5.55)$$

The transformation functions Eq. (5.54) and (5.55) are analyzed in Herrera-Viedma et al. (2007) and the transformation functions Eq. (5.53) has been proposed in Queiroz (2009).

The selection of an adequate transformation function among Eqs. (5.53), (5.54), and (5.55) depends on whether the ratio $NR(X_k, X_l)/NR(X_l, X_k)$ or the difference $NR(X_k, X_l) - NR(X_l, X_k)$ is meaningful (Pedrycz et al. 2011). In particular, the transformation function Eq. (5.53) may be utilized when the difference $NR(X_k, X_l) - NR(X_l, X_k)$ is meaningful. At the same time, the transformation functions Eqs. (5.54) and (5.55) are applicable when the ratio $NR(X_k, X_l)/NR(X_l, X_k)$ is meaningful. The authors of Pedrycz et al. (2011) note that the main difference between Eqs. (5.54) and (5.55) lies in the fact that the strength of $NR(X_k, X_l)$ obtained on the basis of Eq. (5.55) tends be farther from the indifference judgment than the corresponding strength of $NR(X_k, X_l)$ calculated on the basis of Eq. (5.54).

Furthermore, it is necessary to indicate that the additive reciprocal fuzzy preference relations constructed with the use of Eqs. (5.54) or (5.55) satisfy multiplicative transitivity if the corresponding nonreciprocal fuzzy preference relations also satisfy multiplicative transitivity.

At the same time, the additive reciprocal fuzzy preference relations obtained on the basis of applying Eq. (5.53) satisfy additive transitivity only if the corresponding nonreciprocal fuzzy preference relations satisfy the following condition (Pedrycz et al. 2011):

$$\begin{aligned}(NR(X_k,X_j)-NR(X_j,X_k))+(NR(X_j,X_l)-NR(X_l,X_j))\\+(NR(X_k,X_l)-NR(X_l,X_k))=0\end{aligned} \tag{5.56}$$

Example 5.19 Let us consider the nonreciprocal fuzzy preference relation constructed in Example 5.6. Taking into account that differences $NR(X_k, X_l)-NR(X_l, X_k)$ can be considered significant, it is possible to apply Eq. (5.53) to transform Eq. (5.23) into the additive reciprocal fuzzy preference relation. In particular, the result of the use of Eq. (5.52) is the following:

$$RR=\begin{bmatrix}0.50 & 1 & 0.85 & 0.50\\ 0 & 0.50 & 0.40 & 0\\ 0.15 & 0.60 & 0.50 & 0.15\\ 0.50 & 1 & 0.85 & 0.50\end{bmatrix} \tag{5.57}$$

The reader can see that Eq. (5.57) is well compatible with Eq. (5.23).

Example 5.20 Let us transform the following nonreciprocal fuzzy preference relation:

$$NR=\begin{bmatrix}1 & 0.4 & 0.2\\ 1 & 1 & 1\\ 1 & 0.5 & 1\end{bmatrix} \tag{5.58}$$

Taking into account that, for instance, the ratio $NR(X_2, X_1)/NR(X_1, X_2)$ is high, the transformation is to be made by Eqs. (5.54) or (5.55). In particular, the use of Eq. (5.54) leads to

$$RR=\begin{bmatrix}0.50 & 0.29 & 0.17\\ 0.71 & 0.50 & 0.67\\ 0.83 & 0.33 & 0.50\end{bmatrix} \tag{5.59}$$

At the same time, the application of Eq. (5.55) provides

$$RR=\begin{bmatrix}0.50 & 0.14 & 0.04\\ 0.86 & 0.50 & 0.80\\ 0.96 & 0.20 & 0.50\end{bmatrix} \tag{5.60}$$

5.4.5 Transformation of the Additive Reciprocal Fuzzy Preference Relation into the Nonreciprocal Fuzzy Preference Relation

In Pedrycz et al. (2011), the following transformation functions are considered that permit one to convert the additive reciprocal fuzzy preference relation into the nonreciprocal fuzzy preference relation:

$$NR(X_k, X_l) = \begin{cases} 1 + RR(X_k, X_l) - RR(X_l, X_k) & \text{if } RR(X_k, X_l) < 0.5 \\ 1 & RR(X_k, X_l) \geq 0.5 \end{cases} \tag{5.61}$$

$$NR(X_k, X_l) = \begin{cases} \dfrac{RR(X_k, X_l)}{RR(X_l, X_k)} & \text{if } RR(X_k, X_l) < 0.5 \\ 1 & \text{if } RR(X_k, X_l) \geq 0.5 \end{cases} \tag{5.62}$$

$$NR(X_k, X_l) = \begin{cases} \left(\dfrac{RR(X_k, X_l)}{RR(X_l, X_k)}\right)^{0.5} & \text{if } RR(X_k, X_l) < 0.5 \\ 1 & \text{if } RR(X_k, X_l) \geq 0.5 \end{cases} \tag{5.63}$$

These expressions (the construction of Eq. (5.62) was proposed in Queiroz 2009) represent the transformation functions that permit one to reverse conversions of Eqs. (5.53)–(5.55). The recommendations on the use of Eqs. (5.61)–(5.63) are the following (Pedrycz et al. 2011).

Equation (5.61) may be utilized when the additive reciprocal fuzzy preference relation is defined in such a way that the difference $RR(X_k, X_l) - RR(X_l, X_k)$ has a sense. The transformation functions Eqs. (5.62) and (5.63) may utilized when the additive reciprocal fuzzy preference relation is defined in such a way that the ratio $RR(X_k, X_l)/RR(X_l, X_k)$ has sense, indicating how many times X_k is preferred to X_l. The main difference between Eqs. (5.62) and (5.63) is associated with the fact that each pairwise judgment $R(X_k, X_l)$ produced by Eq. (5.62) tend to be closer to the indifference judgment than each corresponding pairwise judgment produced with the application of Eq. (5.63), as can be confirmed in Example 5.21.

Example 5.21 The nonreciprocal fuzzy preference relation Eq. (5.23) constructed in Example 5.6 has been transformed in Example 5.19 in the additive reciprocal fuzzy preference relation Eq. (5.57). Let us transform Eq. (5.57) in the nonreciprocal fuzzy preference relations, applying Eqs. (5.61)–(5.63).

It is not difficult to verify that the use of Eq. (5.61) generates the nonreciprocal fuzzy preference relations Eq. (5.23) that is natural.

The use of Eq. (5.62) to convert Eq. (5.57) generates the following nonreciprocal fuzzy preference relation:

$$NR = \begin{bmatrix} 1 & 1 & 1 & 1 \\ 0 & 1 & 0.67 & 0 \\ 0.18 & 1 & 1 & 0.18 \\ 1 & 1 & 1 & 1 \end{bmatrix} \tag{5.64}$$

At the same time, the use of Eq. (5.63) permits one to obtain

$$NR = \begin{bmatrix} 1 & 1 & 1 & 1 \\ 0 & 1 & 0.45 & 0 \\ 0.03 & 1 & 1 & 0.03 \\ 1 & 1 & 1 & 1 \end{bmatrix} \tag{5.65}$$

The transformation functions presented next permit one to convert the preference information expressed in terms of the different formats directly into the nonreciprocal fuzzy preference relation (Pedrycz et al. 2011) by substituting into Eqs. (5.61), (5.62), or (5.63), the expressions presented before for conversion from the different preference formats into the additive reciprocal fuzzy preference relation.

5.4.6 Transformation of the Ordered Array into the Nonreciprocal Fuzzy Preference Relation

In order to convert preferences expressed in terms of the ordered array into the nonreciprocal fuzzy preference relation, it is possible to use the following transformation function (Pedrycz et al. 2011):

$$NR(X_k, X_l) = \begin{cases} \frac{1}{2} + \frac{O(X_l) - O(X_k)}{2(n-1)} & \text{if } O(X_k) > O(X_l) \\ 1 & \text{if } O(X_k) \le O(X_l) \end{cases} \tag{5.66}$$

which is obtained by substituting Eq. (5.40) into Eq. (5.61). The advantage of applying Eq. (5.40) combined with Eq. (5.61) and not with Eqs. (5.62) or (5.63) is associated with the need to preserve the meaning of the differences $O(X_l) \le O(X_k)$ among the positions of two alternatives in the conversion from the ordered array into the nonreciprocal fuzzy preference relation (Pedrycz et al. 2011).

Example 5.22 The application of Eq. (5.66) to convert the ordered array considered in Example 5.10 permits one to form the following nonreciprocal fuzzy preference relation:

$$NR = \begin{bmatrix} 1 & 0.17 & 0.33 & 1 \\ 1 & 1 & 1 & 1 \\ 1 & 0.33 & 1 & 1 \\ 0.33 & 0 & 0.17 & 1 \end{bmatrix} \tag{5.67}$$

At the same time, Example 5.15 includes the result Eq. (5.43) of converting the same ordered array of Example 5.10 to the additive reciprocal fuzzy preference relation on the basis of Eq. (5.40). Considering this, it is possible to observe that the application of the transformation function Eq. (5.53) to the nonreciprocal fuzzy preference relation Eq. (5.67) generates the additive reciprocal fuzzy preference relation coinciding with Eq. (5.43).

5.4.7 Transformation of the Utility Values into the Nonreciprocal Fuzzy Preference Relation

The preferences presented as a vector of utilities defined on a ratio scale and normalized in the interval [0, 1], can be converted into a nonreciprocal fuzzy preference relation on the basis of one of the following expressions (Pedrycz et al. 2011):

$$NR(X_k, X_l) = \begin{cases} \dfrac{U(X_k)}{U(X_l)} & \text{if } U(X_k) < U(X_l) \\ 1 & \text{if } U(X_k) \geq U(X_l) \end{cases} \tag{5.68}$$

$$NR(X_k, X_l) = \begin{cases} \left(\dfrac{U(X_k)}{U(X_l)}\right)^{0.5} & \text{if } U(X_k) < U(X_l) \\ 1 & \text{if } U(X_k) \geq U(X_l) \end{cases} \tag{5.69}$$

$$NR(X_k, X_l) = \begin{cases} \left(\dfrac{U(X_k)}{U(X_l)}\right)^{2} & \text{if } U(X_k) < U(X_l) \\ 1 & \text{if } U(X_k) \geq U(X_l) \end{cases} \tag{5.70}$$

The expressions Eqs. (5.68) and (5.69) have been obtained by substituting Eq. (5.44) into Eqs. (5.62) and (5.63), respectively. At the same time, the expression Eq. (5.70) has been obtained by the substituting Eq. (5.45) into Eq. (5.62). It is necessary to indicate that the substitution Eq. (5.45) into Eq. (5.63) also corresponds to Eq. (5.62). In order to preserve the meaning of the ratio $U(X_k)/U(X_l)$ in the conversion from the utility values into a nonreciprocal fuzzy preference relation, substitution of Eqs. (5.44) or (5.45) into Eqs. (5.62) or (5.63) is preferable to substitution of Eq. (5.44) or Eq. (5.45) into Eq. (5.61) (Pedrycz et al. 2011).

Characterizing the transformation functions Eqs. (5.68)–(5.70), it is important to indicate that Eq. (5.69) produces pairwise judgments tending somewhat more to an indifference judgment than the other transformation functions do

(Pedrycz et al. 2011). At the same time, Eq. (5.70) produces pairwise judgments tending somewhat more to a strict preference than the other transformation functions do. Finally, Eq. (5.68) can be considered an intermediate case between the extreme cases, Eqs. (6.46) and (6.47), respectively. This is a consequence of the fact that the value of $R(X_k, X_l)$, when $U_p(X_k) < U_p(X_l)$, can be given by one of the ratios satisfying the relationships

$$\left(\frac{U(X_k)}{U(X_l)}\right)^2 \leq \frac{U(X_k)}{U(X_l)} \leq \left(\frac{U(X_k)}{U(X_l)}\right)^{0.5} \tag{5.71}$$

The example from Pedrycz et al. (2011) given next confirms the validity of Eq. (5.71).

Example 5.23 By applying Eqs. (5.68)–(5.70) to convert the utility array $U = \{0.2\ 0.4\ 0.1\ 0.3\}$ from Example 5.16 into a nonreciprocal fuzzy preference relation, we obtain

$$NR = \begin{bmatrix} 1 & 0.5 & 1 & 0.67 \\ 1 & 1 & 1 & 1 \\ 0.5 & 0.25 & 1 & 0.33 \\ 1 & 0.75 & 1 & 1 \end{bmatrix} \tag{5.72}$$

$$NR = \begin{bmatrix} 1 & 0.71 & 1 & 0.82 \\ 1 & 1 & 1 & 1 \\ 0.71 & 0.5 & 1 & 0.58 \\ 1 & 0.87 & 1 & 1 \end{bmatrix} \tag{5.73}$$

$$NR = \begin{bmatrix} 1 & 0.25 & 1 & 0.44 \\ 1 & 1 & 1 & 1 \\ 0.25 & 0.06 & 1 & 0.11 \\ 1 & 0.56 & 1 & 1 \end{bmatrix} \tag{5.74}$$

respectively. It is possible to observe that the obtained nonreciprocal fuzzy preference relations are coherent with the analysis of the transformation functions Eqs. (5.68)–(5.70).

Furthermore, the authors of Pedrycz et al. (2011) analyze the possibility of converting the preferences given in terms of a vector of utilities defined on an interval scale into a nonreciprocal fuzzy preference relation as follows:

$$NR(X_k, X_l) = \begin{cases} 1 + U(X_k) - U(X_l) & \text{if } U(X_k) < U(X_l) \\ 1 & \text{if } U(X_k) \geq U(X_l) \end{cases} \tag{5.75}$$

5.4.8 Transformation of the Multiplicative Preference Relation into the Nonreciprocal Fuzzy Preference Relation

One of the possible ways of transforming the multiplicative preference relation into a nonreciprocal preference relation (Pedrycz et al. 2011) is associated with the use of

$$NR(X_k, X_l) = \begin{cases} 1 + \frac{1}{2}\log_m \frac{M(X_k, X_l)}{M(X_l, X_k)} & \text{if } \log_m M(X_k, X_l) < 0 \\ 1 & \text{if } \log_m M(X_k, X_l) \geq 0 \end{cases} \tag{5.76}$$

The expression has been obtained by substituting Eq. (5.50) into Eq. (5.61). The advantage of considering Eq. (5.61) rather than Eqs. (5.62) or (5.63) lies in the fact that the meaning of the ratio $M(X_k, X_l)/M(X_l, X_k)$ is to some extent preserved, as can be seen in Eq. (5.76) (Pedrycz et al. 2011).

Example 5.24 Let us convert the multiplicative preference relation Eq. (5.36), constructed in Example 6.13, into a nonreciprocal fuzzy preference relation. The use of Eq. (5.76), considering that $m = 9$, generates

$$MR = \begin{bmatrix} 1 & 1 & 1 & 1 \\ 0.685 & 1 & 1 & 1 \\ 0.369 & 0.685 & 1 & 1 \\ 0.053 & 0.369 & 0.685 & 1 \end{bmatrix} \tag{5.77}$$

which is coherent with Eq. (5.36).

5.4.9 Transformation of the Quantitative Information into the Fuzzy Preference Relation

Assume we have an objective function $F(X)$, which is to be maximized, and we are to construct an additive reciprocal fuzzy preference relation for two alternatives X_k and X_l. Then, it is possible (Zhukovin 1988; Ekel 2002) to construct the following expression:

$$RR(X_k, X_l) = \alpha[F(X_k) - F(X_l)] + \beta \tag{5.78}$$

The compliance with the condition Eq. (5.20) for the additive reciprocal preference relations leads to $\beta = 0.5$. It permits one to obtain

$$\alpha\left[\max_{X \in L} F(X_k) - \min_{X \in L} F(X_l)\right] + 0.5 = 1 \tag{5.79}$$

and

$$\alpha = \frac{1}{2\left[\max_{X \in L} F(X_k) - \min_{X \in L} F(X_l)\right]} \tag{5.80}$$

leading to

$$RR(X_k,X_l) = \frac{F(X_k)-F(X_l)}{2\left[\max\limits_{X\in L} F(X) - \min\limits_{X\in L} F(X)\right]} + 0.5 \qquad (5.81)$$

It is not difficult to understand that for $F_p(X)$, which is to be minimized, it is possible to write the following expression:

$$RR(X_k,X_l) = \frac{F(X_l)-F(X_k)}{2\left[\max\limits_{X\in L} F(X) - \min\limits_{X\in L} F(X)\right]} + 0.5 \qquad (5.82)$$

The application of Eq. (5.61) to (5.81) or (5.82) permits one to construct the corresponding nonreciprocal fuzzy preference relations.

5.5 Optimization Problems with Fuzzy Coefficients and Their Analysis

In Chapter 1, the general issues related to the necessity of setting up and solving multicriteria problems have been discussed. In particular, one of the classes of situation, which needs the application of a multicriteria approach, is associated with problems that may be solved on the basis of a single criterion or several criteria. However, if information uncertainty does not permit one to derive unique solutions, it is possible to transform these problems, applying additional criteria including criteria of qualitative character, to reduce the decision uncertainty regions. Taking this into account, let us consider a model, which includes fuzzy coefficients in an objective function and constraints. There exist diverse problems related to system design, planning, operation, and control, which can be formalized within the framework of this type of model. Besides, although there are diverse formulations of optimization problems with fuzziness (Dubois and Prade 1980; Orlovsky 1981; Delgado et al. 1994; Zimmermann 1996; Zimmermann 2008, for instance), in the opinion of the authors Orlovsky (1981) and Pedrycz and Gomide (1998), the problems with fuzzy coefficients in objective functions and constraints are to be considered as a general problem of fuzzy mathematical programming. The problem can be formulated as follows:

$$\text{maximize}\, F(x_1,x_2,\ldots,x_n) \qquad (5.83)$$

subject to constraints

$$G_j(x_1,x_2,\ldots,x_n) \subseteq B_j,\quad j = 1,2,\ldots,m \qquad (5.84)$$

where the objective function Eq. (5.83) and constraints in Eq. (5.84) include fuzzy coefficients.

Let us consider the question of considering constraints of different nature and, primarily, the functional constraints. For simplicity, we can begin from only a single constraint of the following form (Pedrycz et al. 2011):

$$\sum_{i=1}^{n} G_i x_i \subseteq B \tag{5.85}$$

where G_i, $i = 1, 2, \ldots, n$, and B are fuzzy numbers with membership functions $\mu(G_i)$, $i = 1, 2, \ldots, n$, and $\mu(B)$, respectively.

An approach to handling constraints of the form Eq. (5.85) has been proposed in Negoita and Ralescu (1975). In particular, if certain conditions are satisfied (specifically, with regard to the convexity of the fuzzy coefficients G_i, $i = 1, 2, \ldots, n$, and B), and we assume the possibility of ordering

$$0 \le \alpha_1 < \ldots < \alpha_k < \ldots < \alpha_K \le \min\left\{ \min_{1 \le i \le n} \sup \mu(G_i), \mu(B) \right\} \tag{5.86}$$

then the constraint Eq. (5.85) can be modified to obtain the following system of deterministic inclusions:

$$\sum_{i=1}^{n} G_{i,\alpha_k} x_i \subseteq B_{\alpha_k}, \quad k = 1, 2, \ldots, K \tag{5.87}$$

where G_{i,α_k} and B_{α_k}, $k = 1, 2, \ldots, K$, are sets of the α_k-level of G_i, $i = 1, 2, \ldots, n$, and B, respectively.

Considering the definition of sets of a α_k-level (see Chapter 2), it is possible to obtain from Eq. (5.87):

$$\sum_{i=1}^{n} [g_{i_1,\alpha_k}, g_{i_2,\alpha_k}] x_i \subseteq [b_{1,\alpha_k}, b_{2,\alpha_k}], \quad k = 1, 2, \ldots, K \tag{5.88}$$

which means that

$$\sum_{i=1}^{n} g_{i_2,\alpha_k} x_i \le b_{2,\alpha_k}, \quad k = 1, 2, \ldots, K \tag{5.89}$$

and

$$\sum_{i=1}^{n} g_{i_1,\alpha_k} x_i \ge b_{1,\alpha_k}, \quad k = 1, 2, \ldots, K \tag{5.90}$$

Using the principle of explicit domination, realized on the basis of Eqs. (5.94) and (5.95) given next, we can reduce the dimensionality of the sets of inequalities Eqs. (5.89) and (5.90). As a result of normalization (Ekel et al. 1998), carried out in accordance with the expression

$$h_{i,\alpha_k} = g_{i,\alpha_k} \frac{b}{b_{\alpha_k}}, \quad k = 1, 2, \ldots, K, \quad i = 1, 2, \ldots, n \tag{5.91}$$

we can consider, instead of Eqs. (5.89) and (5.90), the sets of constraints

$$\sum_{i=1}^{n} h_{i_2,\alpha_k} x_i \le b,\ \ k = 1,2,\ldots,K \tag{5.92}$$

and

$$\sum_{i=1}^{n} h_{i_1,\alpha_k} x_i \ge b,\ \ k = 1,2,\ldots,K \tag{5.93}$$

respectively. In the Eqs. (5.91)–(5.93), $b > 0$ is a normalizing factor.

If, as a result of analyzing the set of constraints in Eq. (5.92) with $h_{i_2,\alpha_k} \ge 0$, it turns out that

$$h_{i_2,\alpha_q} \le h_{i_2,\alpha_p},\quad q \ne p,\quad i = 1,2,\ldots,n \tag{5.94}$$

then the pth constraint, for a purposeful increase in the variables x_i, $i = 1, 2, \ldots, n$, is disturbed earlier than the qth constraint. For this reason, the qth constraint can be eliminated from further consideration.

In a similar way, the condition of eliminating the qth constraint from consideration in the case of analyzing the set of constraints in Eq. (5.93) is the following:

$$h_{i_1,\alpha_q} \ge h_{i_1,\alpha_p},\quad q \ne p,\quad i = 1,2,\ldots,n \tag{5.95}$$

According to the essence of the optimization problem, one may replace constraints in Eq. (5.84) with constraints

$$g_j(x_1,\ldots,x_n) \le b_j,\quad j = 1,2,\ldots,m' \ge m \tag{5.96}$$

or constraints

$$g_j(x_1,\ldots,x_n) \ge b_j,\quad j = 1,2,\ldots,m'' \ge m \tag{5.97}$$

In this way, for the problem with constraints containing fuzzy coefficients, one can construct an equivalent nonfuzzy analog of the problem whose dimension is reduced by using the principle of explicit domination Eqs. (5.94) or (5.95).

The solution of problems containing fuzzy coefficients in objective functions alone is possible by a modification of traditional mathematical programming methods (Ekel et al. 1998; Ekel 2002).

When applying optimization methods for fuzzy problems, one needs to compare solutions on the levels of the objective function (in essence, to compare or rank corresponding fuzzy numbers to choose the largest or smallest one). If we talk about a problem of linear programming and the corresponding modification of the simplex method for its solution, it is necessary to compare coefficients of nonbasic variables with zero at any cycle of the optimization process.

Taking into consideration the previous discussion, we can apply Eqs. (5.27) and (5.28) (or Eqs. (5.29) and (5.30)) for this comparison. However, we have to keep in mind that, if the membership functions of the solutions (fuzzy numbers) $F(X_k)$ and/or $F(X_l)$ compared are trapezoidal or flat fuzzy numbers, these solutions can be indistinguishable considering the condition Eq. (5.31). In such situations, algorithms based on the modification of traditional optimization methods do not allow one to obtain unique solutions because they "stop" when conditions like Eq. (5.31) arise (Ekel et al. 1998; Galperin and Ekel 2005; Pedrycz et al. 2011). This is natural because a combination of the uncertainty and the relative stability of optimal solutions can produce decision uncertainty regions. This is illustrated by the following simple example (Galperin and Ekel 2005; Pedrycz et al. 2011) where appropriate modification of the simplex method of linear programming is applied.

Example 5.25 Consider the following problem:

$$\text{maximize}\, F(x_1, x_2) = C_1 x_1 + C_2 x_2 \tag{5.98}$$

subject to

$$G_{11} x_1 + G_{12} x_2 \subseteq B_1 \tag{5.99}$$

$$G_{21} x_1 + G_{22} x_2 \subseteq B_2 \tag{5.100}$$

$$x_1 \geq 0, \quad x_2 \geq 0, \quad \text{(numeric)} \tag{5.101}$$

where all coefficients in Eqs. (5.98)–(5.100) are trapezoidal fuzzy numbers defined as $\mu(C_1) = \{1.2, 1.3, 1.6, 1.7\}$, $\mu(C_2) = \{2.1, 2.2, 2.7, 2.8\}$, $\mu(G_{11}) = \{9, 10, 11, 12\}$, $\mu(G_{12}) = \{5, 6, 8, 9\}$, $\mu(B_1) = \{24, 29, 49, 53\}$, $\mu(G_{21}) = \{6, 7, 9, 10\}$, $\mu(G_{22}) = \{6, 7, 9, 10\}$, and $B_2 = \{25, 29, 48, 52\}$.

Considering that G_{11}, G_{12}, B_1, G_{21}, G_{22}, and B_2 are trapezoidal fuzzy numbers, it is sufficient to consider constraints in Eqs. (5.99) and (5.100) for $\alpha_1 = 0$ and $\alpha_2 = \alpha_K = 1$. For this reason, using Eqs. (5.92) and (5.93), it is possible to rewrite Eq. (5.99) as follows:

$$12x_1 + 9x_2 \leq 53 \tag{5.102}$$

$$11x_1 + 8x_2 \leq 49 \tag{5.103}$$

and

$$9x_1 + 5x_2 \geq 24 \tag{5.104}$$

$$10x_1 + 6x_2 \geq 29 \tag{5.105}$$

At the same time, the constraint Eq. (5.100) can be presented by the following inequalities:

$$10x_1 + 10x_2 \leq 52 \tag{5.106}$$

$$9x_1 + 9x_2 \leq 48 \tag{5.107}$$

and

$$6x_1 + 6x_2 \geq 25 \tag{5.108}$$

$$7x_1 + 7x_2 \geq 29 \tag{5.109}$$

Taking into consideration that we have to maximize the objective function with positive coefficients, it is possible to ignore Eqs. (5.104) and (5.105) as well as Eqs. (5.108) and (5.109). The principle of explicit domination Eq. (5.94) applied to Eqs. (5.102) and (5.103) allows us to eliminate Eq. (5.103) from further consideration. At the same time, the application of the principle of explicit domination Eqs. (5.94)–(5.106) and (5.107) permits us to eliminate Eq. (5.107).

Finally, introducing the slack variables $x_3 \geq 0$ and $x_4 \geq 0$, we transform Eqs. (5.102) and (5.106) to

$$12x_1 + 9x_2 + x_3 \leq 53 \tag{5.110}$$

$$10x_1 + 10x_2 + x_4 \leq 52 \tag{5.111}$$

respectively.

To apply the modification of the version of the simplex method given in Rao (1996), we have to consider the minimization problem instead of Eq. (5.98). Thus, our problem is

$$\text{minimize } [-F(x_1,x_2)] = -C_1x_1 - C_2x_2 \tag{5.112}$$

subject to Eqs. (5.110), (5.111), and

$$x_i \geq 0, \ i = 1,\ldots,4 \tag{5.113}$$

Using the previously mentioned modification of the simplex method with executing necessary operations on the fuzzy coefficients, discussed in Chapter 3, of the objective function, for the first cycle we obtain: $x_1 = 4.42$, $x_4 = 7.85$ (basic variables) and $x_2 = 0$, $x_3 = 0$ (nonbasic variables) with $C_2 = \{-1.90, -1.73, -1.00, -0.83\}$ and $C_3 = \{0.10, 0.11, 0.13, 0.14\}$. Since $C_2 < 0$, we are to continue to perform the second cycle: $x_1 = 2.07$, $x_2 = 3.13$ (basic variables) and $x_3 = 0$, $x_4 = 0$ (nonbasic variables) with $C_3 = \{-0.53, -0.47, -0.20, -0.13\}$, $C_4 = \{0.33, 0.40, 0.69, 0.76\}$. Since $C_3 < 0$, we are to continue, obtaining in the third cycle: $x_2 = 5.20$, $x_3 = 6.22$ (basic variables) and $x_1 = 0$, $x_4 = 0$ (nonbasic variables) with $C_1 = \{0.40, 0.60, 1.40, 1.60\}$, $C_4 = \{-0.15, -0.02, 0.51, 0.64\}$. Taking into account that in comparison of C_4 with zero, the situation Eq. (6.31) takes place, the simplex method "stops"; namely it "cannot identify" whether the optimal solution has been obtained or not.

The possible way to overcome this type of situation or, at least, to contract the decision uncertainty regions to the highest extent, is associated with the approach based on formulating and solving one and the same problem within the framework of mutually related models (Ekel et al. 1998; Ekel 2002; Pedrycz et al. 2011). In particular, the problem Eq. (5.83) with the constraints in Eq. (5.84) approximated by Eq. (5.96) and the problem

$$\text{minimize}\, F(x_1,...,x_n) \tag{5.114}$$

subject to the same constraints in Eq. (5.84) approximated by Eq. (5.97) can serve as a mutually related model.

This approach is applicable for solving problems with continuous as well as discrete variables. To understand its essence, let us proceed with an analysis of a certain discrete optimization problem.

The desirability of allowing for constraints on the discreteness of variables in the form of discrete sequences

$$x_{s_i}, \alpha_{s_i}, \beta_{s_i}, ..., \quad s_i = 1, 2, ..., r_i \tag{5.115}$$

has been validated in Zorin and Ekel (1980); here $\alpha_{s_i}, \beta_{s_i}, ...,$ are technical and economic characteristics required for formation of objective functions, constraints, and their increments that correspond to the sth standard value of the variable x_i

It is expedient to utilize discrete sequences of the type in Eq. (5.115) because the characteristics $\alpha_{s_i}, \beta_{s_i}, ...,$ cannot always be fitted closely to the analytical relationships in terms of x_{s_i}, but in discrete sequences of the type in Eq. (5.115) these characteristics may be taken to be exact. Besides, a flexible formalization of combinatorial type problems is possible on the basis of the discrete sequences because they can be different for different variables. Examples of this flexible application of the discrete sequences are given in Zorin and Ekel (1980) and Ekel and Schuffner Neto (2006).

Taking this into account with respect to the expediency of using discrete sequences and by analogy with the problem of Eqs. (5.83) and (5.84), the maximization problem can be formulated as follows.

Assume we are given discrete sequences of the type in Eq. (5.115) (depending on the problem statement the sequences could be increasing or decreasing). From these sequences it is necessary to choose elements such that the objective

$$\text{maximize}\, F\left(x_{s_1}, \alpha_{s_1}, \beta_{s_1}, ..., ..., x_{s_n}, \alpha_{s_n}, \beta_{s_n}, ...\right) \tag{5.116}$$

is met while satisfying the constraints

$$G_j\left(x_{s_1}, \alpha_{s_1}, \beta_{s_1}, ..., ..., x_{s_n}, \alpha_{s_n}, \beta_{s_n}, ...\right) \subseteq B_j, \quad j = 1, 2, ..., m \tag{5.117}$$

Given a maximization problem of the type in Eqs. (5.115)–(5.117) considered before and by analogy with Eq. (5.114), we can formulate a mutually related problem with the objective

$$\text{minimize}\, F\left(x_{s_1}, \alpha_{s_1}, \beta_{s_1}, ..., ..., x_{s_n}, \alpha_{s_n}, \beta_{s_n}, ...\right) \tag{5.118}$$

while satisfying the constraints in Eq. (5.117).

Taking this into account, the constraints in Eq. (5.117) may be reduced to the set of nonfuzzy (numeric) constraints

$$g_j\left(x_{s_1}, \alpha_{s_1}, \beta_{s_1}, \ldots, \ldots, x_{s_n}, \alpha_{s_n}, \beta_{s_n}, \ldots\right) \leq b_j, \quad j = 1, 2, \ldots, m' \geq m \tag{5.119}$$

and

$$g_j\left(x_{s_1}, \alpha_{s_1}, \beta_{s_1}, \ldots, \ldots, x_{s_n}, \alpha_{s_n}, \beta_{s_n}, \ldots\right) \geq b_j, \quad j = 1, 2, \ldots, m'' \geq m \tag{5.120}$$

Next, we consider an example from Ekel et al. (1998) and Pedrycz et al. (2011) to demonstrate the analysis of the mutually related models Eqs. (5.116) and (5.119) (with the increasing (decreasing) discrete sequences Eq. (5.115)), and (5.118) and (5.120) (with the decreasing (increasing) discrete sequences Eq. (5.115)) and the results based on its application.

Example 5.26 Assume we are given the discrete sequence

$$\begin{array}{lccc} & x_{s_i} & \alpha_{s_i} & \beta_{s_i} \\ s_i = 1: & 0, & 0, & 26 \\ s_i = 2: & 1, & 3, & 25 \\ s_i = 3: & 2, & 6, & 23 \\ s_i = 4: & 3, & 9, & 19 \\ s_i = 5: & 4, & 12, & 14 \\ s_i = 6: & 5, & 15, & 9 \end{array} \tag{5.121}$$

From this sequence, it is necessary to choose elements that maximize the objective function

$$F(x_1, x_2) = \left[\left(C_1\alpha_{s_1} + \beta_{s_1}\right) + \left(C_2\alpha_{s_2} + \beta_{s_2}\right)\right] \tag{5.122}$$

subject to the following constraints:

$$G_{11}x_{s_1} + G_{12}x_{s_2} \subseteq B_1 \tag{5.123}$$

$$G_{21}x_{s_1} + G_{22}x_{s_2} \subseteq B_2 \tag{5.124}$$

where coefficients in Eqs. (5.122)–(5.124) are trapezoidal fuzzy numbers defined as $C_1 = \{1.2, 1.4, 1.5, 1.7\}$, $C_2 = \{2.1, 2.4, 2.5, 2.8\}$, $G_{11} = \{5, 6, 8, 9\}$, $G_{12} = \{9, 10, 11, 12\}$, $B_1 = \{24, 29, 49, 53\}$, $G_{21} = \{6, 7, 9, 10\}$, $G_{22} = \{6, 7, 9, 10\}$, and $B_2 = \{25, 29, 48, 52\}$.

As in Example 5.20, it is sufficient to consider the constraints in Eqs. (5.123) and (5.124) for $\alpha_1 = 0$ and $\alpha_K = \alpha_2 = 1$. Considering this, for Eq. (5.123) we can write

$$12x_{s_1} + 9x_{s_2} \leq 53 \tag{5.125}$$

$$11x_{s_1} + 8x_{s_2} \leq 49 \tag{5.126}$$

and

$$9x_{s_1} + 5x_{s_2} \geq 24 \tag{5.127}$$

$$10x_{s_1} + 6x_{s_2} \geq 29 \tag{5.128}$$

Similarly, it is possible to go from the constraints in Eq. (5.124) to the following inequalities:

$$10x_{s_1} + 10x_{s_2} \leq 52 \tag{5.129}$$

$$9x_{s_1} + 9x_{s_2} \leq 48 \tag{5.130}$$

and

$$6x_{s_1} + 6x_{s_2} \geq 25 \tag{5.131}$$

$$7x_{s_1} + 7x_{s_2} \geq 29 \tag{5.132}$$

The principle of explicit domination in Eq. (5.94) applied to Eqs. (5.125) and (5.126) permits one to eliminate Eq. (5.126) from further consideration. The application of the principle of explicit domination in Eqs. (5.94)–(5.129) and (5.130) allows us to eliminate Eq. (5.130).

In the same way, applying the principle of explicit domination in Eqs. (5.95)–(5.127) and (5.128), we can eliminate Eq. (5.127) from further consideration. The application of the principle of explicit domination in Eqs. (5.95)–(5.131) and (5.132) permits one to eliminate Eq. (5.132).

Thus, the problem consists of the maximization of Eq. (5.122) subject to the constraints in Eqs. (5.128) and (5.131). At the same time, the mutually related problem consists of minimizing

$$F(x_1, x_2) = \left[-\left(C_1\alpha_{s_1} + \beta_{s_1}\right) - \left(C_2\alpha_{s_2} + \beta_{s_2}\right)\right] \tag{5.133}$$

subject to the constraints in Eqs. (5.128) and (5.131) by applying the discrete sequence that decreases on s_i:

$$\begin{array}{llll} & x_{s_i} & \alpha_{s_i} & \beta_{s_i} \\ s_i = 1: & 5, & 15, & 26 \\ s_i = 2: & 4, & 12, & 14 \\ s_i = 3: & 3, & 9, & 19 \\ s_i = 4: & 2, & 6, & 23 \\ s_i = 5: & 1, & 3, & 25 \\ s_i = 6: & 0, & 0, & 26 \end{array} \tag{5.134}$$

All steps of the process of solving the problem in Eqs. (5.121), (5.122), (5.125), and (5.129) on the basis of modifying the generalized algorithms of discrete optimization (Ekel et al. 1998; Ekel and Schuffner Neto 2006) are given in Ekel et al. (1998). In particular, this process "stops" when we meet a situation of the

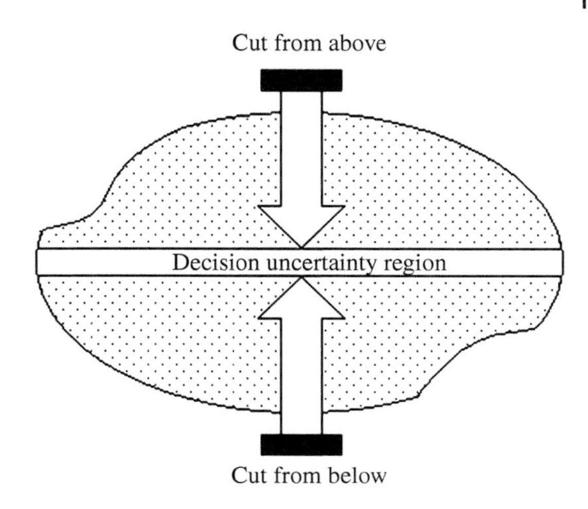

Figure 5.7 Cutting dominated alternatives.

impossibility to distinguish two solutions $X_1 = \{x_1 = 2, x_2 = 3\}$ and $X_2 = \{x_1 = 1, x_2 = 4\}$. At the same time, the solution of the mutually related problem Eqs. (5.134), (5.133), (5.128), and (5.131) leads to the solution $X_3 = \{x_1 = 0, x_2 = 5\}$. As shown in Ekel et al. (1998), there are no more solutions that are competitive. Thus, the decision uncertainty region $X = \{X_1, X_2, X_3\}$ is a formal solution of the problem in Eqs. (5.121)–(5.124).

Figure 5.7 demonstrates the essence of the approach discussed before: the solutions dominated by the initial objective function are cut off from below as well as from above to the greatest degree.

Thus, it was demonstrated that the uncertainty of information, particularly reflected by fuzzy coefficients in objective function and constraints of monocriteria problems, generates the decision uncertainty regions. Their contraction, as indicated before, is possible on the basis of reducing the problem to multicriteria decision-making with applying additional criteria, including criteria of qualitative character. It is natural that the problem of the evaluation, comparison, choice, and/or ordering of alternatives can be initially stated as a multicriteria problem.

5.6 <*X*, *R*> Models of Multicriteria Decision-Making

The problem of multicriteria (multiattribute) analysis of alternatives in a fuzzy environment can be presented as follows.

We are given a set X of alternatives from the decision uncertainty region or predetermined alternatives, which are to be examined by q criteria of quantitative and/or qualitative character. That is, indices $F_p(X_k)$, $p = 1, 2, \ldots, q$, $X_k \in X$, are to be analyzed to make a selection among alternatives. The problem of decision-making, as indicated in Chapter 1 may be presented as a pair <*X*, *R*> where $R = \{R_1, R_2, \ldots, R_p, \ldots, R_q\}$ is a vector fuzzy preference relation (Orlovsky 1981; Fodor and Roubens 1994b; Pedrycz et al. 2011). In this case we have

$$R_p = \left[X \times X, R_p(X_k, X_l)\right], \quad p = 1, 2, \ldots, q, \quad X_k, X_l \in X \tag{5.135}$$

where $R_p(X_k, X_l)$ is a membership function of the *p*th fuzzy preference relation.

Previously, we discussed the use of different preference formats for the presentation of initial information for decision-making and the rationality of using fuzzy preference relations for a uniform preference representation as well as ways of converting diverse preference formats to fuzzy preference relations. Considering this, next, we consider techniques for carrying out the evaluation, comparison, choice, prioritization, and/or ordering of alternatives on the basis of information presented by Eq. (5.135).

Next, we discuss five techniques of multiattribute analysis of alternatives in a fuzzy environment (techniques of analyzing <*X*, *R*> models). The first three techniques are directly based on the notion of the Orlovsky choice function (Orlovsky 1978, 1981). The fourth technique is also based on applying this notion. However, this technique allows a DM to present information related to the importance of fuzzy preference relations in a fuzzy form, particularly in the form of nonreciprocal fuzzy preference relations. The fifth techniques can be considered to be the generalization of the Orlovsky choice function on the basis of applying the ordered weighted average (OWA) operator, discussed in Chapter 4.

Next, we use *R* to denote fuzzy nonstrict preference relations. As we only make use of nonreciprocal fuzzy preference relations here, we do not use any particular notation to indicate whether the fuzzy preference relation under consideration is an additive reciprocal fuzzy preference relation or a nonreciprocal fuzzy preference relation.

5.7 Techniques for Analyzing <*X*, *R*> Models

In this section, we consider techniques of the analysis of <*X*, *R*> models, which are based on the application of the notion of the Orlovsky choice function. This notion was introduced by Orlovsky (1978, 1981) and has since been studied by many researchers. In particular, its axiomatic characterization is given, for instance, in Banerjee (1993); Bouyssou (1997); and Sengupta (1998). The authors of Barrett et al. (1990) demonstrated many interesting and desirable properties of the Orlovsky choice function.

Let us consider the situation of setting up a single fuzzy nonstrict preference relation R given in one of the following forms (Orlovsky 1981):

- $(X_k, X_l) \in R$ or $X_k \succeq X_l$ that means "X_k is not worse than X_l";
- $(X_l, X_k) \in R$ or $X_l \succeq X_k$ that means "X_l is not worse than X_k";
- $(X_k, X_l) \notin R$ or $(X_l, X_k) \notin R$ that means "X_k and X_l are not comparable."

A fuzzy nonstrict preference relation R can be processed to obtain a fuzzy strict preference relation P. In particular, $(X_k, X_l) \in P$ means that X_k is strictly better than X_l (or X_k dominates X_l, i.e. $X_k \succ X_l$). It is natural that the alternative $X_k \in X$ is nondominated in $<X, R>$ if $(X_k, X_l) \in P$ for any $X_l \in X$.

As discussed previously, it is possible to construct the fuzzy strict preference relation P only in terms of the fuzzy nonstrict preference relation R. In particular, it is possible to do it on the basis of applying Eq. (5.13), which can be formally obtained (Orlovsky 1978, 1981) as follows:

$$P = R \setminus R^{-1} \tag{5.136}$$

The expression Eq. (5.13) plays an important role in this section: it can serve as the basis for the evaluation, comparison, choice, prioritization, and/or ordering of alternatives. In particular, it is possible to observe that $P(X_l, X_k), \forall X_k \in X$ is the membership function of the fuzzy set of all X_k that are strictly dominated by X_l. It is natural that that the complementary relation $P^c(X_l, X_k) = 1 - P(X_l, X_k), \forall X_k \in X$ generates the fuzzy set of alternatives that are not dominated by X_l. Considering this, in order to obtain the set of alternatives from X that are not dominated by other alternatives, it is possible to find the fuzzy preference relation that corresponds to the intersection of all $P^c(X_l, X_k), \forall X_k \in X$, on all $X_l \in X$ (Orlovsky 1978, 1981). This intersection, which corresponds to the fuzzy set of nondominated alternatives, can be presented in the following form:

$$ND(X_k) = \inf_{X_l \in X} [1 - P(X_l, X_k)] = 1 - \sup_{X_l \in X} P(X_l, X_k) \tag{5.137}$$

The use of Eq. (5.137) permits one to reflect the level of nondominance of each alternative X_k. Considering this, it is natural to choose alternatives providing the highest level of nondominance X^{ND} as follows:

$$X^{ND} = \left\{ X_k^{ND} \mid X_k^{ND} \in X, ND\left(X_k^{ND}\right) = \sup_{X_k \in X} ND(X_k) \right\} \tag{5.138}$$

Example 5.27 Consider the following nonstrict fuzzy preference relation:

$$R = \begin{bmatrix} 1 & 0.2 & 0.3 & 0.1 \\ 0.5 & 1 & 0.2 & 0.6 \\ 0.1 & 0.6 & 1 & 0.3 \\ 0.6 & 0.1 & 0.5 & 1 \end{bmatrix} \tag{5.139}$$

defined on a set of alternatives $X = \{X_1, X_2, X_3, X_4\}$ to order them and to select the best one.

Applying Eqs. (5.13)–(5.139), it is possible to obtain the membership function of the fuzzy strict preference relation

$$P = \begin{bmatrix} 0 & 0 & 0.2 & 0 \\ 0.3 & 0 & 0 & 0.5 \\ 0 & 0.4 & 0 & 0 \\ 0.5 & 0 & 0.2 & 0 \end{bmatrix} \tag{5.140}$$

The use of Eqs. (5.137) generates the following membership function of the fuzzy set of nondominated alternatives:

$$ND = [0.5\ \ 0.6\ \ 0.8\ \ 0.5] \tag{5.141}$$

It permits us to determine $X_3 \succ X_2 \succ X_1 \sim X_4$ and concur that $X^{ND} = \{X_3\}$.

Orlovsky (1981) introduced the notion of a set of nonfuzzy nondominated alternatives. In particular, if $\sup_{X_k \in X} ND(X_k) = 1$, then the alternatives

$$X^{NFND} = \{X_k^{NFND} \mid X_k^{NFND} \in X, ND(X_k^{NFND}) = 1\} \tag{5.142}$$

are nonfuzzy nondominated. These alternatives can be considered as a nonfuzzy solution to the fuzzy problem.

Example 5.28 Assume there is a set of alternatives $X = \{X_1, X_2, X_3, X_4\}$ and assigned to them the following nonstrict fuzzy preference relation:

$$R = \begin{bmatrix} 1 & 0.8 & 0.5 & 0.5 \\ 0 & 1 & 0 & 0.2 \\ 0.6 & 0.9 & 1 & 0.6 \\ 0 & 0 & 0 & 1 \end{bmatrix} \tag{5.143}$$

Applying Eqs. (5.13)–(5.143), we obtain the membership function of the fuzzy strict preference relation

$$P = \begin{bmatrix} 0 & 0.8 & 0 & 0.5 \\ 0 & 0 & 0 & 0.2 \\ 0.1 & 0.9 & 0 & 0.6 \\ 0 & 0 & 0 & 0 \end{bmatrix} \tag{5.144}$$

The use of Eq. (5.137) permits us to generate the following membership function of the fuzzy set of nondominated alternatives:

$$ND = [0.9\ \ 0.1\ \ 1\ \ 0.4] \tag{5.145}$$

It allows us to obtain $X^{NFND} = \{X_3\}$.

If the fuzzy preference relation R satisfies weak transitivity, then $X^{NFND} \neq \emptyset$. Taking this into account, it should be noted that when the preferences of a DM are expressed as the ordering of alternatives, utility values, or fuzzy estimates, and subsequently converted into fuzzy preference relations with the use of the corresponding transformation functions, then $X^{NFND} \neq \emptyset$, since these transformation functions guarantee weak transitivity of the constructed fuzzy preference relation (Pedrycz et al. 2011). However, it is possible to obtain $X^{NFND} = \emptyset$ if a DM provides the preferences as a multiplicative preference relation or a fuzzy preference relation that does not satisfy weak transitivity, because the corresponding transformation functions transmit existent inconsistencies to the fuzzy preference relations. Thus, the fact of $X^{NFND} = \emptyset$ permits one to detect contradictions in the experts' estimates (Ekel 2002; Pedrycz et al. 2011).

The expressions Eqs. (5.13), (5.137), and (5.138) may be applied to solve choice problems as well as other problems, related to the evaluation, comparison, prioritization, and/or ordering of alternatives with a single fuzzy preference relation. These expressions may also be utilized when R is a vector of fuzzy preference relations.

Let us consider the *first technique* for dealing with a vector fuzzy preference relation R (Orlovsky 1981). The expressions Eqs. (5.13), (5.137), and (5.138) are applicable if we take $R = \cap_{p=1}^{q} R_p$ with the membership function

$$R(X_k,X_l) = \min_{1 \le p \le q} R_p(X_k,X_l), X_k, X_l \in X \tag{5.146}$$

The use of an intersection operator in Eq. (5.146) is associated with the need to satisfy all criteria simultaneously. When using Eq. (5.146), the set X^{ND} fulfills the role of a Pareto set (Orlovsky 1981). It is possible to contract it on the basis of differentiating the importance of R_p, $p = 1, \ldots, q$ with the use of the following convolution:

$$T(X_k,X_l) = \sum_{p=1}^{q} \lambda_p R_p(X_k,X_l), X_k, X_l \in X \tag{5.147}$$

where $\lambda_p \ge 0, p = 1, 2, \ldots, q$, are importance factors for the corresponding criteria, defined as in Eqs. (4.15) and (4.16).

The construction of $T(X_k, X_l)$, $X_k, X_l \in X$, allows one to obtain the membership function $ND'(X_k)$ of the fuzzy set of nondominated alternatives according to an expression similar to Eq. (5.137). The intersection

$$Q(X_k) = \min\{ND(X_k), ND'(X_k)\}, X_k \in X \tag{5.148}$$

provides us with

$$X^{ND} = \left\{ X_k^{ND} \mid X_k^{ND} \in X, Q\left(X_k^{ND}\right) = \sup_{X_k \in X} Q(X_k) \right\} \tag{5.149}$$

The expressions Eqs. (5.137) and (5.138) can serve as the basis for building the *second technique.* This technique is of a lexicographic character and is directed at step-by-step application of criteria for comparing alternatives. The technique permits one (Ekel et al. 1998; Ekel 2001, 2002) to construct a sequence $X^1, X^2, \ldots, X^q$ so that $X \supseteq X^1 \supseteq X^2 \supseteq \ldots \supseteq X^q$ using the following expressions:

$$ND^p(X_k) = \inf_{X_l \in X^{p-1}} \left[1 - P_p(X_l, X_k)\right] = 1 - \sup_{X_l \in X^{p-1}} P_p(X_l, X_k), p = 1, 2, \ldots, q \tag{5.150}$$

$$X^p = \left\{ X_k^{ND,p} \mid X_k^{ND,p} \in X^{p-1}, ND^p\left(X_k^{ND,p}\right) = \sup_{X_l \in X^{p-1}} ND^p(X_k) \right\} \tag{5.151}$$

Characterizing the *second technique,* it should be noted that if R_p satisfies weak transitivity, it is possible to bypass the pairwise comparison of alternatives at the pth step. In this situation, the comparison can be conducted on a serial basis (directly on the basis of Eqs. (5.29) and (5.30)) by memorizing the best alternatives (Ekel 2002).

The *third technique* is based on the use of Eq. (5.137), which can be represented in the following form:

$$ND(X_k) = 1 - \sup_{X_l \in X} P_p(X_l, X_k), p = 1, 2, \ldots, q \tag{5.152}$$

In this way, we can construct the membership functions of the fuzzy sets of nondominated alternatives for each fuzzy preference relation.

The membership functions $ND_p(X_k)$, $p = 1, 2, \ldots, q$ can be considered as the membership functions replacing objective functions $F_p(x)$, $p = 1, 2, \ldots, q$ in analyzing $\langle X, F \rangle$models (Ekel and Schuffner Neto 2006). Therefore, it is possible to construct

$$ND(X_k) = \min_{1 \le p \le q} ND_p(X_k) \tag{5.153}$$

to obtain X^{ND}. If necessary, to differentiate the importance of different preference relations, it is possible to transform Eq. (5.153) to

$$ND(X_k) = \min_{1 \le p \le q} \left[ND_p(X_k)\right]^{\lambda_p} \tag{5.154}$$

The use of Eq. (5.154) does not require the normalization of $\lambda_p \ge 0, p = 1, 2, \ldots, q$ in the same way as Eq. (4.16).

Example 5.29 Assume we are given a set of alternatives $X = \{X_1, X_2, X_3, X_4\}$ and assigned to them the following nonstrict fuzzy preference relations:

$$R_1 = \begin{bmatrix} 1 & 0.8 & 0.5 & 0.1 \\ 1 & 1 & 0.8 & 0.6 \\ 1 & 1 & 1 & 1 \\ 1 & 1 & 1 & 1 \end{bmatrix} \tag{5.155}$$

$$R_2 = \begin{bmatrix} 1 & 1 & 1 & 1 \\ 0.4 & 1 & 0.6 & 0.8 \\ 1 & 1 & 1 & 1 \\ 1 & 1 & 1 & 1 \end{bmatrix} \quad (5.156)$$

$$R_3 = \begin{bmatrix} 1 & 1 & 1 & 1 \\ 1 & 1 & 1 & 0.7 \\ 1 & 1 & 1 & 0.8 \\ 1 & 1 & 1 & 1 \end{bmatrix} \quad (5.157)$$

Let us consider the solution of the problem on the basis of the *first technique*. The intersection of the fuzzy nonstrict preference relations Eqs. (5.155)–(5.157), applying Eq. (5.146), generates

$$R = \begin{bmatrix} 1 & 0.8 & 0.5 & 0.1 \\ 0.4 & 1 & 0.6 & 0.6 \\ 1 & 1 & 1 & 0.8 \\ 1 & 1 & 1 & 1 \end{bmatrix} \quad (5.158)$$

The utilization of Eq. (5.13) permits us to construct the following fuzzy strict preference relation:

$$P = \begin{bmatrix} 0 & 0.4 & 0 & 0 \\ 0 & 0 & 0 & 0 \\ 0.5 & 0.4 & 0 & 0 \\ 0.9 & 0.4 & 0.2 & 0 \end{bmatrix} \quad (5.159)$$

The application of Eq. (5.137) generates

$$ND = [0.1 \;\; 0.6 \;\; 0.8 \;\; 1] \quad (5.160)$$

and $X^{ND} = \{X_4\}$.

Let us consider the application of the *second technique*, assuming that the criteria are arranged, for example, in the following order of importance: $p = 1$, $p = 2$, and $p = 3$.

Consistently applying Eqs. (5.13), (5.137), and (5.138), we obtain

$$P_1 = \begin{bmatrix} 0 & 0 & 0 & 0 \\ 0.2 & 0 & 0 & 0 \\ 0.5 & 0.2 & 0 & 0 \\ 0.9 & 0.4 & 0 & 0 \end{bmatrix} \quad (5.161)$$

$$ND^1 = [0.1 \;\; 0.6 \;\; 1 \;\; 1] \quad (5.162)$$

and $X^1 = \{X_3, X_4\}$.

For the second step, we can construct the fuzzy nonstrict preference relation, the fuzzy strict preference relation, and the fuzzy set of nondominated alternatives, considering only alternatives X_3 and X_4 as follows:

$$R^2 = \begin{bmatrix} 1 & 1 \\ 1 & 1 \end{bmatrix} \tag{5.163}$$

$$P_2 = \begin{bmatrix} 0 & 0 \\ 0 & 0 \end{bmatrix} \tag{5.164}$$

$$ND^2 = \begin{bmatrix} 1 & 1 \end{bmatrix} \tag{5.165}$$

and $X^2 = \{X_3, X_4\}$. Thus, the second step does not allow us to narrow down the decision uncertainty region.

Let us construct the fuzzy nonstrict preference relation, the fuzzy strict preference relation, and the fuzzy set of nondominated alternatives for the third step as follows:

$$R^3 = \begin{bmatrix} 1 & 0.8 \\ 1 & 1 \end{bmatrix} \tag{5.166}$$

$$P_3 = \begin{bmatrix} 0 & 0 \\ 0.2 & 0 \end{bmatrix} \tag{5.167}$$

$$ND^3 = \begin{bmatrix} 0.8 & 1 \end{bmatrix} \tag{5.168}$$

and $X^2 = \{X_4\}$.

Let us consider the application of the *third technique*. The membership function ND_1 of the set of nondominated alternatives for the first fuzzy nonstrict R_1 is Eq. (5.162). The analysis of the fuzzy nonstrict preference relation Eq. (5.156) generates the following membership function of nondominated alternatives:

$$ND_2 = \begin{bmatrix} 1 & 0.4 & 1 & 1 \end{bmatrix} \tag{5.169}$$

The analysis of the fuzzy nonstrict preference relation Eq. (5.157) leads to

$$ND_2 = \begin{bmatrix} 1 & 0.7 & 0.8 & 1 \end{bmatrix} \tag{5.170}$$

Thereby, all three methods allowed us to obtain the same result. Taking this into account, it should be noted that the application of the *second technique* may lead to solutions different from the results obtained on the basis of the *first technique*. However, solutions based on the *first technique* and the *third technique*, which share a single generic basis, may in some cases also differ from each other (Ekel and Schuffner Neto 2006). At the same time, the *third technique* is more preferential from the substantial point of view. In particular, the use of the *first technique* can lead to choosing alternatives with the degree of nondominance equal to one, though these alternatives are not the best ones from the point of view of all preference relations. The *third technique* can generate this result only for alternatives that are the best solutions from the point of view of all fuzzy preference relations. It should be stressed that the possibility to obtain different solutions on the basis of different approaches (as it is demonstrated by an example in Section 5.9) is to be considered natural, and the choice of the approach is a prerogative of DM.

The described techniques have been implemented within the framework of interactive decision-making system for multicriteria analysis of alternatives in a fuzzy environment, named MDMS (Pedrycz et al. 2011). These techniques are of a universal nature and are already being used to solve problems in power engineering (Canha et al. 2007; Ekel et al. 2016), naval engineering (Botter and Ekel 2005), and management (Berredo 2005).

All three techniques aimed at analyzing the <*X*, *R*> models described before require the explicit direct or indirect ordering of the criteria. Considering this, it is necessary to distinguish the results of Orlovsky (1981), which allow one to present information related to the importance of the criteria as a nonreciprocal fuzzy preference relation:

$$\Lambda = \left[\lambda \times \lambda, \Lambda\left(\lambda_p, \lambda_t\right)\right], \quad p, t = 1, 2, \ldots, q \tag{5.171}$$

Using the membership functions of the fuzzy sets of nondominated alternatives for all preference relations Eq. (5.137), it is possible to construct the following fuzzy preference relation induced by the preference relations Eq. (5.137) and (5.171):

$$R_\Lambda(X_k, X_l) = \sup_{\lambda_p, \lambda_t \in \Lambda} \min_{X_k, X_l \in X} \left\{ND_p(X_k),\ ND_t(X_l),\ \Lambda\left(\lambda_p, \lambda_t\right)\right\}, \quad p, t = 1, 2, \ldots, q \tag{5.172}$$

The fuzzy preference relations in Eq. (5.172) can be considered (Ekel et al. 2006) as a result of aggregating the family of R_p, $p = 1, 2, \ldots, q$, with the use of information reflecting the relative importance of criteria given in the form Eq. (5.171).

Applying Eqs. (5.13) and (5.137) to (5.172), it is possible to construct the fuzzy set of nondominated alternatives $ND'_\Lambda(X_k)$. As shown in (Orlovsky 1981), the set $ND'_\Lambda(X_k)$ is to be modified in accordance with the following relationship:

$$ND_\Lambda(X_k) = \min\left\{ND'_\Lambda(X_k), R_\Lambda(X_k, X_k)\right\} \tag{5.173}$$

To better understand the fourth technique, consider an example from Pedrycz et al. (2011).

Example 5.30 We are given a set of alternatives $X = \{X_1, X_2, X_3\}$ that are to be compared applying three criteria. The corresponding nonstrict fuzzy preference relations are the following:

$$R_1 = \begin{bmatrix} 1 & 1 & 1 \\ 0.8 & 1 & 0.6 \\ 0.4 & 0.6 & 1 \end{bmatrix} \tag{5.174}$$

$$R_2 = \begin{bmatrix} 1 & 0.4 & 0.3 \\ 1 & 1 & 0.7 \\ 0.9 & 0.6 & 1 \end{bmatrix} \tag{5.175}$$

$$R_3 = \begin{bmatrix} 1 & 0.8 & 0.7 \\ 0.4 & 1 & 0.8 \\ 0.3 & 0.6 & 1 \end{bmatrix} \tag{5.176}$$

The information related to the importance of criteria is presented as follows:

$$\Lambda = \begin{bmatrix} 1 & 0.8 & 0.7 \\ 1 & 1 & 0.5 \\ 0.9 & 0.7 & 1 \end{bmatrix} \tag{5.177}$$

The application of Eqs. (5.13) to (5.174)–(5.176), permits one to construct the following fuzzy strict preference relations:

$$P_1 = \begin{bmatrix} 0 & 0.2 & 0.6 \\ 0 & 0 & 0 \\ 0 & 0 & 0 \end{bmatrix} \tag{5.178}$$

$$P_2 = \begin{bmatrix} 0 & 0 & 0 \\ 0.6 & 0 & 0.1 \\ 0.6 & 0 & 0 \end{bmatrix} \tag{5.179}$$

and

$$P_3 = \begin{bmatrix} 0 & 0.4 & 0.4 \\ 0 & 0 & 0.2 \\ 0 & 0 & 0 \end{bmatrix} \tag{5.180}$$

for the first, second, and third criteria, respectively. Applying Eqs. (5.137) to (5.178)–(5.180), we obtain the following membership functions of the fuzzy sets of nondominated alternatives:

$$ND_1 = [1 \;\; 0.8 \;\; 0.4] \tag{5.181}$$

$$ND_2 = [0.4 \quad 1 \quad 0.9] \tag{5.182}$$

and

$$ND_3 = [1 \quad 0.6 \quad 0.6] \tag{5.183}$$

for the first, second, and third criteria, respectively.

The use of Eq. (5.172) for the processing of Eqs. (5.181)–(5.183) together with Eq. (5.177), permits us to obtain

$$R_\Lambda = \begin{bmatrix} 1 & 0.8 & 0.8 \\ 1 & 1 & 0.9 \\ 0.9 & 0.9 & 0.4 \end{bmatrix} \tag{5.184}$$

The use of Eq. (5.13) permits us to construct the corresponding fuzzy strict preference relation

$$P_\Lambda = \begin{bmatrix} 0 & 0 & 0 \\ 0.2 & 0 & 0 \\ 0.1 & 0 & 0 \end{bmatrix} \tag{5.185}$$

which, in accordance with Eq. (5.137), leads to

$$ND'_\Lambda = [0.8 \quad 1 \quad 1] \tag{5.186}$$

Finally, the application of Eq. (5.173), taking into account that

$$R_\Lambda(X_k, X_k) = [1 \quad 1 \quad 0.4] \tag{5.187}$$

generates

$$ND_\Lambda = [0.8 \quad 1 \quad 0.4] \tag{5.188}$$

Finally, applying the results discussed in Section 4.6 as well as Grabisch et al. (1998); Pedrycz et al. (2011); and Ekel et al. (2016), we can speak about the analysis of <*X*, *R*> models with applying the OWA operator. In particular, in this case, Eq. (4.39) is transformed in the following expression:

$$\mathrm{OWA}\left(R_1(X_k, X_l), R_2(X_k, X_l), \ldots, R_q(X_k, X_l)\right) = \sum_{i=1}^{q} w_i b_i \tag{5.189}$$

Examples of applying the OWA operator for multicriteria analysis of alternatives in a fuzzy environment can be found in Pedrycz et al. (2011).

5.8 Practical Examples of Analyzing <X, R> Models

The example given next demonstrates three techniques for analyzing <*X*, *R*> models based on the concept of the strict fuzzy preference relation (Pedrycz et al. 2011).

Example 5.31 The problem of substation planning in power systems considering the uncertainty of information is analyzed in Fontoura Filho et al. (1994). Its practical application is associated with a group of substations 138/13.8 kV of a power utility. In particular, careful analysis has been executed to select a solution from three alternatives on the basis of their total costs where the uncertainty of interest rates is modeled as trapezoidal membership functions. The details about the membership functions of alternative costs are given in Table 5.7 and are also illustrated in Figure 5.8.

It is evident that the selection of the most preferable alternative is hampered: the difference between the alternatives 1 and 2 is equal to 0.38% for the left bounds of the corresponding membership functions for a certainty of 70% that has been accepted in (Fontoura Filho et al. 1994), but does not give grounds to proceed with a convincing decision. This may also be illustrated by analyzing a nonstrict fuzzy preference relation

$$R_1 = \begin{bmatrix} 1 & 1 & 1 \\ 1 & 1 & 1 \\ 1 & 0.912 & 1 \end{bmatrix} \tag{5.190}$$

constructed on the basis of Figure 5.8. Applying Eqs. (5.13)–(5.190), we obtain the following strict fuzzy preference relation:

Table 5.7 Total costs (US$ 1000) of alternatives.

Alternative	1	2	3	4
1	20 291	22 007	22 769	27 054
2	21 058	21 831	22 378	25 865
3	21 977	22 749	23 098	24 276

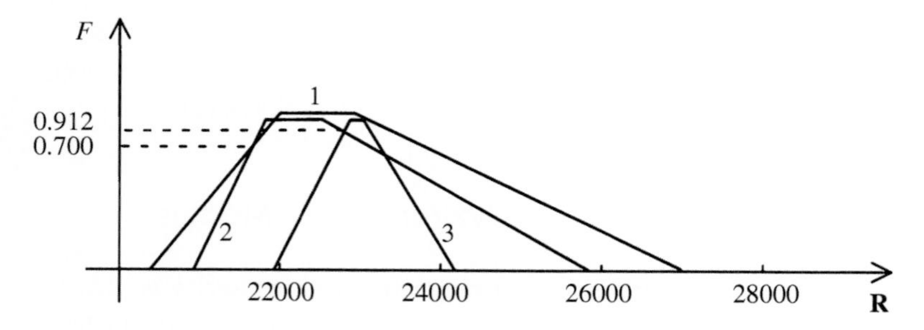

Figure 5.8 Total alternative costs.

$$P_1 = \begin{bmatrix} 0 & 0 & 0 \\ 0 & 0 & 0.088 \\ 0 & 0 & 0 \end{bmatrix} \quad (5.191)$$

Using Eq. (5.137), we can obtain the membership function of the fuzzy set of nondominated alternatives

$$ND = \begin{bmatrix} 1 & 1 & 0.912 \end{bmatrix} \quad (5.192)$$

which indicates that the alternatives 1 and 2 are indistinguishable.

Taking this into account, it is possible to consider indices "*Flexibility of Development*" and "*Damage to Agriculture*" as additional criteria denoted by $F_2(X_k)$ and $F_3(X_k)$. The membership functions corresponding to the normalized fuzzy values $S(F)$= (*very small, small, middle, large,* and *very large*) of the linguistic variables *Flexibility of Development* and *Damage to Agriculture*, which can be used to estimate $F_2(X_k)$ and $F_3(X_k)$, are given in Figure 5.9.

Assume that the alternatives have received the following estimates: $F_2(X_1)$ = *large*, $F_2(X_2)$ = *large*, and $F_2(X_3)$ = *very large* for the second criterion and $F_3(X_1)$ = *small*, $F_3(X_2)$ = *middle*, and $F_3(X_3)$ = *large* for the third criterion. Considering this as well as the necessity to maximize $F_2(X_k)$ and to minimize $F_3(X_k)$, it is possible to construct the matrices of the nonstrict fuzzy preference relations

$$R_2 = \begin{bmatrix} 1 & 1 & 0.909 \\ 1 & 1 & 0.909 \\ 1 & 1 & 1 \end{bmatrix} \quad (5.193)$$

and

$$R_3 = \begin{bmatrix} 1 & 1 & 1 \\ 0.938 & 1 & 1 \\ 0.625 & 0.938 & 1 \end{bmatrix} \quad (5.194)$$

for the second and third criterion, respectively.

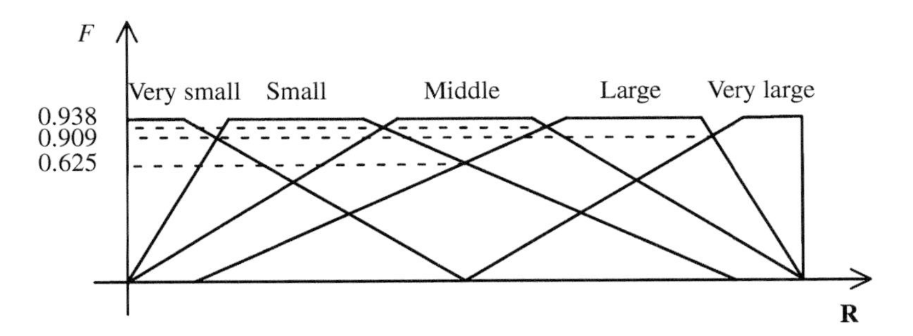

Figure 5.9 Membership functions for normalized fuzzy values.

Applying the *first technique* presented in Section 5.8, it is possible to obtain the intersection of Eqs. (5.190), (5.193), and (5.194) as follows:

$$R = \begin{bmatrix} 1 & 1 & 0.909 \\ 0.938 & 1 & 0.909 \\ 0.625 & 0.912 & 1 \end{bmatrix} \tag{5.195}$$

Applying Eq. (5.13) to Eq. (5.195), one can construct the following strict fuzzy preference relation:

$$P = \begin{bmatrix} 0 & 0.062 & 0.274 \\ 0 & 0 & 0 \\ 0 & 0.03 & 0 \end{bmatrix} \tag{5.196}$$

that permits us, using Eq. (5.137), to obtain the membership function of the fuzzy set of nondominated alternatives

$$ND = [1 \;\; 0.938 \;\; 0.716] \tag{5.197}$$

The alternative one has the maximum degree of nondominance and it is natural to consider it as the solution; that is, $X^{ND} = \{X_1\}$. Thus, we have obtained the solution without applying the convolution Eq. (5.147).

Let us consider the application of the *second technique* if the criteria are arranged, for example, in the following order of their importance: $p = 1$, $p = 2$, and $p = 3$.

Applying Eqs. (5.13), (5.150), and (5.151), we obtain, from Eq. (5.190), the result coinciding with Eq. (5.192). This is obvious, and $X^1 = \{X_1, X_2\}$. Thus, the alternatives X_1 and X_2 are to be considered for a sequent analysis, and from Eq. (5.193) we can proceed with the second step:

$$R^2 = \begin{bmatrix} 1 & 1 \\ 1 & 1 \end{bmatrix} \tag{5.198}$$

that leads to

$$ND^2 = [1 \;\; 1] \tag{5.199}$$

and $X^2 = \{X_1, X_2\}$. The second step does not allow us to narrow down the decision uncertainty region.

From Eq. (5.194), we can proceed with the third step

$$R^3 = \begin{bmatrix} 1 & 1 \\ 0.938 & 1 \end{bmatrix} \tag{5.200}$$

Then

$$ND^3 = [1 \;\; 0.938] \tag{5.201}$$

leading to $X^3 = \{X_1\}$.

It is natural that if the criteria are arranged in another order, it is possible to obtain another solution. For instance, it is not difficult to verify if $p = 2$, $p = 3$, and $p = 1$, then $X^1 = \{X_3\}$.

Finally, let us apply the *third technique*. The membership function of the set of nondominated alternatives for the first fuzzy preference relation $ND_1(X_k)$ is Eq. (5.192). Processing Eq. (5.193), it is possible to build

$$ND_2 = \begin{bmatrix} 0.909 & 0.909 & 1 \end{bmatrix} \tag{5.202}$$

In an analogous way, the processing of Eq. (5.194) leads us to

$$ND_3 = \begin{bmatrix} 1 & 0.938 & 0.625 \end{bmatrix} \tag{5.203}$$

The intersection of Eqs. (5.192), (5.202), and (5.203) produces

$$ND = \begin{bmatrix} 0.909 & 0.909 & 0.625 \end{bmatrix} \tag{5.204}$$

and generates $X^{ND} = \{X_1, X_2\}$.

In this way, the *third technique* does not permit one to choose a unique alternative. It allowed us only to exclude alternative X_3 from further considerations: information given by Eqs. (5.190), (5.193), and (5.194) is not sufficient to choose a unique alternative.

Therefore, the *first technique* allows one to choose the alternative X_1. The *second technique* indicates the alternative X_1 (for the order of importance: $p = 1, p = 2$, and $p = 3$) as well. The *third technique* only permits one to eliminate the alternative X_3.

The example given next demonstrates the applicability of the results presented in Section 5.8 to analyze problems associated with nonfuzzy preference relations as well (Borisov et al. 1990).

Example 5.32 The direct development of an integrated project by an enterprise presents considerable difficulties. Several ways exist around these difficulties:

1) Training proper professionals;
2) Inviting new professionals capable of developing a project;
3) Contracting another enterprise with the necessary profile.

The decision of a manager is to be based on applying the following criteria:

1) Project development duration;
2) Financial expenditures;
3) Project development quality.

Thus, we have three alternatives X_1, X_2, and X_3, which are to be analyzed from the point of view of the indicated previous criteria. Let us apply the *first technique* given in Section 5.8 by taking into account the following importance factors: $\lambda_1 = 0.6$, $\lambda_2 = 0.2$, and $\lambda_3 = 0.2$.

From the point of view of the first criterion, a manager has an opinion that X_1 is as good as X_2 and X_3 is better than X_2. This information permits one to construct the following nonstrict preference relation:

$$R_1 = \begin{bmatrix} 1 & 1 & 0 \\ 1 & 1 & 0 \\ 0 & 1 & 1 \end{bmatrix} \tag{5.205}$$

The preferences expressed from the point of view of the second criterion are the following: X_1 is better than X_2 and X_3, and X_2 is better than X_3. These preferences permit one to construct the second nonstrict preference relation

$$R_2 = \begin{bmatrix} 1 & 1 & 1 \\ 0 & 1 & 1 \\ 0 & 0 & 1 \end{bmatrix} \tag{5.206}$$

Finally, the preferences from the point of view of the third criterion are presented as follows: X_1 is as good as X_2 and X_3 is better than X_1. It allows one to construct the third nonstrict preference relation

$$R_3 = \begin{bmatrix} 1 & 1 & 0 \\ 1 & 1 & 0 \\ 1 & 0 & 1 \end{bmatrix} \tag{5.207}$$

The intersection of Eqs. (5.205)–(5.207) leads to the formation of the following nonstrict preference relation:

$$R = \begin{bmatrix} 1 & 1 & 0 \\ 0 & 1 & 0 \\ 0 & 0 & 1 \end{bmatrix} \tag{5.208}$$

The use of Eq. (5.13) to process Eq. (5.208) permits one to construct the following strict preference relation:

$$P = \begin{bmatrix} 0 & 0 & 0 \\ 1 & 0 & 0 \\ 0 & 0 & 0 \end{bmatrix} \tag{5.209}$$

providing

$$ND = [0 \ \ 1 \ \ 1] \tag{5.210}$$

on the basis of Eq. (5.137).

In such a manner, we have to introduce the convolution Eq. (5.147) into consideration by applying the importance factors given previously as follows:

$$T = 0.6\begin{bmatrix} 1 & 1 & 0 \\ 1 & 1 & 0 \\ 0 & 1 & 1 \end{bmatrix} + 0.2\begin{bmatrix} 1 & 1 & 1 \\ 0 & 1 & 1 \\ 0 & 0 & 1 \end{bmatrix} + 0.2\begin{bmatrix} 1 & 1 & 0 \\ 1 & 1 & 0 \\ 1 & 0 & 1 \end{bmatrix} = \begin{bmatrix} 1 & 1 & 0.2 \\ 0.8 & 1 & 0.2 \\ 0.2 & 0.6 & 1 \end{bmatrix} \tag{5.211}$$

It is not difficult to understand that Eq. (5.211) generates the following strict preference relation:

$$P = \begin{bmatrix} 0 & 0.2 & 0 \\ 0 & 0 & 0 \\ 0 & 0.4 & 0 \end{bmatrix} \tag{5.212}$$

The membership function of the set of nondominated alternatives, which corresponds to Eq. (5.212), is the following:

$$ND' = [1 \;\; 0.6 \;\; 1] \tag{5.213}$$

Finally, the intersection of Eqs. (5.210) and (5.213) realized in accordance with Eq. (5.148) leads to

$$Q = [0 \;\; 0.6 \;\; 1] \tag{5.214}$$

Given this, we define $X^{ND} = \{X_3\}$. It means that the suitable alternative is to recommend to contract another enterprise with the necessary profile to develop the project.

The example given next demonstrates the analysis of the <*X*, *R*> models with fuzzy ordering of criteria (Orudjev 1983).

Example 5.33 The problem of choosing a local reactive power source at a power system bus with reactive power shortage is considered in Orudjev (1983) and Pedrycz et al. (2011). The following alternatives are considered:

1) Controlled thyristor reactor with constantly connected capacitor banks;
2) Controlled thyristor reactor with capacitor banks connected through the reactor;
3) Synchronous compensator;
4) Capacitor banks with smooth thyristor control.

The decision is to be made on the basis of using the following criteria:

1) Reliability;
2) Investment;
3) Control rapidity.

The nonstrict fuzzy preference relations corresponding to the first, second, and third criteria are the following:

$$R_1 = \begin{bmatrix} 1 & 0.7 & 0.4 & 0.8 \\ 0 & 1 & 0.2 & 1 \\ 0.5 & 0.3 & 1 & 0.1 \\ 0.8 & 0.4 & 0.2 & 1 \end{bmatrix} \tag{5.215}$$

$$R_2 = \begin{bmatrix} 1 & 0.1 & 0.5 & 0.8 \\ 0 & 1 & 0.8 & 0.6 \\ 0.7 & 0.4 & 1 & 0.7 \\ 0.4 & 0.8 & 0.2 & 1 \end{bmatrix} \tag{5.216}$$

$$R_3 = \begin{bmatrix} 1 & 0.9 & 0.12 & 0.3 \\ 0.3 & 1 & 0.8 & 0.5 \\ 0.3 & 0.15 & 1 & 0.7 \\ 0.9 & 0.6 & 0.2 & 1 \end{bmatrix} \tag{5.217}$$

respectively.

The information related to the importance of considered criteria is presented in the following form:

$$\Lambda = \begin{bmatrix} 1 & 0.8 & 0.6 \\ 0.5 & 1 & 0.7 \\ 0.2 & 0.1 & 1 \end{bmatrix} \tag{5.218}$$

The application of Eq. (5.13) to Eqs. (5.125–5.217) to construct the corresponding strict fuzzy preference relations and, then, the utilization of Eq. (5.137) yields the following membership functions of the fuzzy sets of nondominated alternatives:

$$ND_1 = [0.9 \;\; 0.3 \;\; 0.9 \;\; 0.4] \tag{5.219}$$

$$ND_2 = [0.8 \;\; 0.8 \;\; 0.6 \;\; 0.5] \tag{5.220}$$

and

$$ND_3 = [0.4 \;\; 0.4 \;\; 0.35 \;\; 0.5] \tag{5.221}$$

for the first, second, and third criterion, respectively.

The application of Eq. (5.172) to process Eqs. (5.219)–(5.221) together with Eq. (5.218) leads to the following relation:

$$R_\Lambda = \begin{bmatrix} 1 & 0.8 & 0.9 & 0.5 \\ 0.8 & 0.9 & 0.5 & 0.5 \\ 0.9 & 0.8 & 0.9 & 0.8 \\ 0.5 & 0.5 & 0.5 & 0.5 \end{bmatrix} \tag{5.222}$$

The application of Eq. (5.13) to Eq. (5.222) to construct the corresponding strict fuzzy preference relations and, then, the use of Eq. (5.173) forms the following membership function of the fuzzy sets of nondominated alternatives:

$$ND' = [1 \quad 0.7 \quad 1 \quad 0.7] \tag{5.223}$$

Finally, the application of Eq. (5.173), considering that

$$R_{\Lambda} = [1 \quad 0.9 \quad 0.9 \quad 0.5] \tag{5.224}$$

leads to

$$ND_{\Lambda} = [1 \quad 0.7 \quad 0.9 \quad 0.5] \tag{5.225}$$

Thus, a controlled thyristor reactor with constantly connected capacitor banks should be selected.

The last example demonstrates the analysis of the <*X*, *R*> models formed on the basis of applying diverse preference formats and transformation functions to convert these formats to the nonstrict fuzzy preference relations providing homogeneous information for the use of decision-making procedures (Ekel et al. 2019).

Example 5.34 The expansion of electric transmission systems is fundamental to meet the observable growing demands for electricity. Such expansion is generally characterized by technical complexity, the necessity of considering socio-environmental factors, and the need for significant investments. Moreover, the process of their analysis has to be capable to prioritize the projects in accordance with a wide range of other considerations, including strategic goals of concessionaires and investors. Thus, there is a problem of managing the process of analyzing several investment alternatives in electric transmission, carried out by concessionaires or investors, which plan to participate in auctions of new transmission projects. The necessity of applying the express analysis is associated with the following considerations.

The granting authorities (usually, regulatory agencies) make available, together with auction announcements on bidding for electric transmission projects, reports containing information on the items that are part of each lot. In addition to the importance for future implantation, this information is related to decisions of competitors on participation or not participation in the auction, and on the composition of their proposals on lots of interest. However, the previous indicated reports are usually published on a date close to the auction (for instance, one month of antecedence in Brazil). This does not allow one to effectively analyze all the interest. Thus, it is advantageous to carry out preliminarily studies in order to prepare the best proposals, with greater chances to get the corresponding lots.

To carry out efficient studies without "waste" of resources invested in the analysis of ventures that prove to be very risky or generally unattractive, it is advisable to prioritize the projects to be studied. Given that, at this point, there is still insufficient information on the final definition of the corridor and the preferential routes of the lines, it is necessary to create a methodology for the convincing use of the information already available. Considering this, the results of Ekel et al. (2019) are related to decision-making procedures for supporting the evaluation, comparison, choice, prioritization, and/or ordering of electric transmission projects in the prospecting stage, aimed at greater efficiency and effectiveness in the advanced preparation for auctions.

The process of elaborating decisions on transmission projects proposed in (Ekel et al. 2019) includes two stages. The first one is associated with studies related to the criteria relevant to the decision, which allow one to form the most rational estimates of the considered lots. These criteria often have a spatial nature. Considering this, the proposed models assume the use of Geographic Information Systems (GISs) that play an important role in supporting the analysis of such class of problems (Malczewski 2006; Shafiullah et al. 2016). Another important aspect is the need to apply the techniques of multiobjective analysis under conditions of uncertainty. In particular, it is proposed in Ekel et al. (2019) to use the generalization of the well-known Dijkstra's algorithm (Dijkstra 1959) (related to graph optimization, allowing the construction of transmission line routes) to analyze multiobjective problems. This generalization is complemented by the consideration of the uncertainty factor, applying a possibilistic approach. In particular, a general scheme of decision-making under conditions of uncertainty is used, discussed in Chapter 7, which permits one to construct solutions, including robust solutions, to multiobjective problems.

The second stage is directed at defining a portfolio with the most appropriate and favorable areas for implanting the transmission lines from the set of lots analyzed at the first stage. This stage is based on applying techniques of preference modeling in a fuzzy environment discusses in the present chapter. Their application allows one to adequately consider quantitative as well as qualitative criteria whose estimates are based on knowledge, experience, and intuition of a DM (individual or group).

In Ekel et al. (2019), the first stage is based on applying the thematic maps that reflect technical-economic cost c^E and socio-environmental cost c^S.

Let us consider four auction lots to demonstrate the most important aspects of the two-stage multicriteria analysis of new transmission lines. Without discussing the execution of the first stage, its results are given in Table 5.8.

Let us consider the procedures of the second stage for the following situations:

1) Elaboration of the recommendation on the basis of applying the additional criterion "Expected Profit";

Table 5.8 Rational estimates of the analyzed lots.

Lot	$F_1(X_k)=c^E$	$F_2(X_k)=c^S$
X_1	141.07	12.10
X_2	133.03	13.56
X_3	138.05	13.66
X_4	123.98	12.73

2) Elaboration of the recommendation on the basis of applying the criteria "Expected Profit" and "Time of Load Time Achievement."

The alternative estimates from the point of view of the criterion "Expected Profit" can be constructed in the quantitative form. At the same time, the alternative estimates from the point of view of the criterion "Time of Planned Load Achievement" can be obtained in the qualitative form.

Taking into account that $F_1(X_k)$ and $F_2(X_k)$ are to be minimized, it is possible to apply Eq. (5.82) to construct the following additive reciprocal fuzzy preference relations:

$$RR_1 = \begin{bmatrix} 0.5000 & 0.2648 & 0.4175 & 0.0000 \\ 0.7352 & 0.5000 & 0.6527 & 0.2352 \\ 0.5825 & 0.3473 & 0.5000 & 0.0825 \\ 1.0000 & 0.7648 & 0.9173 & 0.5000 \end{bmatrix} \tag{5.266}$$

for $F_1(X_k)$ and

$$RR_2 = \begin{bmatrix} 0.5000 & 0.9679 & 1.0000 & 0.7019 \\ 0.0321 & 0.5000 & 0.5321 & 0.2340 \\ 0.0000 & 0.4679 & 0.5000 & 0.2019 \\ 0.2981 & 0.7660 & 0.7981 & 0.5000 \end{bmatrix} \tag{5.267}$$

for $F_2(X_k)$, respectively.

Applying the transformation function Eq. (5.83), it is possible to convert these additive reciprocal fuzzy preference relations to the following nonreciprocal fuzzy preference relations:

$$NR_1 = \begin{bmatrix} 1.0000 & 0.5296 & 0.8323 & 0.0000 \\ 1.0000 & 1.0000 & 1.0000 & 0.4704 \\ 1.0000 & 0.6946 & 1.0000 & 0.1652 \\ 1.0000 & 1.0000 & 1.0000 & 1.0000 \end{bmatrix} \tag{5.268}$$

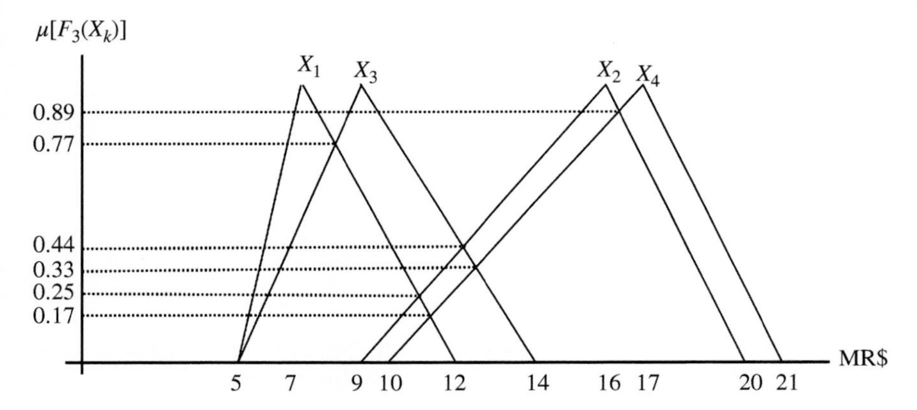

Figure 5.10 Membership functions of fuzzy values of estimates for the criterion "Expected Profit."

and

$$NR_2 = \begin{bmatrix} 1.0000 & 1.0000 & 1.0000 & 1.0000 \\ 0.0642 & 1.0000 & 1.0000 & 0.4680 \\ 0.0000 & 0.9358 & 1.0000 & 0.4038 \\ 0.5962 & 1.0000 & 1.0000 & 1.0000 \end{bmatrix} \tag{5.269}$$

respectively.

Figure 5.10 presents the estimates (membership functions of fuzzy values) for the alternatives from the point of view of the criterion "Expected Profit."

Applying Eqs. (5.29) and (5.30), defined for $F_p(X_k) \geq F_p(X_l)$ and $F_p(X_l) \geq F_p(X_k)$, respectively, it is possible to obtain the following nonreciprocal fuzzy preference relation:

$$NR_3 = \begin{bmatrix} 1.00 & 0.25 & 0.77 & 0.17 \\ 1.00 & 1.00 & 1.00 & 0.89 \\ 1.00 & 0.44 & 1.00 & 0.33 \\ 1.00 & 1.00 & 1.00 & 1.00 \end{bmatrix} \tag{5.270}$$

Using the *first technique* for processing of NR_1, NR_2, and NR_3, it is possible to obtain

$$R = \begin{bmatrix} 1.0000 & 0.2500 & 0.7700 & 0.0000 \\ 0.0642 & 1.0000 & 1.0000 & 0.4680 \\ 0.0000 & 0.4400 & 1.0000 & 0.1652 \\ 0.5962 & 1.0000 & 1.0000 & 1.0000 \end{bmatrix} \tag{5.271}$$

on the basis of Eq. (5.146). Applying Eq. (5.13), one can build the fuzzy strict preference relation

$$P = \begin{bmatrix} 0.0000 & 0.1858 & 0.7700 & 0.0000 \\ 0.0000 & 0.0000 & 0.5600 & 0.0000 \\ 0.0000 & 0.0000 & 0.0000 & 0.0000 \\ 0.5962 & 0.5320 & 0.8348 & 0.0000 \end{bmatrix} \quad (5.272)$$

generating the set of nondominated alternatives with the following membership function:

$$ND = [0.4038 \quad 0.4680 \quad 0.1652 \quad 1.0000] \quad (5.273)$$

Therefore, the analysis on the basis of applying the three criteria permits us to elaborate the following recommendation: $X_4 \succ X_2 \succ X_1 \succ X_3$.

Now, the criterion "Time of Planned Load Achievement," which is of a qualitative character, is added. The membership functions of fuzzy values that can be used for estimating the alternatives from the point of view of the criterion "Time of Planned Load Achievement" are presented in Figure 5.11 (S – Small, M – Medium, and L – Large).

The alternatives have received the following estimates: $X_1 - S$, $X_2 - L$, $X_3 - M$, and $X_4 - L$. Applying Eqs. (5.29) and (5.30) for these estimates, it is possible to construct the following nonreciprocal fuzzy preference relation:

$$NR_4 = \begin{bmatrix} 1.00 & 1.00 & 1.00 & 1.00 \\ 0.30 & 1.00 & 0.75 & 1.00 \\ 0.75 & 1.00 & 1.00 & 1.00 \\ 0.30 & 1.00 & 0.75 & 1.00 \end{bmatrix} \quad (5.274)$$

Using the *first technique* for NR_1, NR_2, NR_3, and NR_4, it is possible to obtain

$$R = \begin{bmatrix} 1.0000 & 0.2500 & 0.7700 & 0.0000 \\ 0.0642 & 1.0000 & 0.7500 & 0.4680 \\ 0.0000 & 0.4400 & 1.0000 & 0.1652 \\ 0.3000 & 1.0000 & 0.7500 & 1.0000 \end{bmatrix} \quad (5.275)$$

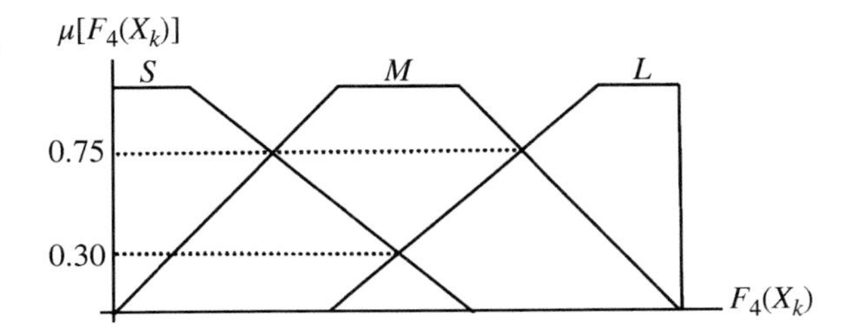

Figure 5.11 Membership functions of fuzzy values of estimates for the criterion "Time of Planned Load Achievement."

The application of Eq. (5.13) provides

$$P = \begin{bmatrix} 0.0000 & 0.1858 & 0.7700 & 0.0000 \\ 0.0000 & 0.0000 & 0.3100 & 0.0000 \\ 0.0000 & 0.0000 & 0.0000 & 0.0000 \\ 0.3000 & 0.5320 & 0.5848 & 0.0000 \end{bmatrix} \tag{5.276}$$

generating the set of nondominated alternatives with the following membership function:

$$ND = [0.7000 \quad 0.4680 \quad 0.2300 \quad 1.0000]$$

Thus, the introduction of the fourth criterion leads to the following recommendation: $X_4 \succ X_1 \succ X_2 \succ X_3$, which is different from the recommendation obtained on the basis of applying three criteria. However, still the best alternative is X_4.

5.9 Conclusions

In this chapter, we have considered some selected issues related to constructing and analyzing <X, R> models serving for preference modeling through binary fuzzy preference relations. The properties of the binary fuzzy preference relations as well as conditions of their transitivity have been discussed.

In dealing with <X, R> models, a fundamental question arises as to how can one build fuzzy preference relations? The corresponding techniques were considered. Much attention has been paid to a natural and convincing approach to constructing fuzzy preference relations based on the ordering of fuzzy quantities and, in particular, on the use of the fuzzy number ranking index, generated by the concept of the membership function of the generalized preference relation.

The input of the preference information plays a fundamental role in decision-making processes, as the recommendations are derived from the models, which are constructed in accordance with the provided information. Considering that this information is often subjective, vague, and uncertain, it is of high importance to provide involved experts with the means to articulate their preferences as a maximum truthful and accurate manner. Otherwise, if any expert is forced to express his/her preferences applying a preference format with which he/she does not feel comfortable, the input of the preference information can become a critical step in multicriteria analysis. Taking this into account, we have considered five main types of preference format. The presence of different preference formats requires their conversion to a unique format to provide homogeneous information for decision-making procedures. Considering this, so-called transformation functions are introduced. They permit one to reduce different preference formats to fuzzy preference relations as well as quantitative information to fuzzy preference relations.

We have discussed the questions of the emergence and importance of problems of multicriteria evaluation, comparison, choice, prioritization, and/or ordering of alternatives. Two types of situation exist that give rise to these problems. The first one is related to the direct statement of multiattribute decision-making problems when the consequences associated with solutions to problems cannot be estimated with a single criterion. The second class is related to problems that may be solved on the basis of a single criterion or on the basis of several criteria; however, if the uncertainty of information does not permit to obtain a unique solution (as it is demonstrated in the chapter by solving continuous and discrete problems of mathematical programming with fuzzy coefficients), it is possible to include additional criteria and thereby convert these problems into multiattribute tasks. Diverse techniques of multiattribute analysis in a fuzzy environment are discussed. These techniques can lead to different solutions. However, this is to be considered natural and intuitively appealing. The choice of a specific technique is a prerogative of the DM. This selection has to be based on the essence of the problem and the possible sources of information and its uncertainty.

Although these techniques are directly related to individual decision-making, they can be applied to procedures of group decision-making. The use of the presented results is illustrated by solving practical problems coming from different areas. One of these problems demands an integration of $<X, M>$ and $<X, R>$ models.

Exercises

5.1 Verify whether the fuzzy relation

$$R = \begin{bmatrix} 1 & 0.5 & 0.7 & 0.7 \\ 0.5 & 1 & 0.5 & 0.5 \\ 0.6 & 0.6 & 1 & 0.6 \\ 0.5 & 0.5 & 0.5 & 1 \end{bmatrix}$$

satisfies the conditions of (i) min-transitivity, (ii) product transitivity, and (iii) Lukasiewicz-transitivity.

5.2 Verify whether the multiplicative preference relation

$$MR = \begin{bmatrix} 1 & 5 & 6 & 7 \\ 0.2 & 1 & 4 & 6 \\ 0.17 & 0.25 & 1 & 4 \\ 0.14 & 0.17 & 0.25 & 1 \end{bmatrix}$$

satisfies the condition of multiplicative transitivity.

5.3 Applying Eqs. (5.13) and (5.14), construct the fuzzy strict preference relation and fuzzy indifference relation from the fuzzy nonstrict preference relation of Problem 5.1.

5.4 Applying Eqs. (5.29) and (5.30), construct the nonreciprocal fuzzy preference relations from the comparison of the alternatives X_1, X_2, and X_3 that are to be evaluated from the point of view of a minimized criterion F. The corresponding levels of membership functions are given in Table 5.9.

Table 5.9 Membership functions of $F(X_1)$, $F(X_2)$, and $F(X_3)$.

R	2	4	6	8	10	12	14	16
$\mu[F(X_1)]$	0	0.5	1	0.5	0.2	0	0	0
$\mu[F(X_2)]$	0	0.20	0.6	0.8	1	1	0.6	0
$\mu[F(X_3)]$	0	0	0.4	1	0.8	0.6	0.4	0

5.5 Convert the ordered array $O = \{2\ 1\ 3\ 4\ 5\}$ into an additive reciprocal fuzzy preference relation.

5.6 Convert the array of utility values $U = \{1\ 0\ 0.6\ 0.4\}$ into nonreciprocal fuzzy preference relations, applying (i) consistently to Eqs. (5.48) and (5.61) and (ii) Eq. (5.75), and compare the obtained results.

5.7 Convert the multiplicative preference relation of Problem Eq. (5.2) into nonreciprocal fuzzy preference relations, applying (i) consistently to Eqs. (5.51) and (5.61) and (ii) Eq. (5.76) (with $m = 9$), and compare the obtained results.

5.8 Applying results given in Subsection 5.4.9, construct nonreciprocal fuzzy preference relations related to indicators "Investment Level" and "Expected Profit Level" for the four industrial projects presented in Table 5.10.

Table 5.10 Indicators of industrial projects (US$ 1000).

Project	Investment Level	Expected Profit Level
X_1	11 340.00	2600.00
X_2	8450.00	1950.00
X_3	9980.00	2080.00
X_4	10 760.00	2230.00

5.9 Construct the membership function of the fuzzy set of nondominated alternatives for the following fuzzy nonstrict preference relation:

$$R = \begin{bmatrix} 1 & 0.5 & 1 & 0.4 \\ 1 & 1 & 0.8 & 1 \\ 0.6 & 1 & 1 & 1 \\ 1 & 0.8 & 0.5 & 1 \end{bmatrix}$$

5.10 Verify the existence of the nonfuzzy solution in the problem described by the following fuzzy nonstrict preference relation:

$$R = \begin{bmatrix} 1 & 1 & 1 & 0.6 \\ 0.6 & 1 & 0.8 & 0 \\ 0.8 & 1 & 1 & 0.2 \\ 1 & 1 & 1 & 1 \end{bmatrix}$$

5.11 Apply the *first technique* for analyzing <X, R> models to solve the problem that includes the following fuzzy nonstrict preference relations:

$$R_1 = \begin{bmatrix} 1 & 1 & 1 \\ 0.8 & 1 & 1 \\ 0.8 & 1 & 1 \end{bmatrix}$$

$$R_2 = \begin{bmatrix} 1 & 1 & 1 \\ 1 & 1 & 1 \\ 0.8 & 0.8 & 1 \end{bmatrix}$$

$$R_3 = \begin{bmatrix} 1 & 0.8 & 0.6 \\ 1 & 1 & 0.8 \\ 1 & 1 & 1 \end{bmatrix}$$

Take into account, if necessary, the following: $\lambda_1 = 0.4$, $\lambda_2 = 0.4$, and $\lambda_3 = 0.2$.

5.12 Utilize the *third technique* for analyzing <X, R> models to solve the problem that includes the fuzzy nonstrict preference relations given in Problem 5.11. If necessary, apply $\lambda_1 = 1$, $\lambda_2 = 2$, and $\lambda_3 = 1$.

5.13 Apply the *second technique* for analyzing <X, R> models to solve the problem that includes the fuzzy nonstrict preference relations given in Problem 5.11 following that order of importance.

5.14 Use the third technique for analyzing <*X*, *R*> models to solve a problem that includes the fuzzy nonstrict preference relations given in Problem 5.11. If necessary, apply $\lambda_1 = 1$, $\lambda_2 = 2$, and $\lambda_3 = 1$.

5.15 Solve the problem formulated in Example 5.31 if the alternatives have the following estimates:

$F_1(X_1)$= *middle*, $F_1(X_2)$= *middle*, and $F_1(X_3)$= *large*
$F_2(X_1)$= *small*, $F_2(X_2)$= *middle*, and $F_2(X_3)$= *large*
$F_3(X_1)$= *middle*, $F_3(X_2)$= *middle*, and $F_3(X_3)$= *large*

5.16 Verify the possibility of changing the solution (alternative X_1) of the problem defined by Example 5.31, if the information related to the importance of criteria is presented in the following form:

$$\Lambda = \begin{bmatrix} 1 & 0.5 & 0.2 \\ 0.8 & 1 & 0.1 \\ 0.7 & 0.8 & 1 \end{bmatrix}$$

References

Abbasbandy, S. and Asady, B. (2006). Ranking of fuzzy numbers by sign distance. *Information Sciences* 176 (16): 2405–2416.

Abbasbandy, S. and Hajjari, T. (2009). A new approach for ranking of trapezoidal fuzzy numbers. *Computers and Mathematics with Applications* 57 (3): 413–419.

Alsina, C. (1985). On a family of connectives for fuzzy sets. *Fuzzy Sets and Systems* 16 (3): 231–235.

Baas, S.M. and Kwakernaak, H. (1977). Rating and raking of multi-aspect alternatives using fuzzy sets. *Automatica* 13 (1): 47–58.

Baldwin, J.F. and Guild, N.C.F. (1979). Comparison of fuzzy sets on the same decision space. *Fuzzy Sets and Systems* 2 (3): 213–231.

Banerjee, A. (1993). Rational choice under fuzzy preferences: the Orlovsky choice function. *Fuzzy Sets and Systems* 54 (3): 295–299.

Barrett, C.R., Patanalk, P.K., and Salles, M. (1990). On choosing rationally when preferences are fuzzy. *Fuzzy Sets and Systems* 34 (2): 197–212.

Belton, V. (1999). Multiple-criteria problem structuring and analysis in a value theory framework. In: *Multicriteria Decision Making: Advances in MCDM Models, Algorithms, Theory* (eds. T. Gal, T. Stewart and T. Hanne), 12-2–12-29. Dordrecht: Kluwer.

Berredo, R.C., Ekel, P. Ya., Galperin, E.A., and Sant'anna, A.S. (2005) Fuzzy preference modeling and its management applications. *Proceedings of the International Conference on Industrial Logistics*, Montevideo, pp. 41–50.

Borisov, O.N., Krumberg, O.A., and Fedorov, I.P. (1990). *Decision Making on the Basis of Fuzzy Models: Utilization Examples*. Riga (in Russian): Zinatne.

Botter, R.C. and Ekel, P. Ya. (2005) Fuzzy preference relations and their naval engineering applications. *Proceedings of the XIX Congress of Pan-American Institute of Naval Engineering*, Guayaquil, Paper 7–8.

Bouyssou, D. (1997). Acyclic fuzzy preference and the Orlovski choice function: a note. *Fuzzy Sets and Systems* 89 (1): 107–111.

Burfardi, A. (1998). On the construction of fuzzy preference structures. *Journal of Multi-criteria Decision Analysis* 7 (3): 169–175.

Canha, L., Ekel, P., Queiroz, J., and Schuffner Neto, F. (2007). Models and methods of decision making in fuzzy environment and their applications to power engineering problems. *Numerical Linear Algebra with Applications* 14 (3): 369–390.

Chen, C. and Klein, C.M. (1997). An efficient approach to solving fuzzy MADM problems. *Fuzzy Sets and Systems* 88 (1): 51–67.

Chen, L.H. and Lu, H.W. (2002). The preference order on fuzzy numbers. *Computers and Mathematics with Applications* 44 (10–11): 1455–1465.

Chen, S.J. and Hwang, C.L. (1992). *Fuzzy Multiple Attribute Decision Making: Methods and Applications*. Berlin: SpringerVerlag.

Chen, S.M. and Chen, J.H. (2009). Fuzzy risk analysis based on ranking generalized fuzzy numbers with different heights and different spreads. *Expert Systems with Applications* 36 (2): 6833–6842.

Cheng, C.H. (1998). A new approach for ranking fuzzy numbers by distance methods. *Fuzzy Sets and Systems* 95 (3): 307–313.

Chiclana, F., Herrera, F., and Herrera-Viedma, E. (1998). Integrating three representation models in fuzzy multipurpose decision making based on fuzzy preference relations. *Fuzzy Sets and Systems* 97 (1): 33–48.

Chiclana, F., Herrera, F., and Herrera-Viedma, E. (2001). Integrating multiplicative preference relations in a multipurpose decision-making model base don fuzzy preference relations. *Fuzzy Sets and Systems* 122 (2): 277–291.

Chiclana, F., Herrera-Viedma, E., and Herrera, F. (2004) T-additive and T-multiplicative transitivity: new consistency properties for providing preference relations. *Proceedings of the Tenth International Conference on Information Processing and Management of Uncertainty in Knowledge-based Systems*, Perugia, pp. 403–410.

Chu, T.C. and Tsao, C.T. (2002). Ranking fuzzy numbers with an area between the centroid point and the original point. *Computers and Mathematics with Applications* 43 (1): 111–117.

De Baets, B. and Fodor, J.C. (1997). Twenty years of fuzzy preference structures. *Rivista di Matematica per le Scienze Economiche e Sociali* 20 (1): 45–66.

Delgado, M., Kacprzyk, J., Verdegay, J.L., and Vila, M.A. (eds.) (1994). *Fuzzy Optimization: Recent Advances*. Heidelberg: Physica-Verlag.

Destercke, S. and Couso, I. (2015). Ranking of fuzzy intervals seen through the imprecise probabilistic lens. *Fuzzy Sets and Systems* 278 (1): 20–39.

Dijkstra, E.W. (1959). A note on two problems in connexion with graphs. *Numerische Mathematik* 1 (1): 269–271.

Dubois, D. and Prade, H. (1980). *Fuzzy Sets and Systems: Theory and Applications*. New York: Academic Press.

Dubois, D. and Prade, H. (1983). Ranking fuzzy numbers in the setting of possibility theory. *Information Sciences* 30 (3): 183–224.

Dubois, D. and Prade, H. (1999) A unified view of ranking techniques for fuzzy numbers. *Proceedings of the 1999 International Conference on Fuzzy Numbers*, Seoul, pp. 1328–1333.

Dyer, J.S. (2005). MAUT – multiattribute utility theory. In: *Multiple Criteria Decision Analysis: State of the Art Surveys* (eds. J. Figueira, S. Greco and M. Ehrgott), 265–292. New York: Springer-Verlag.

Ekel, P., Kokshenev, I., Parreiras, R. et al. (2016). Multiobjective and multiattribute decision making in a fuzzy environment and their power engineering applications. *Information Sciences* 360-361 (1): 100–119.

Ekel, P., Pedrycz, W., and Schinzinger, R. (1998). A general approach to solving a wide class of fuzzy optimization problems. *Fuzzy Sets and Systems* 97 (1): 49–66.

Ekel, P., Queiroz, J., Parreiras, R., and Palhares, R. (2009). Fuzzy set based models and methods of multicriteria group decision making. *Nonlinear Analysis: Theory, Methods and Applications* 71 (12): 409–419.

Ekel, P.Y. (2001). Methods of decision making in fuzzy environment and their applications. *Nonlinear Analysis: Theory, Methods and Applications* 47 (5): 979–990.

Ekel, P.Y. (2002). Fuzzy sets and models of decision making. *Computers and Mathematics with Applications* 44 (7): 863–875.

Ekel, P.Y., Lisboa, A.C., Pereira, J.G. Jr. et al. (2019). Two-stage multicriteria georeferenced express analysis of new electric transmission line projects. *International Journal of Electrical Power & Energy Systems* 108: 415–431.

Ekel, P.Y. and Popov, V.A. (1985). Consideration of the uncertainty factor in problems of modelling and optimizing electrical networks. *Power Engineering* 23 (2): 45–52.

Ekel, P.Y. and Schuffner Neto, F.H. (2006). Algorithms of discrete optimization and their application to problems with fuzzy coefficients. *Information Sciences* 176 (19): 2846–2868.

Ekel, P.Y., Silva, M.R., Schuffner Neto, F., and Palhares, R.M. (2006). Fuzzy preference modeling and its application to multiobjective decision making. *Computers and Mathematics with Applications* 52 (1–2): 179–196.

Facchinetti, G. (2002). Ranking functions induced by weighted average of fuzzy numbers. *Fuzzy Optimization and Decision Making* 1 (3): 313–327.

Farquhar, P.H. and Keller, L.R. (1989). Preference intensity measurement. *Annals of Operations Research* 19 (1–4): 205–217.
Fodor, J.C. and De Baets, B. (2008). Fuzzy preference modeling: fundamentals and recent advances. In: *Fuzzy Sets and Their Extensions: Representation, Aggregation and Models* (eds. H. Bustince, F. Herrera and J. Montero), 207–217. New York: Springer-Verlag.
Fodor, J.C. and Roubens, M. (1993) Fuzzy strict preference relations in decision making. *Proceedings of the Second IEEE International Conference on Fuzzy Systems*, San Francisco, pp. 1145–1149.
Fodor, J.C. and Roubens, M. (1994a). Valued preference structures. *European Journal of Operational Research* 79 (2): 277–286.
Fodor, J.C. and Roubens, M. (1994b). *Fuzzy Preference Modelling and Multicriteria Decision Support.* Boston: Kluwer.
Fodor, J.C. and Rudas, I. (2006) Two functional equations in fuzzy preference modeling, INES'06. *Proceedings of the International Conference on Intelligent Engineering Systems*, London, pp. 17–23.
Fontoura Filho, R.N., Ales, J.C.O., and Tortelly, D.L.S. (1994) Uncertainty models applied to the substation planning. *Technical Papers of the 4th International Conference on Probabilistic Methods Applied to Power Systems*, Rio de Janeiro, pp. 1–7.
Fortemps, P. and Roubens, M. (1996). Ranking and defuzzification methods based on area compensation. *Fuzzy Sets and Systems* 82 (3): 319–330.
Freeling, S. (1980). Fuzzy sets and decision analysis. *IEEE Transactions on Systems, Man, and Cybernetics* 10 (3): 341–354.
Galperin, E.A. and Ekel, P.Y. (2005). Synthetic realization approach to fuzzy global optimization via gamma algorithm. *Mathematical and Computer Modelling* 41 (13): 1457–1468.
Grabisch, M., Orlovski, S.A., and Yager, R.R. (1998). Fuzzy aggregation of numerical preferences. In: *Fuzzy Sets in Decision Analysis, Operations Research and Statistics, the Handbooks of Fuzzy Sets Series*, Vol. 4, 31–68. Boston: Kluwer.
Harker, P.T. and Vargas, L.G. (1987). The theory of ratio scale estimation Saaty's analytic hierarchy process. *Management Science* 33 (11): 1383–1402.
Herrera-Viedma, E., Alonso, S., Chiclana, F., and Herrera, F. (2007). A consensus model for group decision making with incomplete fuzzy preference relations. *IEEE Transactions on Fuzzy Systems* 15 (5): 863–877.
Herrera-Viedma, E., Herrera, F., Chiclana, F., and Luque, M. (2004). Some issues on consistency of fuzzy preference relations. *European Journal of Operational Research* 154 (1): 98–109.
Horiuchi, K. and Tamura, N. (1998). VSOP fuzzy numbers and their fuzzy ordering. *Fuzzy Sets and Systems* 93 (2): 197–210.
Jain, R. (1976). Decision-making in the presence of fuzzy variables. *IEEE Transactions on Systems, Man, and Cybernetics* 6 (10): 698–703.

Keeney, R.L. and Raiffa, H. (1976). *Decisions with Multiple Objectives: Preferences and Values Tradeoffs*. New York: Wiley.

Kulshreshtha, P. and Shekar, B. (2000). Interrelationships among fuzzy preference-based choice functions and significance of rationality conditions: a taxonomic and intuitive perspective. *Fuzzy Sets and Systems* 109 (3): 429–445.

Lee, E.S. and Li, R.L. (1988). Comparison of fuzzy numbers based on the probability measure of fuzzy events. *Computers and Mathematics with Applications* 15 (3): 887–896.

Lee-Kwang, H. (1999). A method for ranking fuzzy numbers and its application to decision-making. *IEEE Transactions on Fuzzy Systems* 7 (1): 677–685.

Liu, X.W. and Han, S.L. (2005). Ranking fuzzy numbers with preference weighting function expectations. *Computers and Mathematics with Applications* 49 (11–12): 1731–1753.

Llamazares, B. (2003). Characterization of fuzzy preference structures through Lukasiewicz triplets. *Fuzzy Sets and Systems* 136 (2): 217–235.

Malczewski, J. (2006). GIS-based multicriteria decision analysis: a survey of the literature. *International Journal of Geographical Information Science* 20 (7): 703–726.

Modarres, M. and Sadi-Nezhad, S. (2001). Ranking fuzzy numbers by preference ratio. *Fuzzy Sets and Systems* 118 (3): 429–436.

Negoita, C.V. and Ralescu, D.A. (1975). *Application of Fuzzy Sets to Systems Analysis*. Basel: Birkhäuser.

Orlovsky, S.A. (1978). Decision-making with a fuzzy preference relation. *Fuzzy Sets and Systems* 1 (3): 155–167.

Orlovsky, S.A. (1981). *Problems of Decision Making with Fuzzy Information*. Moscow (in Russian): Nauka.

Orudjev, F.D. (1983). Expert estimates and fuzzy set theory in investigating electrical systems. *Elektrichestvo* 85 (4): 7–11 (in Russian).

Ovchinnikov, S. (1981). Structure of fuzzy binary relations. *Fuzzy Sets and Systems* 6 (2): 169–195.

Ovchinnikov, S. and Roubens, M. (1991). On strict preference relations. *Fuzzy Sets and Systems* 43 (3): 319–326.

Ovchinnikov, S. and Roubens, M. (1992). On fuzzy strict preference, indifference and incomparability relations. *Fuzzy Sets and Systems* 49 (1): 15–20.

Öztürk, M., Tsoukiàs, A., and Vincke, P. (2005). Preference modelling. In: *Multiple Criteria Decision Analysis: State of the Art Surveys* (eds. J. Figueira, S. Greco and M. Ehrgott), 265–292. New York: Springer.

Parreiras, P., Ekel, P., and Bernandes, F. Jr. (2012a). A dynamic consensus scheme based on a nonreciprocal fuzzy preference relation modeling. *Information Sciences* 211 (1): 1–17.

Parreiras, R.O., Ekel, P.Y., Martini, J.S.C., and Palhares, R.M. (2010). A flexible consensus scheme for multicriteria group decision making under linguistic assessments. *Information Sciences* 180 (7): 1075–1089.

Parreiras, R.O., Ekel, P.Y., and Morais, D.C. (2012b). Fuzzy set based consensus schemes for multicriteria group decision making applied to strategic planning. *Group Decision and Negotiation* 21 (2): 153–183.

Pedrycz, W., Ekel, P., and Parreiras, R. (2011). *Fuzzy Multicriteria Decision-Making: Models, Methods, and Applications*. Chichester: Wiley.

Pedrycz, W. and Gomide, F. (1998). *An Introduction to Fuzzy Sets: Analysis and Design*. Cambridge, MA: MIT Press.

Popov, V.A. and Ekel, P.Y. (1987). Fuzzy set theory and problems of controlling the design and operation of electric power systems. *Soviet Journal of Computer and System Sciences* 25 (4): 92–99.

Queiroz, J.C.B. (2009) *Models and methods of decision making support for strategic management*. Ph.D. thesis (in Portuguese). Federal University of Minas Gerais, Belo Horizonte.

Raj, P.A. and Kumar, D.N. (1999). Ranking alternatives with fuzzy weights using maximizing set and minimizing set. *Fuzzy Sets and Systems* 105 (3): 365–375.

Rao, S. (1996). *Engineering Optimization: Theory and Practice*. New York: Wiley.

Roubens, M. (1989). Some properties of choice functions based on valued binary relations. *European Journal of Operational Research* 40 (3): 309–321.

Saadi-Nezhad, S. and Shahnazari-Shahrezaei, P. (2013). Ranking fuzzy numbers using preference ratio: a utility function approach. *Decision Science Letters* 2 (3): 149–162.

Saaty, T. (1980). *The Analytic Hierarchy Process*. New York: McGraw-Hill.

Salo, A.A. and Hämälainen, R.P. (1997). On the measurement of preferences in the analytic hierarchy process. *Journal of Multi-Criteria Decision Analysis* 6 (6): 309–319.

Sengupta, K. (1998). Fuzzy preference and Orlovsky choice procedure. *Fuzzy Sets and Systems* 93 (2): 231–234.

Shafiullah, M., Rahman, S.M., Mortoja, M.G., and Al-Ramadan, B. (2016). Role of spatial analysis technology in power system industry: an overview. *Renewable and Sustainable Energy Reviews* 66 (4): 584–595.

Tanino, T. (1984). Fuzzy preference orderings in group decision making. *Fuzzy Sets and Systems* 12 (1): 117–131.

Tseng, T.Y. and Klein, C.M. (1989). New algorithm for the ranking procedure in fuzzy decision making. *IEEE Transactions on Systems, Man, and Cybernetics* 19 (5): 1289–1296.

Von Winterfeldt, D. and Edwards, W. (1986). *Decision Analysis and Behavioral Research*. Cambridge: Cambridge University Press.

Wang, X. and Kerre, E.E. (2001). Reasonable properties for the ordering of fuzzy quantities (I) and (II). *Fuzzy Sets and Systems* 118 (3): 387–405.

Wang, Y.J. and Lee, H.S. (2008). The revised method of ranking fuzzy numbers with an area between the centroidand original points. *Computers and Mathematics with Applications* 55 (9): 2033–2042.

Wang, Y.M. and Luo, Y. (2009). Area ranking of fuzzy numbers based on positive and negative ideal points. *Computers and Mathematics with Applications* 58 (9): 1767–1779.

Yager, R.R. (1981). A procedure for ordering fuzzy sets of the unit interval. *Information Sciences* 24 (2): 143–161.

Zeshui, X. and Cuiping, W. (1999). A consistency improving method in the analytic hierarchy process. *European Journal of Operational Research* 116 (2): 443–449.

Zhang, Q., Chen, J.C.H., and Chong, P.P. (2004). Decision consolidation: criteria weight determination using multiple preference formats. *Decision Support Systems* 38 (2): 247–258.

Zhang, Q., Wang, Y., and Yang, Y. (2007) Fuzzy multiple attribute decision making with eight types of preference information on alternatives. *Proceedings of the 2007 IEEE Symposium on Computational Intelligence in Multicriteria Decision Making*, Honolulu, pp. 288–293.

Zhukovin, V.E. (1988). *Fuzzy Multicriteria Models of Decision Making*. Tbilisi (in Russian): Metsniereba.

Zimmermann, H.J. (1996). *Fuzzy Set Theory and Its Application*. Boston, MA: Kluwer.

Zimmermann, H.J. (2008). *Fuzzy Sets, Decision Making, and Expert Systems*. Boston, MA: Kluwer.

Zorin, V.V. and Ekel, P.Y. (1980). Discrete-optimization methods for electrical supply systems. *Power Engineering* 18 (5): 19–30.

6

Dealing with Uncertainty of Information

A Classic Approach

The classic approach encountered when considering uncertainty, which comes with a broad range of practical applications (Kaufman 1961; Belyaev 1977), is associated with the analysis carried out for the given solution alternatives (strategies) and the given representative combinations of initial data, states of nature, or scenarios realized on the basis of constructing the corresponding payoff matrices. This analysis is based on applying so-called choice criteria. The use of certain choice criteria leads to robust solutions, which are considered in the present chapter. The basic phases to support the application of the classic approach when dealing with the uncertainty factor are presented. The questions of constructing representative combinations of initial data, states of nature, or scenarios are also discussed. The use of the presented results is illustrated by solving a monocriteria problem in the presence of uncertainty.

6.1 Characterization of the Classic Approach to Dealing with Uncertainty of Information

The classic approach (Luce and Raiffa 1957; Raiffa 1968; Webster 2003) to dealing with the uncertainty of information is based on the assumption that the analysis is carried out for the given solution alternatives (strategies) X_k, $k = 1, 2, \ldots, K$ and the given representative combinations of initial data, states of nature, or scenarios Y_s, $s = 1, 2, \ldots, S$. Making use of alternatives and scenarios, we associate them with the corresponding payoff matrix in the form shown in Table 6.1.

The payoff matrix reflects effects (or consequences) of one or other solution alternative X_k, $k = 1, 2, \ldots, K$ for the corresponding combinations of initial data, states of nature, or scenarios Y_s, $s = 1, 2, \ldots, S$.

Multicriteria Decision-Making under Conditions of Uncertainty: A Fuzzy Set Perspective, First Edition. Petr Ekel, Witold Pedrycz, and Joel Pereira, Jr.

Table 6.1 Payoff matrix.

	Y_1	...	Y_s	...	Y_S
X_1	$F(X_1, Y_1)$	...	$F(X_1, Y_s)$	...	$F(X_1, Y_S)$
...	...	...	...	...	...
X_k	$F(X_k, Y_1)$	...	$F(X_k, Y_s)$	...	$F(X_k, Y_S)$
...	...	...	...	...	...
X_K	$F(X_K, Y_1)$	...	$F(X_K, Y_s)$	...	$F(X_K, Y_S)$

Belyaev (1977) defines the following basic phases to support the application of the classic approach when dealing with the uncertainty factor:

- mathematical formulation of the problem;
- selection of the representative combinations of initial data (selection of the states of nature or scenarios);
- determination and preliminary analysis of solution alternatives;
- construction of the payoff matrix;
- analysis of the payoff matrix and the choice of the rational solution alternatives;
- selection of the final solution.

Without the detailed discussion of the phases identified here, it is worth emphasizing that an analysis of the payoff matrices and the choice of the rational solution alternatives are based on the use of the corresponding choice criteria (Luce and Raiffa 1957; Raiffa 1968; Webster 2003). Those being utilized most frequently and exhibiting a general character are the criteria of Wald, Laplace, Savage, and Hurwicz. At the same time, it is necessary to indicate that other choice criteria are available in the literature, for example, the criteria of Hodges and Lehmann (Hodges and Lehmann 1952), Bayes (Trukhaev 1981), maximal probability (Trukhaev 1981), and so on. However, the use of these criteria presumes the availability of the certain information (usually, probabilistically described) about the states of nature.

6.2 Payoff Matrices and Characteristic Estimates

To better understand the nature of the criteria of Wald, Laplace, Savage, and Hurwicz, the matrix presented in Table 6.1 is now extended and shown in Table 6.2, in which we take into account recommendations presented in (Belyaev 1977).

Table 6.2 Payoff matrix with characteristic estimates.

	Y_1	…	Y_s	…	Y_S	$F^{max}(X_k)$	$F^{min}(X_k)$	$\bar{F}(X_k)$	$r^{max}(X_k)$
X_1	$F(X_1, Y_1)$	…	$F(X_1, Y_s)$	…	$F(X_1, Y_S)$	$F^{max}(X_1)$	$F^{min}(X_1)$	$\bar{F}(X_1)$	$r^{max}(X_1)$
…	…	…	…	…	…	…	…	…	…
X_k	$F(X_k, Y_1)$	…	$F(X_k, Y_s)$	…	$F(X_k, Y_S)$	$F^{max}(X_k)$	$F^{min}(X_k)$	$\bar{F}(X_k)$	$r^{max}(X_k)$
…	…	…	…	…	…	…	…	…	…
X_K	$F(X_K, Y_1)$	…	$F(X_K, Y_s)$	…	$F(X_K, Y_S)$	$F^{max}(X_K)$	$F^{min}(X_K)$	$\bar{F}(X_K)$	$r^{max}(X_K)$
$F^{max}(Y_s)$	$F^{max}(Y_1)$	…	$F^{max}(Y_s)$	…	$F^{max}(Y_S)$				

This extension is associated with the incorporation of the following estimates (Pedrycz et al. 2011):

- The objective function maximum level

$$F^{\max}(X_k) = \max_{1 \le s \le S} F(X_k, Y_s) \tag{6.1}$$

This level is determined for the given solution alternative and, as the name stipulates, is the most optimistic estimate when the objective function is to be maximized or the most pessimistic estimate if the objective function is to be minimized for the considered solution alternative.

- The objective function minimum level

$$F^{\min}(X_k) = \min_{1 \le s \le S} F(X_k, Y_s) \tag{6.2}$$

computed for the given solution alternative. It is the most pessimistic estimate when the objective function is to be maximized, or is treated as the most optimistic estimate if the objective function is to be minimized for the considered solution alternative.

- The objective function

$$\bar{F}(X_k) = \frac{1}{S}\sum_{s=1}^{S} F(X_k, Y_s) \tag{6.3}$$

determined for the given solution alternative.

- The risk (regret) maximum level

$$r^{\max}(X_k) = \max_{1 \le s \le S} r(X_k, Y_s) \tag{6.4}$$

where $r(X_k, Y_s)$ is an over-expenditure that takes place under combination of the representative combination of initial data, state of nature, or scenario Y_s, and the choice of the solution alternative X_k instead of the solution alternative that is locally optimal for the given Y_s. The estimates of over-expenditures provide a certain description of the situation as they show a relative difference of the objective function values under the choice of one solution alternative in place of another. In fact, the over-expenditures characterize a damage level associated with the uncertainty of the situation itself.

To determine risks (regrets) $r(X_k, Y_s)$, it is necessary to define the maximum value of the objective function (if it is to be maximized, as it is considered in Table 6.2) for each representative combination of initial data, state of nature or scenario Y_s (for each column of the payoff matrix):

$$F^{\max}(Y_s) = \max_{1 \le k \le K} F(X_k, Y_s) \tag{6.5}$$

It is evident that if the objective function is to be minimized, it is necessary to define its minimum for each representative combination of initial data, state of nature, or scenario Y_s (for each column of the payoff matrix):

Table 6.3 Risk matrix.

	Y_1	...	Y_s	...	Y_S	$r^{max}(X_k)$
X_1	$r(X_1, Y_1)$	...	$r(X_1, Y_s)$	...	$r(X_1, Y_S)$	$r^{max}(X_1)$
...	...	...	...	...	...	...
X_k	$r(X_k, Y_1)$	...	$r(X_k, Y_s)$	...	$r(X_k, Y_S)$	$r^{max}(X_k)$
...	...	...	...	...	...	...
X_K	$r(X_K, Y_1)$	...	$r(X_K, Y_s)$	...	$r(X_K, Y_S)$	$r^{max}(X_K)$

$$F^{\min}(Y_s) = \min_{1 \le k \le K} F(X_k, Y_s) \tag{6.6}$$

The risk associated with any solution alternative X_k and any representative combination of initial data, state of nature, or scenario Y_s can be evaluated as follows:

$$r(X_k, Y_s) = F^{\max}(Y_s) - F(X_k, Y_s) \tag{6.7}$$

if the objective function is to be maximized, or

$$r(X_k, Y_s) = F(X_k, Y_s) - F^{\min}(Y_s) \tag{6.8}$$

if the objective function is to be minimized.

Carrying out calculations on the basis of Eqs. (6.7) or (6.8) for all X_k, $k = 1, 2, \ldots, K$ and Y_s, $s = 1, 2, \ldots, S$, we obtain the risk (regret) matrix shown in Table 6.3. Note that any column of this matrix includes at least a single zero element $r(X_k, Y_s) = 0$.

Example 6.1 Consider the payoff matrix presented in Table 6.4. If the solved problem is associated with maximization, then applying Eq. (6.7), it is possible to construct the risk matrix presented in Table 6.5.

In the case of the problem of minimization, applying Eq. (6.8), we can build the risk matrix shown in Table 6.6.

Finally, based on the data presented in Table 6.4, it is possible to use Eqs. (6.1)–(6.4) to build the payoff matrix with characteristic estimates, as presented in Table 6.7. Note that the Table 6.7 is based on the assumption that the objective function is to be maximized. For a minimization function, the values of the column $r^{max}(X_k)$ are based on the corresponding column in Table 6.6, as shown in Table 6.8.

Table 6.4 Example 6.1: Payoff matrix.

	Y_1	Y_2	Y_3	Y_4
X_1	13	7	9	12
X_2	13	8	14	10
X_3	10	9	7	12
X_4	10	8	11	13

Table 6.5 Example 6.1: Risk matrix for the problem of maximization.

	Y_1	Y_2	Y_3	Y_4	$r^{max}(X_k)$
X_1	0	2	5	1	5
X_2	0	1	0	3	3
X_3	3	0	7	1	7
X_4	3	1	3	0	3

Table 6.6 Example 6.1: Risk matrix for the problem of minimization.

	Y_1	Y_2	Y_3	Y_4	$r^{max}(X_k)$
X_1	3	0	2	2	3
X_2	3	1	7	0	7
X_3	0	2	0	2	2
X_4	0	1	4	3	4

Table 6.7 Example 6.1: Payoff matrix with characteristic estimates for the problem of maximization.

	Y_1	Y_2	Y_3	Y_4	$F^{max}(X_k)$	$F^{min}(X_k)$	$\bar{F}(X_k)$	$r^{max}(X_k)$
X_1	13	7	9	12	13	7	10.25	5
X_2	13	8	14	10	14	8	11.25	3
X_3	10	9	7	12	12	7	9.50	7
X_4	10	8	11	13	13	8	10.50	3

Table 6.8 Example 6.1: Payoff matrix with characteristic estimates for the problem of minimization.

	Y_1	Y_2	Y_3	Y_4	$F^{max}(X_k)$	$F^{min}(X_k)$	$\bar{F}(X_k)$	$r^{max}(X_k)$
X_1	13	7	9	12	13	7	10.25	3
X_2	13	8	14	10	14	8	11.25	7
X_3	10	9	7	12	12	7	9.50	2
X_4	10	8	11	13	13	8	10.50	4

6.3 Choice Criteria and Their Application

The choice criteria of Wald, Laplace, Savage, and Hurwicz are based on the use of the characteristic estimates $F^{max}(X_k)$, $F^{min}(X_k)$, $\bar{F}(X_k)$, and $r^{max}(X_k)$, defined by Eqs. (6.1)–(6.4). In this section, we discuss the particular characteristics of each choice criterion.

The criterion of Wald utilizes the estimates $F^{min}(X_k)$ or $F^{max}(X_k)$. The Wald criterion stipulates that one has to choose the solution alternative X^W, for which, for the maximization objective function, the estimate $F^{min}(X_k)$ attains maximum, that is,

$$\max_{1 \le k \le K} F^{min}(X_k) = \max_{1 \le k \le K} \min_{1 \le s \le S} F(X_k, Y_s) \tag{6.9}$$

or, for a minimization objective function, the estimate $F^{max}(X_k)$ attains minimum, that is

$$\min_{1 \le k \le K} F^{max}(X_k) = \min_{1 \le k \le K} \max_{1 \le s \le S} F(X_k, Y_s) \tag{6.10}$$

The use of this criterion generates solution alternatives, assuming the most unfavorable combination of initial data. It guarantees that the objective function level is not greater than a certain value at any possible future conditions (in the case of the maximized objective function) or is not lesser than a certain value at any possible future conditions (in the case of the minimized objective function). This is its dignity (Belyaev 1977). On the other hand, the orientation toward the most unfavorable combination of initial data is extremely cautious (pessimistic or conservative) (Belyaev 1977).

The criterion of Laplace uses the estimate $\bar{F}(X_k)$ and is oriented to choose the solution alternative X^L, for which this estimate attains its maximum:

$$\max_{1 \le k \le K} \bar{F}(X_k) = \max_{1 \le k \le K} \frac{1}{S} \sum_{s=1}^{S} F(X_k, Y_s) \tag{6.11}$$

for a maximization objective function, or its minimum

$$\min_{1 \le k \le K} \bar{F}(X_k) = \min_{1 \le k \le K} \frac{1}{S} \sum_{s=1}^{S} F(X_k, Y_s) \tag{6.12}$$

for a minimization objective function.

This criterion corresponds to the principle of "insufficient reason" (Belyaev 1977), that is, it is based upon the assumption that we have no basis to distinguish one or another combination of initial data. Thus, it is necessary to act as they are equally probable. This is its drawback. However, the average score is sufficiently important.

The criterion of Savage is associated with the use of the estimate $r^{max}(X_k)$ and allows one to choose the solution alternative X^S, for which this estimate reaches minimum:

$$\min_{1 \le k \le K} r^{\max}(X_k) = \min_{1 \le k \le K} \max_{1 \le s \le S} r(X_k, Y_s) \tag{6.13}$$

As in the case of the Wald choice criterion, the use of Eq. (6.10) is based on the *minimax* principle. Therefore, the Savage choice criterion can also be considered conservative. However, experience (Belyaev 1977) shows that the recommendations based on applying Eq. (6.13) are mismatched with the decisions obtained with the use of Eq. (6.10). Operating with values of $r^{max}(X_k)$, we obtain a slightly different evaluation of the situation, which could lead to more "daring" (less conservative) recommendations.

Finally, the criterion of Hurwicz utilizes a linear combination of the estimates $F^{min}(X_k)$ and $F^{max}(X_k)$ chooses the solution alternative X^H for which this combination attains the maximum:

$$\begin{aligned} &\max_{1 \le k \le K} \left[\alpha F^{\min}(X_k) + (1-\alpha) F^{\max}(X_k)\right] \\ &= \max_{1 \le k \le K} \left[\alpha \min_{1 \le s \le S} F(X_k, Y_s) + (1-\alpha) \max_{1 \le s \le S} F(X_k, Y_s)\right] \end{aligned} \tag{6.14}$$

for a maximization objective function, or the minimum:

$$\begin{aligned} &\min_{1 \le k \le K} \left[\alpha F^{\max}(X_k) + (1-\alpha) F^{\min}(X_k)\right] \\ &= \min_{1 \le k \le K} \left[\alpha \max_{1 \le s \le S} F(X_k, Y_s) + (1-\alpha) \min_{1 \le s \le S} F(X_k, Y_s)\right] \end{aligned} \tag{6.15}$$

for a minimization problem.

In Eqs. (6.14) and (6.15), $\alpha \in [0, 1]$ is the index "pessimism-optimism" whose magnitude is defined by a DM. If $\alpha = 1$, the Hurwicz choice criterion is turned into the Wald choice criterion, and if $\alpha = 0$, Eq. (6.15) is turned into the "extreme optimism" (*minmin*) criterion for which the most favorable combination of initial data is assumed. When $0 < \alpha < 1$, we obtain something that is an average and this is the attractiveness of the Hurwicz criterion. Belyaev (1977) recommends choosing a range from 0.5 to 1.

Generally, we can talk about obtaining so-called robust solutions (Roy 2010) in the case of applying the Wald and Savage choice criteria. Besides, the use of the Hurwicz choice criterion with the index "pessimism-optimism" α = 1 also provides robust solutions.

Example 6.2 The application of the Wald, Laplace, Savage, and Hurwicz choice criteria to the data presented in Table 6.7 leads to the following results for the maximized objective function:

- $X^W = \{X_2, X_4\}$ since $\min_{1 \le s \le 4} \{X_2\} = \min_{1 \le s \le 4} \{X_4\} = 8$
- $X^L = \{X_2\}$ since $\frac{1}{4}\sum_{s=1}^{4} F(X_2, Y_s) = 11.25$
- $X^S = \{X_2, X_4\}$ since $\max_{1 \le s \le 4} r\{X_2\} = \max_{1 \le s \le 4} r\{X_4\} = 3$
- $X^H = \{X_2\}$ since $\alpha \min_{1 \le s \le 4} F(X_2, Y_s) + (1 - \alpha) \max_{1 \le s \le 4} F(X_2, Y_s) = 9.50$

The last result has been obtained for α = 0.75.

Thus, the use of the different choice criteria does not lead to a unique solution alternative.

Let us consider the case of the minimized objective function. The results of the application of the choice criteria to the data presented in Table 6.8 are the following:

- $X^W = \{X_3\}$ since $\max_{1 \le s \le 4} \{X_3\} = 12$
- $X^L = \{X_3\}$ since $\frac{1}{4}\sum_{s=1}^{4} F(X_2, Y_s) = 9.50$
- $X^S = \{X_3\}$ since $\max_{1 \le s \le 4} r\{X_3\} = 2$
- $X^H = \{X_3\}$ since $\alpha \max_{1 \le s \le 4} F(X_3, Y_s) + (1 - \alpha) \min_{1 \le s \le S} F(X_3, Y_s) = 10.75$

The last result has also been obtained for α = 0.75.

6.4 Elements of Constructing Representative Combinations of Initial Data, States of Nature, or Scenarios

The first stage in the decision-making process is concerned with the construction of a payoff matrix for all combinations of the solution alternatives X_k, $k = 1, 2, \ldots, K$ and the representative states of nature Y_s, $s = 1, 2, \ldots, S$. Taking this into account, it is necessary to indicate that the questions related to constructing representative combinations of initial data, states of nature, or scenarios, including the definition of their number, are of a general nature and the corresponding answers vary from area to area (Amer et al. 2012). Generally, the main condition in building representative combinations of initial data, states of nature, or scenarios is that each scenario is to be the representation of a plausible reality (Durbach and Stewart 2012).

Given the experience of (Pedrycz et al. 2011; Pereira et al. 2015) in the present work, so-called LP_τ-sequences (proposed in Sobol' 1966, 1979 and classified in Niederreiter 1978) are also used to construct the representative combinations of initial data, states of nature, or scenarios. The sequences of points constructed on the basis of Sobol' (1966, 1979) fill the multidimensional cube in a very uniform mode. These sequences have superior characteristics of uniformity among other uniformly distributed sequences (Sobol' 1979).

The results of Sobol' (1966, 1979) allow one to determine points Q_s, $s = 1, 2, \ldots, S$ with coordinates q_{st}, $t = 1, 2, \ldots, T$ in the corresponding unit hypercube Q^T. For instance, if we have $T = 5$ and it is necessary to create $S = 6$ representative combinations of initial data, states of nature, or scenarios, the coordinates of Q_s, $s = 1, 2, \ldots, 6$ for $t = 1, 2, \ldots, 5$ determined on the basis of Sobol' (1966, 1979) are presented in Table 6.9.

In reality, the selection of representative combinations of initial data, states of nature, or scenarios is reduced to the formation of points of a uniformly distributed sequence in Q^T and their transformation to the hypercube C^T, defined by the lower c'_t and upper c''_t bounds of the corresponding uncertain coefficient of the analyzed objective function. Taking this into account, if points Q_s, $s = 1, 2, \ldots, S$ with coordinates q_{st}, $t = 1, 2, \ldots, T$ form a uniformly distributed sequence in Q^T, then points C_s, $s = 1, 2, \ldots, S$ with the coordinates are expressed as follows:

$$c_{st} = c'_t + \left(c''_t - c'_t\right) q_{st}, \quad t = 1, 2, \ldots, T \tag{6.16}$$

Example 6.3 Applying LP_τ-sequences, let us construct five representative combinations of initial data, states of nature, or scenarios for an objective function, which includes the following three interval-valued coefficients:

- $c_1 = [2, 4]$
- $c_2 = [4, 7]$
- $c_3 = [3, 5]$

Using part of Table 6.9 (corresponding to $S = 5$ and $T = 3$) presented in Table 6.10 and applying Eq. (6.16), one can build coordinates of representative combinations of initial data, states of nature, or scenarios for an objective function

The data in Table 6.10 permit us to construct coordinates of representative combinations of initial data, states of nature, or scenarios presented in Table 6.11.

Table 6.9 Points of the LP_τ-sequences in Q^5.

s	t = 1	t = 2	t = 3	t = 4	t = 5
1	0.500	0.500	0.500	0.500	0.500
2	0.250	0.750	0.250	0.750	0.250
3	0.750	0.250	0.750	0.250	0.750
4	0.125	0.625	0.875	0.875	0.625
5	0.625	0.125	0.375	0.375	0.125
6	0.375	0.375	0.625	0.125	0.875

Table 6.10 Example 6.3: Points of the LP_τ-sequences in Q^3.

s	t = 1	t = 2	t = 3
1	0.500	0.500	0.500
2	0.250	0.750	0.250
3	0.750	0.250	0.750
4	0.125	0.625	0.875
5	0.625	0.125	0.375

Table 6.11 Example 6.3: Representative combinations of initial data.

s	t = 1	t = 2	t = 3
1	3.000	5.500	4.000
2	2.500	6.250	3.500
3	3.500	4.750	4.500
4	2.250	5.875	4.750
5	3.250	4.375	3.750

6.5 Application Example

Let us consider the following problem with interval coefficients present in the objective function:

$$F(x) = [3,7]x_1 + [4,9]x_2 + [2,7]x_3 + [5,8]x_4 \rightarrow \min \tag{6.17}$$

subject to the following constraints:

$$0 \le x_1 \le 30 \tag{6.18}$$

$$0 \le x_2 \le 50 \tag{6.19}$$

$$0 \le x_3 \le 35 \tag{6.20}$$

$$0 \le x_4 \le 25 \tag{6.21}$$

$$x_1 + x_2 + x_3 + x_4 = 100 \tag{6.22}$$

Table 6.12 includes points of the LP_τ-sequences in Q^4 in accordance with the number of coefficients in the objective function Eq. (6.17). These points can serve for the generation of $S = 7$ representative combinations of initial data, states of nature, or scenarios.

The application of Eq. (6.16) helps one to process Eq. (6.17) by using points of the LP_τ-sequences of Table 6.12. In particular, one can construct seven optimization problems with deterministic coefficients

$$F(x) = 5.000x_1 + 6.500x_2 + 4.500x_3 + 6.500x_4 \to \min \tag{6.23}$$

$$F(x) = 4.000x_1 + 7.750x_2 + 3.250x_3 + 7.250x_4 \to \min \tag{6.24}$$

$$F(x) = 6.000x_1 + 5.250x_2 + 5.750x_3 + 5.750x_4 \to \min \tag{6.25}$$

$$F(x) = 3.500x_1 + 7.125x_2 + 6.375x_3 + 7.625x_4 \to \min \tag{6.26}$$

$$F(x) = 5.500x_1 + 4.625x_2 + 3.875x_3 + 6.125x_4 \to \min \tag{6.27}$$

$$F(x) = 4.500x_1 + 5.875x_2 + 5.125x_3 + 5.375x_4 \to \min \tag{6.28}$$

$$F(x) = 6.500x_1 + 8.375x_2 + 2.625x_3 + 6.875x_4 \to \min \tag{6.29}$$

which are subject to the same constraints in Eqs. (6.18)–(6.22).

The solutions to these optimization problems are the following:

- $s = 1$: $x_1^0 = 30$, $x_2^0 = 35$, $x_3^0 = 35$, $x_4^0 = 0$ for Eq. (6.23)
- $s = 2$: $x_1^0 = 30$, $x_2^0 = 10$, $x_3^0 = 35$, $x_4^0 = 25$ for Eq. (6.24)
- $s = 3$: $x_1^0 = 0$, $x_2^0 = 50$, $x_3^0 = 35$, $x_4^0 = 15$ for Eq. (6.25)

Table 6.12 Application example: Points of the LP_τ-sequences in Q^4.

s	$t = 1$	$t = 2$	$t = 3$	$t = 4$
1	0.500	0.500	0.500	0.500
2	0.250	0.750	0.250	0.750
3	0.750	0.250	0.750	0.250
4	0.125	0.625	0.875	0.875
5	0.625	0.125	0.375	0.375
6	0.375	0.375	0.625	0.125
7	0.875	0.875	0.125	0.625

- $s = 4$: $x_1^0 = 30$, $x_2^0 = 35$, $x_3^0 = 35$, $x_4^0 = 0$ for Eq. (6.26)
- $s = 5$: $x_1^0 = 15$, $x_2^0 = 50$, $x_3^0 = 35$, $x_4^0 = 0$ for Eq. (6.27)
- $s = 6$: $x_1^0 = 30$, $x_2^0 = 10$, $x_3^0 = 35$, $x_4^0 = 25$ for Eq. (6.28)
- $s = 7$: $x_1^0 = 30$, $x_2^0 = 10$, $x_3^0 = 35$, $x_4^0 = 25$ for Eq. (6.29)

In such a way, we can form the following four solution alternatives for the problems in Eqs. (6.17)–(6.22):

- $X_1 = (30, 35, 35, 0)$
- $X_2 = (30, 10, 35, 25)$
- $X_3 = (0, 50, 35, 15)$
- $X_4 = (15, 50, 35, 0)$

Substituting these solutions into Eqs. (6.23)–(6.29), one can construct a payoff matrix presented in Table 6.13.

The use of the data in Table 6.13 and the application of Eq. (6.8) produces the risk matrix presented in Table 6.14.

Now, we can build a matrix that includes characteristic estimates. This matrix is presented in Table 6.15.

Finally, analyzing the data in Table 6.15, on the basis of applying the Wald, Laplace, Savage, and Hurwicz choice criteria, it is possible to choose the rational solution alternatives. The Wald criterion in Eq. (6.10) indicates $X^W = \{X_1\}$. The Laplace choice criterion in Eq. (6.12) leads to $X^L = \{X_1, X_2\}$. The use of the Savage choice criterion in Eq. (6.13) produces $X^S = \{X_1\}$. The Hurwicz choice criterion in Eq. (6.15), applied with $\alpha = 0.75$, also generates $X^H = \{X_1\}$.

Table 6.13 Application example: Payoff matrix.

	Y_1	Y_2	Y_3	Y_4	Y_5	Y_6	Y_7
X_1	535.00	565.00	505.00	520.00	580.00	462.50	577.50
X_2	535.00	577.50	492.50	507.50	542.50	500.00	590.00
X_3	580.00	550.00	610.00	553.75	613.75	458.75	693.75
X_4	557.50	553.75	561.25	540.62	608.12	449.38	631.88

Table 6.14 Application example: Risk matrix.

	Y_1	Y_2	Y_3	Y_4	Y_5	Y_6	Y_7	$r^{max}(X_k)$
X_1	0	15.00	12.50	12.50	37.50	13.12	0	37.50
X_2	0	27.50	0	0	0	50.62	12.50	50.62
X_3	45.00	0	117.50	46.25	71.25	9.37	116.25	117.50
X_4	22.50	3.75	68.75	33.12	65.62	0	54.38	68.75

Table 6.15 Application example: Characteristic estimates.

	$F^{max}(X_k)$	$F^{min}(X_k)$	$\bar{F}(X_k)$	$r^{max}(X_k)$
X_1	580.00	462.50	535.00	37.50
X_2	590.00	492.50	535.00	50.62
X_3	693.75	458.75	580.00	117.50
X_4	631.88	449.38	557.50	68.75

6.6 Conclusions

We have considered the classic approach to taking into account the uncertainty factor in analyzing monocriteria decision-making models. This approach is associated with the construction of so-called payoff matrices, reflecting effects that can be obtained for different combinations of solution alternatives (strategies) and representative combinations of initial data, states of nature, or scenarios. To obtain rational solution alternatives, the payoff matrices are processed with the use of so-called choice criteria, which are discussed. Some of them permit one to construct robust solutions. The basic phases to support the utilization of the classic approach have been presented. The questions of the construction of representative combinations of initial data, states of nature, or scenarios on the basis of applying LP_τ-sequences have been discussed. The LP_τ-sequences have superior characteristics of uniformity among other uniformly distributed sequences. The use of the presented results has been illustrated by solving the monocriteria problem in conditions of uncertainty.

Before processing with the generalization of the classic approach to dealing with the uncertainty factor in multiobjective problems (Chapter 7), it is worth noting that there have been some other models addressing the consideration of uncertainty. For instance, in Yager (1996); Kuchta (2007); and Wen and Iwamura (2008), fuzzy sets were discussed as a viable alternative, and in Ahn and Yager (2014) the OWA operator was used for decision-making under uncertainty. However, all these studies are focused on monocriteria problems.

Exercises

6.1 Using the data in Table 6.10, construct four representative combinations of initial data, states of nature, or scenarios for an objective function that includes the following three interval coefficients:

- $c_1 = [1, 5]$
- $c_2 = [3, 6]$
- $c_3 = [2, 7]$

6.2 Apply the classic approach to considering the uncertainty of information (taking into account $\alpha = 0.75$ for the Hurwicz choice criterion) to analyze the payoff matrix given in Table 6.16. The objective function is to be minimized.

Table 6.16 Problem 6.2: Payoff matrix.

	Y_1	Y_2	Y_3	Y_4
X_1	10	12	12	11
X_2	8	13	11	14
X_3	10	11	12	13
X_4	12	13	12	9

6.3 Apply the classic approach to considering the uncertainty of information (taking into account $\alpha = 0.75$ for the Hurwicz choice criterion) to analyze the payoff matrix given in Table 6.17. The objective function is to be maximized.

Table 6.17 Problem 6.3: Payoff matrix.

	Y_1	Y_2	Y_3	Y_4
X_1	105	112	96	118
X_2	95	115	108	120
X_3	110	98	99	115
X_4	107	111	102	107

6.4 Verify the possibility of changing the solution of the problem, defined by Problems 6.1 and 6.2, for the Hurwicz criterion in the case of setting $\alpha = 0.25$.

6.5 The levels of the characteristic estimates given in Table 6.15 have permitted the selection of the solution alternative X_1 on the basis of applying the Hurwicz choice criterion with setting $\alpha = 0.75$. It is necessary to find a boundary value of α that permits one to select the solution alternative X_4 instead of X_1, applying the Hurwicz choice criterion.

References

Ahn, B.S. and Yager, R.R. (2014). The use of ordered weighted averaging method for decision making under uncertainty. *International Transactions in Operational Research* 21 (2): 247–262.

Amer, M., Daim, T.U., and Jetter, A. (2012). A review of scenario planning. *Futures* 46 (1): 23–40.

Belyaev, L.S. (1977). *A Practical Approach to Choosing Alternate Solutions to Complex Optimization Problems Under Uncertainty.* Laxenburg: IIASA.

Durbach, I.N. and Stewart, T.J. (2012). Modeling uncertainty in multi-criteria decision analysis. *European Journal of Operational Research* 223 (1): 1–14.

Hodges, J.L. and Lehmann, E.L. (1952). The use of previous experience in reaching statistical decisions. *The Annals of Mathematical Statistics* 23 (3): 396–407.

Kaufman, G.M. (1961). *Statistical Decision and Related Techniques in Oil and Gas Exploration.* Englewood Cliffs: Prentice Hall.

Kuchta, D. (2007). Choice of the best alternative in case of a continuous set of states of nature-application of fuzzy numbers. *Fuzzy Optimization and Decision Making* 6 (2): 173–178.

Luce, R.D. and Raiffa, H. (1957). *Games and Decisions.* New York: Wiley.

Niederreiter, H. (1978). Quasi-Monte Carlo methods and pseudo-random numbers. *Bulletin of the American Mathematical Society* 84 (6): 951–1041.

Pedrycz, W., Ekel, P., and Parreiras, R. (2011). *Fuzzy Multicriteria Decision-Making: Models, Methods and Applications.* John Wiley & Sons.

Pereira, J.G. Jr., Ekel, P.Y., Palhares, R.M., and Parreiras, R.O. (2015). On multicriteria decision making under conditions of uncertainty. *Information Sciences* 324: 44–59.

Raiffa, H. (1968). *Decision Analysis.* Reading, MA: Addison-Wesley.

Roy, B. (2010). To better respond to the robustness concern in decision aiding: four proposals based on a twofold observation. In: *Handbook of Multicriteria Analysis* (eds. C. Zopounidis and P.M. Pardalos), 3–24. Berlin: Springer.

Sobol', I.M. (1966). On the distribution of points in a cube and integration grids. *Achievements of Mathematical Sciences* 21 (5): 271–272. (in Russian).

Sobol', I.M. (1979). On the systematic search in a hypercube. *SIAM Journal on Numerical Analysis* 16 (5): 790–793.

Trukhaev, R.I. (1981). *Models of Decision Making in Conditions of Uncertainty.* Moscow (in Russian): Nauka.

Webster, T.J. (2003). *Managerial Economics: Theory and Practice.* London: Academic Press.

Wen, M. and Iwamura, K. (2008). Fuzzy facility location-allocation problem under the Hurwicz criterion. *European Journal of Operational Research* 184 (2): 627–635.

Yager, R.R. (1996) Fuzzy set methods for uncertainty representation in risky financial decisions. *Proceedings of the IEEE/IAFE Conference on Computational Intelligence for Financial Engineering*, New York, NY, pp. 59–65.

7

Generalization of the Classic Approach to Dealing with Uncertainty of Information and General Scheme of Multicriteria Decision-Making under Conditions of Uncertainty

This chapter deals with multicriteria decision-making problems under conditions of uncertainty. One of its main contributions is the consideration of choice criteria of the classic approach to handle information uncertainty in monocriteria decision-making as objective functions within the framework of multiobjective models. Such consideration of choice criteria is of a fundamental character and allows one to modify the generalization, proposed and discussed in Ekel et al. (2008, 2011) and Pedrycz et al. (2011), of the classic approach to handle information uncertainty for solving multicriteria problems. The modification is helpful in overcoming limitations of the indicated generalization (which can lead to contradictory decisions) and chart a general methodology for multicriteria decision-making under conditions of uncertainty. This methodology is based on a possibilistic approach to produce solutions, including robust solutions, in multicriteria analysis. Its usage, in the original form, is instrumental in using available quantitative information to the highest extent to reduce the decision uncertainty regions. If the solving capacity related to quantitative information processing does not permit one to obtain unique solutions, the methodology assumes the use, at the final decision stage, of qualitative information based on knowledge, experience, and intuition of experts involved in the decision-making process, in particular, within the framework of $\langle X, R \rangle$ models. Thus, the general scheme is based on combining $\langle X, F \rangle$ models, $\langle X, R \rangle$ models, and the generalization of the classic approach to dealing with uncertainty of information.

However, increasingly, we encounter problems whose essence requires the consideration of the objectives (investment attractiveness, political effect, maintenance flexibility, etc.) formed on the basis of qualitative information, at all decision process stages. Considering this, the chapter describes the general scheme of multicriteria decision-making under conditions of uncertainty aimed at generating multicriteria solutions, including multicriteria robust solutions,

Multicriteria Decision-Making under Conditions of Uncertainty: A Fuzzy Set Perspective, First Edition. Petr Ekel, Witold Pedrycz, and Joel Pereira, Jr.

by constructing representative combinations of initial data, states of nature, or scenarios with direct using qualitative information (with the possibility for experts to apply diverse preference formats processed by transformation functions) presented along with quantitative information, realizing a process of information fusion within the multiobjective models.

The use of the presented results is illustrated by solving problems coming from the strategic planning area.

7.1 Generalization of the Classic Approach to Dealing with Uncertainty of Information in Multicriteria Decision Problems

As discussed in Chapter 4, the application of the Bellman–Zadeh approach (Bellman and Zadeh 1970) to decision-making in a fuzzy environment when analyzing multicriteria models provides constructive means to derive harmonious solutions on the basis of analyzing associated *maxmin* problems. This circumstance permits one to propose the generalization of the classic approach to deal with information uncertainty by applying the Bellman–Zadeh approach to decision-making in a fuzzy environment (Ekel et al. 2008, 2011; Pedrycz et al. 2011).

The results of Ekel et al. (2008, 2011) and Pedrycz et al. (2011) are related to analyzing payoff matrices, whose construction was discussed in Chapter 6. Naturally, if the considered problem includes q objective functions, then q payoff matrices are to be constructed and analyzed.

Applying Eq. (4.31) to minimized objective functions or Eq. (4.32) to maximized ones, one constructs the normalized or modified payoff matrix for the pth objective function. This payoff matrix is presented in Table 7.1.

The availability of q modified payoff matrices helps us to construct the aggregated payoff matrix presented in Table 7.2 by applying Eq. (4.28).

Table 7.1 Modified payoff matrix for the pth objective function.

	Y_1	…	Y_s	…	Y_S
X_1	$\mu_{A_p}(X_1, Y_1)$	…	$\mu_{A_p}(X_1, Y_s)$	…	$\mu_{A_p}(X_1, Y_S)$
…	…	…	…	…	…
X_k	$\mu_{A_p}(X_k, Y_1)$	…	$\mu_{A_p}(X_k, Y_s)$	…	$\mu_{A_p}(X_k, Y_S)$
…	…	…	…	…	…
X_K	$\mu_{A_p}(X_K, Y_1)$	…	$\mu_{A_p}(X_K, Y_s)$	…	$\mu_{A_p}(X_K, Y_S)$

Table 7.2 Aggregated payoff matrix with characteristic estimates.

	Y_1	...	Y_s	...	Y_S	$\mu_D^{max}(X_k)$	$\mu_D^{min}(X_k)$	$\bar{\mu}_D(X_k)$	$r^{max}(X_k)$
X_1	$\mu_D(X_1, Y_1)$	...	$\mu_D(X_1, Y_s)$	...	$\mu_D(X_1, Y_S)$	$\mu_D^{max}(X_1)$	$\mu_D^{min}(X_1)$	$\bar{\mu}_D(X_1)$	$r^{max}(X_1)$
...	...	...	...	...	...	...	...	...	...
X_k	$\mu_D(X_k, Y_1)$	...	$\mu_D(X_k, Y_s)$	...	$\mu_D(X_k, Y_S)$	$\mu_D^{max}(X_k)$	$\mu_D^{min}(X_k)$	$\bar{\mu}_D(X_k)$	$r^{max}(X_k)$
...	...	...	...	...	...	...	...	...	...
X_K	$\mu_D(X_K, Y_1)$	...	$\mu_D(X_K, Y_s)$	...	$\mu_D(X_K, Y_S)$	$\mu_D^{max}(X_K)$	$\mu_D^{min}(X_K)$	$\bar{\mu}_D(X_K)$	$r^{max}(X_K)$
$\mu_D^{max}(Y_s)$	$\mu_D^{max}(Y_1)$	...	$\mu_D^{max}(Y_s)$	...	$\mu_D^{max}(Y_S)$				

The characteristic estimates of Table 7.2 are the following:

- The membership function maximum level (optimistic estimate)

$$\mu_D^{max}(X_k) = \max_{1 \le s \le S} \mu_D(X_k, Y_s) \tag{7.1}$$

- The membership function minimum level (corresponding to the pessimistic estimate)

$$\mu_D^{min}(X_k) = \min_{1 \le s \le S} \mu_D(X_k, Y_s) \tag{7.2}$$

- The membership function average level

$$\bar{\mu}_D(X_k) = \frac{1}{S}\sum_{s=1}^{S} \mu_D(X_k, Y_s) \tag{7.3}$$

- The risk maximum level, which is defined as Eq. (6.4) with $r(X_k, Y_s) = \mu_D^{max}(Y_s) - \mu_D(X_k, Y_s)$ where $\mu_D^{max}(Y_s) = \max\limits_{1 \le k \le K} \mu_D(X_k, Y_s)$.

In this case, it is possible to construct the aggregated risk matrix (similar to the risk matrix given in Table 6.3) as well. One observes that if the risk matrices constructed for each objective function reflect the particular risks (monocriteria risk estimates), the aggregated risk matrix reflects the aggregated risks (multicriteria risk) in decision-making (Pedrycz et al. 2011).

The characteristic estimates $\mu_D^{max}(X_k)$, $\mu_D^{min}(X_k)$, $\bar{\mu}_D(X_k)$, and $r_D^{max}(X_k)$ considered here can serve as the basis for the modified choice criteria that are to be used under the generalization of the classic approach (Ekel et al. 2008, 2011; Pedrycz et al. 2011).

In particular, the modified Wald criterion assumes the following form:

$$\max_{1\le k\le K}\mu_D(X_k) = \max_{1\le k\le K}\min_{1\le s\le S}\min_{1\le p\le q}\mu_{A_p}(X_k, Y_s) \tag{7.4}$$

The modified Laplace criterion can be presented as follows:

$$\max_{1\le k\le K}\mu_D(X_k) = \max_{1\le k\le K}\frac{1}{S}\sum_{s=1}^{S}\min_{1\le p\le q}\mu_{A_p}(X_k, Y_s) \tag{7.5}$$

The modified Savage criterion comes in the following form:

$$\min_{1\le k\le K} r_D^{\max}(X_k) = \min_{1\le k\le K}\max_{1\le s\le S}\left[\max_{1\le k\le K}\min_{1\le p\le q}\mu_{A_p}(X_k, Y_s) - \min_{1\le p\le q}\mu_{A_p}(X_k, Y_s)\right] \tag{7.6}$$

Finally, the Hurwicz criterion takes on this form:

$$\max_{1\le k\le K}\left[\alpha\min_{1\le k\le K}\mu_D(X_k) + (1-\alpha)\max_{1\le k\le K}\mu_D(X_k)\right]$$
$$= \max_{1\le k\le K}\left[\alpha\min_{1\le k\le K}\min_{1\le p\le q}\mu_{D_p}(X_k, Y_s) + (1-\alpha)\max_{1\le k\le K}\min_{1\le p\le q}\mu_{D_p}(X_k, Y_s)\right] \tag{7.7}$$

Although the generalization of the classic approach (Luce and Raiffa 1957; Raiffa 1968; Webster 2003) to consider the uncertainty of information in multicriteria decision-making is concerned here with the modification of the criteria of Wald, Laplace, Savage, and Hurwicz, the same line of thought can be extended to other types of choice criteria being encountered in the literature. For instance, it is possible to apply the criteria of Hodges and Lehmann, Bayes, maximal probability, and so on (Hodges and Lehmann 1952; Trukhaev 1981). However, the use of these criteria presumes the availability of the certain type of information (usually coming in a probabilistic form) about the representative combination of initial data states of nature or scenarios.

Example 7.1 Let us consider a bicriteria decision-making problem associated with analyzing the solution alternatives X_k, $k = 1, 2, \ldots, 4$ with the presence of the representative combinations of initial data, states of nature, or scenarios Y_s, $s = 1, 2, \ldots, 4$. The corresponding payoff matrices are presented in Tables 7.3 and 7.4.

Assuming that both criteria have to be maximized, we can apply Eq. (4.32) to build the modified payoff matrices. Tables 7.5 and 7.6 include the modified payoff matrices for the first criterion and second criterion, respectively.

Table 7.3 Example 7.1: Payoff matrix for the first criterion.

	Y_1	Y_2	Y_3	Y_4
X_1	14	15	11	16
X_2	9	11	13	14
X_3	11	15	10	12
X_4	16	13	12	14

Table 7.4 Example 7.1: Payoff matrix for the second criterion.

	Y_1	Y_2	Y_3	Y_4
X_1	48	44	38	46
X_2	43	38	39	47
X_3	46	40	38	37
X_4	40	38	39	45

Table 7.5 Example 7.1: Modified payoff matrix for the first criterion.

	Y_1	Y_2	Y_3	Y_4
X_1	0.71	0.86	0.29	1
X_2	0	0.29	0.57	0.71
X_3	0.29	0.86	0.14	0.43
X_4	1	0.57	0.43	0.71

Table 7.6 Example 7.1: Modified payoff matrix for the second criterion.

	Y_1	Y_2	Y_3	Y_4
X_1	1	0.64	0.09	0.82
X_2	0.55	0.09	0.18	0.91
X_3	0.82	0.27	0.09	0
X_4	0.27	0.09	0.18	0.73

Table 7.7 Example 7.1: Aggregated payoff matrix.

	Y_1	Y_2	Y_3	Y_4
X_1	0.71	0.64	0.09	0.82
X_2	0	0.09	0.18	0.71
X_3	0.29	0.27	0.09	0
X_4	0.27	0.09	0.18	0.71

The application of Eq. (4.28) permits one to build the aggregated payoff matrix, which is presented in Table 7.7. The use of Eq. (6.7) to process data of Table 7.7 leads to the aggregated risk matrix given in Table 7.8.

The aggregated payoff matrix with characteristic estimates is presented in Table 7.9. The characteristic estimates permit us to indicate the solution alternatives selected by the modified Wald criterion: $X^W = \{X_1, X_4\}$. The application of the modified Laplace criterion leads to $X^L = \{X_1\}$. The use of the modified Savage criterion leads $X^S = \{X_1\}$ as well. Finally, the use of the modified Hurwicz criterion also indicates $X^H = \{X_1\}$. Thus, it is possible to conclude that the alternatives X_2 and X_3 can be excluded with a high degree of reliability.

Table 7.8 Example 7.1: Aggregated risk matrix.

	Y_1	Y_2	Y_3	Y_4
X_1	0	0	0.09	0
X_2	0.71	0.55	0	0.11
X_3	0.42	0.37	0.09	0.82
X_4	0.44	0.55	0	0.11

Table 7.9 Example 7.1: Aggregated payoff matrix with characteristic estimates.

	Y_1	Y_2	Y_3	Y_4	$\mu_D^{max}(X_k)$	$\mu_D^{min}(X_k)$	$\bar{\mu}_D(X_k)$	$r^{max}(X_k)$
X_1	0.71	0.64	0.09	0.82	0.82	0.09	0.57	0.09
X_2	0	0.09	0.18	0.71	0.71	0	0.25	0.71
X_3	0.29	0.27	0.09	0	0.29	0	0.16	0.82
X_4	0.27	0.09	0.18	0.71	0.71	0.09	0.31	0.55

The approach described here has been used to solve some problems related to multiobjective decision-making under conditions of uncertainty (Ekel et al. 2008, 2011; Pedrycz et al. 2011). However, certain limitations of the generalization of the classic approach have been reported in Pereira Jr. et al. (2015) and Ekel et al. (2016). To demonstrate them, consider Example 7.2.

Example 7.2 Let us analyze a hypothetical problem related to minimizing two objective functions. The corresponding payoff matrices with characteristic estimates (associated with the use of the Laplace choice criterion in the monocriteria case and the modified Laplace choice criterion in the bicriteria case) are presented in Tables 7.10 and 7.11.

The corresponding modified payoff matrices for both objective functions are presented in Tables 7.12 and 7.13, respectively. Finally, the aggregated payoff matrix with characteristic estimates is given in Table 7.14.

It is not difficult to observe that the solution alternative X_2 is more preferable than X_1 from the point of view of both objective functions, when applying the Laplace criterion (Tables 7.10 and 7.11). However, the analysis of the aggregated

Table 7.10 Example 7.2: Payoff matrix with characteristic estimates for the first objective function.

	Y_1	Y_2	$\bar{F}(X_k)$
X_1	9.00	9.00	9.00
X_2	4.20	11.49	7.80
X_3	15.00	7.80	11.40
X_4	3.00	13.80	8.40

Table 7.11 Example 7.2: Payoff matrix with characteristic estimates for the second objective function.

	Y_1	Y_2	$\bar{F}(X_k)$
X_1	8.40	14.80	11.60
X_2	13.20	5.20	9.20
X_3	2.00	18.00	10.00
X_4	11.60	13.20	12.40

Table 7.12 Example 7.2: Modified payoff matrix for the first objective function.

	Y_1	Y_2
X_1	0.50	0.50
X_2	0.90	0.30
X_3	0	0.60
X_4	1	0.10

Table 7.13 Example 7.2: Modified payoff matrix for the second objective function.

	Y_1	Y_2
X_1	0.60	0.20
X_2	0.30	0.90
X_3	1	0
X_4	0.40	0.30

Table 7.14 Example 7.2: Aggregated payoff matrix with characteristic estimates.

	Y_1	Y_2	$\bar{\mu}_D(X_k)$
X_1	0.50	0.20	0.35
X_2	0.30	0.30	0.30
X_3	0	0	0
X_4	0.40	0.10	0.25

payoff matrix (Table 7.14) shows that the solution alternative X_1 is more preferable than X_2.

Although the considered example is associated with using the Laplace choice criterion, the use of other choice criteria quite often leads to similar contradictions. Taking this into consideration, the results of Pereira Jr. et al. (2015) and Ekel et al. (2016), discussed next, are directed at improving the results of Ekel et al. 2008, 2011) and Pedrycz et al. (2011) to overcome the identified contradictions.

7.2 Consideration of Choice Criteria of the Classic Approach to Dealing with Uncertainty of Information as Objective Functions within the Framework of <X, F> Models

The classic approach to dealing with the uncertainty of information is directed at analyzing the problems Eqs. (6.9), (6.11), (6.13), and (6.14), or (6.10), (6.12), (6.13), and (6.15) for a given objective function in an environment with several representative combinations of initial data, states of nature, or scenarios Y_s, $s = 1, 2, \ldots, S$. Therefore, it is possible to consider the Wald, Laplace, Savage, and Hurwicz choice criteria as objective functions.

For the Wald criterion, the objective function is defined as

$$F^W(X_k) = F^{\min}(X_k) = \min_{1 \le s \le S} F(X_k, Y_s) \tag{7.8}$$

for a maximization problem or

$$F^W(X_k) = F^{\max}(X_k) = \max_{1 \le s \le S} F(X_k, Y_s) \tag{7.9}$$

for a minimization problem.

The objective function for the Laplace choice criterion is

$$F^L(X_k) = \bar{F}(X_k) = \frac{1}{S}\sum_{s=1}^{S} F(X_k, Y_s). \tag{7.10}$$

The Savage criterion can be represented as an objective function as follows

$$F^S(X_k) = r^{\max}(X_k) = \max_{1 \le s \le S} r(X_k, Y_s) \tag{7.11}$$

Finally, the following objective function can be defined for the Hurwicz choice criterion:

$$\begin{aligned} F^H(X_k) &= \alpha F^{\min}(X_k) + (1-\alpha) F^{\max}(X_k) \\ &= \alpha \min_{1 \le s \le S} F(X_k, Y_s) + (1-\alpha) \max_{1 \le s \le S} F(X_k, Y_s) \end{aligned} \tag{7.12}$$

for a maximization problem or

$$\begin{aligned} F^H(X_k) &= \alpha F^{\max}(X_k) + (1-\alpha) F^{\min}(X_k) \\ &= \alpha \max_{1 \le s \le S} F(X_k, Y_s) + (1-\alpha) \min_{1 \le s \le S} F(X_k, Y_s) \end{aligned} \tag{7.13}$$

for a minimization problem.

It allows one to construct q problems that, generally, include four or less (if not the all choice criteria are used in the analysis) objective functions as follows:

$$F_{r,p}(X) \rightarrow \underset{X \in L}{\text{extr}}, \quad r = 1,2,\ldots,t \leq 4, p = 1,2,\ldots,q \tag{7.14}$$

where $F_{1,p}(X) = F_p^W(X_k)$, $F_{2,p}(X) = F_p^L(X_k)$, $F_{3,p}(X) = F_p^S(X_k)$, and $F_{4,p}(X) = F_p^H(X_k)$.

Thus, the analysis of the solution alternatives and consequent choice of the rational solution alternatives can be realized within the framework of the <X, F> models. Applying Eqs. (4.31) or (4.32) to construct the corresponding membership functions for $F_{r,p}(X)$, $r = 1, 2, \ldots, t$, $p = 1, 2, \ldots, q$, one can solve the problem Eq. (4.30) for the solution alternatives X_k, $k = 1, 2, \ldots, K$. This method of analysis provides the choice of the rational solution alternatives in accordance with the principle of the Pareto optimality (Pareto 1886) and permits one to overcome the limitations of the generalization of the classic approach to dealing with the uncertainty of information indicated previously. Taking this into account, the payoff matrix with characteristic estimates (Table 6.2) is to be presented as the payoff matrix with the choice criteria estimates for $p = 1, 2, \ldots, q$ (Table 7.15) or for simplicity as the matrix of choice criteria estimates $p = 1, 2, \ldots, q$ (Table 7.16).

The availability of q matrices with choice criteria estimates allows one, applying Eqs. (4.31) or (4.32), to construct q modified matrices of choice criteria estimates, as shown in Table 7.17.

Finally, the presence of q modified matrices of choice criteria estimates permits one to construct the aggregated matrix of choice criteria estimates by applying Eq. (4.28), as shown in Table 7.18. This matrix includes the estimates calculated on the basis of Eq. (4.29) to provide the choice of the rational solution alternatives using Eq. (4.30).

Example 7.3 Returning to Example 7.2, it is not difficult to find that the construction of the modified matrix of estimates for the Laplace choice criterion for the first objective function leads to results shown in Table 7.19. At the same time, the modified matrix of estimates for the Laplace choice criterion for the second objective function is given in Table 7.20. The aggregation of these modified matrices of estimates for the Laplace choice criterion generates the problem solution $X^L = \{X_2\}$ (Table 7.21).

Table 7.15 Payoff matrix with choice criteria estimates for the *p*th objective function.

	Y_1	...	Y_s	...	Y_S	$F_p^W(X_k)$	$F_p^L(X_k)$	$F_p^S(X_k)$	$F_p^H(X_k)$
X_1	$F_p(X_1, Y_1)$	...	$F_p(X_1, Y_s)$	...	$F_p(X_1, Y_S)$	$F_p^W(X_1)$	$F_p^L(X_1)$	$F_p^S(X_1)$	$F_p^H(X_1)$
...	...	...	...	...	...	...	...	...	...
X_k	$F_p(X_k, Y_1)$	...	$F_p(X_k, Y_s)$	...	$F_p(X_k, Y_S)$	$F_p^W(X_k)$	$F_p^L(X_k)$	$F_p^S(X_k)$	$F_p^H(X_k)$
...	...	...	...	...	...	...	...	...	...
X_K	$F_p(X_K, Y_1)$	...	$F_p(X_K, Y_s)$	...	$F_p(X_K, Y_S)$	$F_p^W(X_K)$	$F_p^L(X_K)$	$F_p^S(X_K)$	$F_p^H(X_K)$
						$\min_{1 \le k \le K} F_p^W(X_k)$	$\min_{1 \le k \le K} F_p^L(X_k)$	$\min_{1 \le k \le K} F_p^L(X_k)$	$\min_{1 \le k \le K} F_p^H(X_k)$
						$\min_{1 \le k \le K} F_p^W(X_k)$	$\max_{1 \le k \le K} F_p^L(X_k)$	$\max_{1 \le k \le K} F_p^L(X_k)$	$\max_{1 \le k \le K} F_p^H(X_k)$

Table 7.16 Matrix of choice criteria estimates for the *p*th objective function.

	$F_p^W(X_k)$	$F_p^L(X_k)$	$F_p^S(X_k)$	$F_p^H(X_k)$
X_1	$F_p^W(X_1)$	$F_p^L(X_1)$	$F_p^S(X_1)$	$F_p^H(X_1)$
...	...	...	...	...
X_k	$F_p^W(X_k)$	$F_p^L(X_k)$	$F_p^S(X_k)$	$F_p^H(X_k)$
...	...	...	...	...
X_K	$F_p^W(X_K)$	$F_p^L(X_K)$	$F_p^S(X_K)$	$F_p^H(X_K)$
	$\min_{1 \le k \le K} F_p^W(X_k)$	$\min_{1 \le k \le K} F_p^L(X_k)$	$\min_{1 \le k \le K} F_p^L(X_k)$	$\min_{1 \le k \le K} F_p^H(X_k)$
	$\min_{1 \le k \le K} F_p^W(X_k)$	$\max_{1 \le k \le K} F_p^L(X_k)$	$\max_{1 \le k \le K} F_p^L(X_k)$	$\max_{1 \le k \le K} F_p^H(X_k)$

Table 7.17 Modified matrix of choice criteria estimates for the *p*th objective function.

	$\mu_{A_p}^W(X_k)$	$\mu_{A_p}^L(X_k)$	$\mu_{A_p}^S(X_k)$	$\mu_{A_p}^H(X_k)$
X_1	$\mu_{A_p}^W(X_1)$	$\mu_{A_p}^L(X_1)$	$\mu_{A_p}^S(X_1)$	$\mu_{A_p}^H(X_1)$
...	...	...	...	...
X_k	$\mu_{A_p}^W(X_k)$	$\mu_{A_p}^L(X_k)$	$\mu_{A_p}^S(X_k)$	$\mu_{A_p}^H(X_k)$
...	...	...	...	...
X_K	$\mu_{A_p}^W(X_K)$	$\mu_{A_p}^L(X_K)$	$\mu_{A_p}^S(X_K)$	$\mu_{A_p}^H(X_K)$

Table 7.18 Aggregated payoff matrix of choice criteria estimates.

	$\mu_D^W(X_k)$	$\mu_D^L(X_k)$	$\mu_D^S(X_k)$	$\mu_D^H(X_k)$
X_1	$\mu_D^W(X_1)$	$\mu_D^L(X_1)$	$\mu_D^S(X_1)$	$\mu_D^S(X_1)$
...	...	...	...	...
X_k	$\mu_D^W(X_k)$	$\mu_D^L(X_k)$	$\mu_D^S(X_k)$	$\mu_D^H(X_k)$
...	...	...	...	...
X_K	$\mu_D^W(X_K)$	$\mu_D^L(X_K)$	$\mu_D^S(X_K)$	$\mu_D^H(X_K)$
	$\max_{1 \le k \le K} \mu_D^W(X_k)$	$\max_{1 \le k \le K} \mu_D^L(X_k)$	$\max_{1 \le k \le K} \mu_D^S(X_k)$	$\max_{1 \le k \le K} \mu_D^H(X_k)$

Table 7.19 Example 7.3: Modified matrix of estimates for the Laplace choice criterion for the first objective function.

	$\mu_{A_1}^{L}(X_k)$
X_1	0.67
X_2	1
X_3	0
X_4	0.83

Table 7.20 Example 7.3: Modified matrix of estimates for the Laplace choice criterion for the second objective function.

	$\mu_{A_2}^{L}(X_k)$
X_1	0.25
X_2	1
X_3	0.75
X_4	0

Table 7.21 Example 7.3: Aggregated matrix of choice criteria estimates.

	$\mu_{D}^{L}(X_k)$
X_1	0.25
X_2	1
X_3	0
X_4	0

Example 7.4 Let us consider the following multiobjective problem with interval coefficients present in the objective functions (Pedrycz et al. 2011):

$$F_1(x) = [2.70,3.30]x_1 + [11.70,14.30]x_2 + [7.20,8.80]x_3 \rightarrow \min \tag{7.15}$$

$$F_2(x) = [5.40,6.60]x_1 + [3.60,4.40]x_2 + [4.50,5.50]x_3 \rightarrow \min \tag{7.16}$$

subject to the following constraints:

$$0 \le x_1 \le 10 \tag{7.17}$$

$$0 \le x_2 \le 12 \tag{7.18}$$

$$0 \le x_3 \le 14 \tag{7.19}$$

$$x_1 + x_2 + x_3 = 30 \tag{7.20}$$

The first stage in the decision-making process is concerned with constructing two payoff matrices for all combinations of the solution alternatives X_k, $k = 1,2, \ldots, K$ and all representative combinations of initial data, states of nature, or scenarios Y_s, $s = 1,2, \ldots, S$.

As in Chapter 6, we can apply the LP$_\tau$-sequences to generate Y_s, $s = 1, 2, \ldots, S$.

Assuming that we have to construct $S = 7$ representative combinations of initial data, states of nature, or scenarios, then it is necessary to obtain points Q_s, $s = 1, 2, \ldots, 7$ with coordinates q_{st}, $t = 1, 2\ldots, 6$ (we have six coefficients in Eqs. (7.15) and (7.16)) in the corresponding unit hypercube Q^T. Applying the results of Sobol' (1966, 1979), we can expand Table 6.9 to present q_{st}, $s = 1, 2, \ldots, 7$, $t = 1, 2\ldots, 6$ in Table 7.22.

Applying Eq. (6.16), we can form S representative combinations of initial data, states of nature, or scenarios, presented in Table 7.23.

Table 7.22 Example 7.4: Points of the LP$_\tau$-sequences in Q^6.

s	$t = 1$	$t = 2$	$t = 3$	$t = 4$	$t = 5$	$t = 6$
1	0.500	0.500	0.500	0.500	0.500	0.500
2	0.250	0.750	0.250	0.750	0.250	0.750
3	0.750	0.250	0.750	0.250	0.750	0.250
4	0.125	0.625	0.875	0.875	0.625	0.125
5	0.625	0.125	0.375	0.375	0.125	0.625
6	0.375	0.375	0.625	0.125	0.875	0.875
7	0.875	0.875	0.125	0.625	0.375	0.375

Table 7.23 Example 7.4: Representative combinations of initial data, states of nature, or scenarios Y_s, $s = 1, 2, \ldots, S$.

s	$t = 1$	$t = 2$	$t = 3$	$t = 4$	$t = 5$	$t = 6$
1	3.00	13.00	8.00	6.00	4.00	5.00
2	2.85	13.65	7.60	6.30	3.80	5.25
3	3.15	12.35	8.40	5.70	4.20	4.75
4	2.93	12.68	8.20	5.55	4.30	5.38
5	2.78	13.33	8.60	6.45	4.10	4.63
6	3.08	12.03	7.80	5.85	3.70	5.13
7	3.23	13.98	7.40	6.15	3.90	4.88

The coordinates of points given in Table 7.23 serve as a basis for constructing the following seven (in accordance with the representative combinations of initial data, states of nature, or scenarios) multiobjective problems:

$$F_1(x) = 3.00x_1 + 13.00x_2 + 8.00x_3 \rightarrow \min \quad (7.21)$$

$$F_2(x) = 6.00x_1 + 4.00x_2 + 5.00x_3 \rightarrow \min \quad (7.22)$$

$$F_1(x) = 2.85x_1 + 13.65x_2 + 7.60x_3 \rightarrow \min \quad (7.23)$$

$$F_2(x) = 6.30x_1 + 3.80x_2 + 5.25x_3 \rightarrow \min \quad (7.24)$$

$$F_1(x) = 3.15x_1 + 12.35x_2 + 8.40x_3 \rightarrow \min \quad (7.25)$$

$$F_2(x) = 5.70x_1 + 4.20x_2 + 4.75x_3 \rightarrow \min \quad (7.26)$$

$$F_1(x) = 2.93x_1 + 12.68x_2 + 8.20x_3 \rightarrow \min \quad (7.27)$$

$$F_2(x) = 5.55x_1 + 4.30x_2 + 5.38x_3 \rightarrow \min \quad (7.28)$$

$$F_1(x) = 2.78x_1 + 13.33x_2 + 8.60x_3 \rightarrow \min \quad (7.29)$$

$$F_2(x) = 6.45x_1 + 4.10x_2 + 4.63x_3 \rightarrow \min \quad (7.30)$$

$$F_1(x) = 3.08x_1 + 12.03x_2 + 7.80x_3 \rightarrow \min \quad (7.31)$$

$$F_2(x) = 5.85x_1 + 3.70x_2 + 5.13x_3 \rightarrow \min \quad (7.32)$$

$$F_1(x) = 3.23x_1 + 13.98x_2 + 7.40x_3 \rightarrow \min \quad (7.33)$$

$$F_2(x) = 6.15x_1 + 3.90x_2 + 4.88x_3 \rightarrow \min \quad (7.34)$$

which are subject to the same constraints in Eqs. (7.17)–(7.20).

The solutions to these problems, obtained with the use of the Bellman–Zadeh approach to decision-making in a fuzzy environment for analyzing $<X, F>$ models, are the following:

- $s = 1$: $x_1^0 = 7.00$, $x_2^0 = 9.00$, $x_3^0 = 14.00$ for Eqs. (7.21) and (7.22)
- $s = 2$: $x_1^0 = 8.95$, $x_2^0 = 10.50$, $x_3^0 = 10.55$ for Eqs. (7.23) and (7.24)
- $s = 3$: $x_1^0 = 7.00$, $x_2^0 = 9.00$, $x_3^0 = 14.00$ for Eqs. (7.25) and (7.26)
- $s = 4$: $x_1^0 = 9.95$, $x_2^0 = 10.50$, $x_3^0 = 9.55$ for Eqs. (7.27) and (7.28)
- $s = 5$: $x_1^0 = 7.00$, $x_2^0 = 9.00$, $x_3^0 = 14.00$ for Eqs. (7.29) and (7.30)
- $s = 6$: $x_1^0 = 9.93$, $x_2^0 = 11.35$, $x_3^0 = 8.72$ for Eqs. (7.31) and (7.32)
- $s = 7$: $x_1^0 = 7.00$, $x_2^0 = 9.00$, $x_3^0 = 14.00$ for Eqs. (7.33) and (7.34)

Therefore, we can form the following four solution alternatives for the problem Eqs. (7.15)–(7.20):

$$X_1 = (7.00, 9.00, 14.00)$$

$$X_2 = (8.95, 10.50, 10.55)$$

$$X_3 = (9.95, 10.50, 9.55)$$

$$X_4 = (9.93, 11.35, 8.72)$$

Table 7.24 Example 7.4: Payoff matrix for the first objective function.

	Y_1	Y_2	Y_3	Y_4	Y_5	Y_6	Y_7
X_1	250.00	249.20	250.80	249.43	259.83	239.03	252.03
X_2	243.75	249.01	246.49	245.87	255.58	236.17	253.77
X_3	242.75	244.26	241.24	240.60	249.76	231.45	249.60
X_4	247.10	249.50	244.70	244.52	253.89	235.14	255.27

Substituting these solutions into Eqs. (7.21), (7.23), (7.25), (7.27), (7.29), (7.31), and (7.33), we can construct the payoff matrix for the first objective function (Table 7.24). When substituting them into Eqs. (7.22), (7.24), (7.26), (7.28), (7.30), (7.32), and (7.34), we construct the payoff matrix for the second objective function (Table 7.26).

Let us consider the solution of the monocriteria problem Eq. (7.15) subject to the constraints in Eqs. (7.17)–(7.20), analyzing the payoff matrix given in Table 7.24 (Pereira Jr. et al. 2015). The corresponding matrix of the choice criteria estimates is presented in Table 7.25.

The analysis of Table 7.25 indicates that the use of the choice Wald criterion permits one to choose the solution alternative $X^W = \{X_3\}$. The use of the Laplace choice criterion allows one to find $X^L = \{X_3\}$. The application of the Savage choice criterion leads to $X^S = \{X_3\}$. Finally, the use of the Hurwicz choice criterion with $\alpha = 0.75$ generates $X^H = \{X_3\}$ as well. In such a manner, the solution alternative X_3 is to be considered as the solution to the monocriteria problem Eq. (7.15) with the constraints in Eqs. (7.17)–(7.20) with a high degree of confidence.

Let us consider the solution of the monocriteria problem Eq. (7.16) subject to the constraints in Eqs. (7.17)–(7.20) by analyzing the payoff matrix given in

Table 7.25 Example 7.4: Matrix of choice criteria estimates for the first objective function.

	$F^W(X_k)$	$F^L(X_k)$	$F^S(X_k)$	$F^H(X_k)$
X_1	259.83	250.05	10.07	254.63
X_2	255.58	247.81	5.82	250.73
X_3	249.76	242.81	0.00	245.18
X_4	255.27	247.16	5.24	250.24
$\min_{1 \le k \le K} F(X_k)$	249.76	242.81	0.00	245.18
$\max_{1 \le k \le K} F(X_k)$	259.83	250.05	10.07	254.63

Table 7.26 Example 7.4: Payoff matrix for the second objective function.

	Y_1	Y_2	Y_3	Y_4	Y_5	Y_6	Y_7
X_1	148.00	151.80	144.20	152.87	146.87	146.07	146.47
X_2	148.45	151.67	145.23	151.58	149.62	145.33	147.48
X_3	149.45	152.72	146.18	151.75	151.44	146.05	148.75
X_4	148.58	151.47	145.69	150.83	150.96	144.82	147.89

Table 7.27 Example 7.4: Matrix of choice criteria estimates for the second objective function.

	$F^W(X_k)$	$F^L(X_k)$	$F^S(X_k)$	$F^H(X_k)$
X_1	152.87	148.04	2.04	150.70
X_2	151.67	148.48	2.75	150.06
X_3	152.72	149.48	4.57	151.05
X_4	151.47	148.61	4.09	149.81
$\min_{1 \le k \le K} F(X_k)$	151.47	148.04	2.04	149.81
$\max_{1 \le k \le K} F(X_k)$	152.87	149.48	4.57	151.05

Table 7.26. The corresponding matrix of choice criteria estimates is presented in Table 7.27.

The analysis of Table 7.27 shows that $X^W = \{X_4\}$, $X^L = \{X_1\}$, $X^S = \{X_1\}$, and $X^H = \{X_4\}$. In such a manner, the alternatives X_1 and X_4 are to be considered as the solutions to the monocriteria problem Eq. (7.16) with the constraints in Eqs. (7.17)–(7.20). Formally, these alternatives cannot be distinguished on the basis of information defined by the payoff matrix in Table 7.26.

Let us return to the problem described by expressions Eqs. (7.15)–(7.20). The information in Table 7.25 permits one to construct the modified matrix of choice criteria estimates for the first objective function (Table 7.28). The modified matrix of choice criteria estimates for the second objective function (Table 7.29) has been obtained on the basis of information given in Table 7.27.

The modified matrices of choice criteria estimates presented in Tables 7.28 and 7.29 result in the construction of the aggregated payoff matrix presented in Table 7.30.

Thus, in this case, the use of the choice criteria of Wald, Laplace, and Hurwicz leads to the same solution alternative: $X^W = X^L = X^H = \{X_4\}$. At the same time, the Savage criterion leads to the solution $X^S = \{X_2\}$. (By the way, the application

Table 7.28 Example 7.4: Modified matrix of choice criteria estimates for the first objective function.

	$\mu_{A_1}^W(X_k)$	$\mu_{A_1}^L(X_k)$	$\mu_{A_1}^S(X_k)$	$\mu_{A_1}^H(X_k)$
X_1	0	0	0	0
X_2	0.42	0.31	0.42	0.41
X_3	1	1	1	1
X_4	0.45	0.40	0.48	0.46

Table 7.29 Example 7.4: Modified matrix of choice criteria estimates for the second objective function.

	$\mu_{A_2}^W(X_k)$	$\mu_{A_2}^L(X_k)$	$\mu_{A_2}^S(X_k)$	$\mu_{A_2}^H(X_k)$
X_1	0	1	1	0.28
X_2	0.86	0.69	0.72	0.80
X_3	0.11	0	0	0
X_4	1	0.60	0.19	1

Table 7.30 Example 7.4: Aggregated payoff matrix of choice criteria estimates.

	$\mu_D^W(X_k)$	$\mu_D^L(X_k)$	$\mu_D^S(X_k)$	$\mu_D^H(X_k)$
X_1	0	0	0	0
X_2	0.42	0.31	0.42	0.41
X_3	0.11	0	0	0
X_4	0.45	0.40	0.19	0.46

of the results of Ekel et al. 2008, 2011; Pedrycz et al. 2011, discussed in Section 7.1, lead to other solution alternatives: $X^W = X^L = X^H = \{X_4\}$ and $X^S = \{X_3\}$; Pedrycz et al. 2011.)

Thus, the solution alternatives X_2 and X_4 are the result of the processing of available quantitative information. As indicated previously, if the solving capacity related to quantitative information processing does not permit one to obtain unique solutions, it is necessary to use, at the final decision stage, qualitative information based on knowledge, experience, and intuition of experts involved in the decision-making process; in particular, within the framework of $\langle X, R\rangle$ models.

7.3 Construction of Objectives and Elaboration of Representative Combination of Initial Data, States of Nature, or Scenarios using Qualitative Information

As it was indicated previously, diverse classes of problems exist whose essence requires the consideration of the objectives formed on the basis of qualitative information at all stages of the decision-making process. Taking this into account, the present section mainly reflects the results (Ramalho et al. 2019) that permit one to generate solutions, including robust solutions, to multiobjective decision-making problems on the basis of direct using qualitative information within the framework of the possibilistic approach implemented with the application of the generalization of the classic approach to dealing with the uncertainty of information. The construction of the objective functions on the basis of qualitative information preserves the homogeneous formulation of the objectives, discussed in Chapter 4.

The approach discussed in the present section is associated with the following stages (Ramalho 2017; Ramalho et al. 2019):

1) Elicitation of preferences;
2) Representation of preferences within multiplicative preference relations;
3) Definition of preference vectors on the basis of the analytic hierarchy process (AHP);
4) Aggregation of preferences and generation of representative combination of initial data, states of nature, or scenarios.

This approach is based on applying and combining the results of Yager (1978); Saaty et al. (2003, 2007); Kokshenev et al. (2015); and Ekel et al. (2016). In particular, Yager (1978) uses the Bellman–Zadeh approach to decision-making in a fuzzy environment (Bellman and Zadeh 1970) in conjunction with the AHP (Saaty 1980) for analyzing multiattribute problems that include objectives with unequal importance levels. The construction of a linear programming model, where either coefficients of an objective function or coefficients of constraints are estimated from the AHP, is proposed in Saaty et al. (2003). Some examples of applying the Saaty's results (Saaty et al. 2003) to solve problems of allocating intangible resources through binary linear programming are given in Saaty et al. (2007). Finally, the results by Kokshenev et al. (2015) and Ekel et al. (2016) are directed at the use of the ordered weighted average (OWA) operator to regulate the level of intercriterion compensation or the degree of optimism or risk appetite inherent to the decision attitude (Damodaran 2008; Palomares et al. 2012; Kokshenev et al. 2015) in multiattribute decision-making.

7.3.1 Elicitation of Preferences

The objective functions formed on the basis of qualitative information are to include elements reflecting the preferences of one or more involved experts, expressed by the corresponding preference structures or formats. In this chapter, we consider the application of the following preference formats, described in Chapter 5:

- ordering of the alternatives;
- additive reciprocal fuzzy preference relations;
- nonreciprocal fuzzy preference relations;
- fuzzy estimates;
- multiplicative preference relations.

Taking this into account, it should be noted that the nonreciprocal fuzzy preference relations and fuzzy estimates are equivalent to a certain extent. In particular, if two alternatives $x_k \in X$ and $x_l \in X$ have fuzzy estimates with the membership functions $\mu(x_k)$ and $\mu(x_l)$, then the quantity $NR\ (x_k, x_l)$ is the degree of preference $\mu(x_k) \succeq \mu(x_l)$, while the quantity $NR\ (x_l, x_k)$ is the degree of preference $\mu(x_l) \succeq \mu(x_k)$. Using the concept of a generalized preference relation, discussed in Chapter 5, as well as Eqs. (5.27) and (5.28), the quantities $NR\ (x_k, x_l)$ and $NR\ (x_l, x_k)$ can be evaluated, respectively, as the follows:

$$NR(x_k,x_l) = \sup_{x_k,x_l \in X} \min\{\mu(x_k),\mu(x_l),\mu_R(x_k,x_l)\} \tag{7.35}$$

$$NR(x_l,x_k) = \sup_{x_k,x_l \in X} \min\{\mu(x_k),\mu(x_l),\mu_R(x_l,x_k)\} \tag{7.36}$$

where $\mu_R(x_k, x_l)$ and $\mu_R(x_l, x_k)$ are the membership functions of the corresponding fuzzy preference relations that, respectively, reflect the essence of the preferences of x_k over x_l and of x_l over x_k (for instance, as indicated in Chapter 5, "more attractive," "more flexible," etc.).

When the indicator in terms of which alternatives x_k and x_l are evaluated can be measured on a numerical scale, and if the essence of preference behind relation R is coherent with the natural order ($\leq$) along the axis of measured values of this indicator, then Eqs. (7.35) and (7.36), respectively, are reduced to

$$NR(x_k,x_l) = \sup_{x_k \leq x_l} \min_{x_k,x_l \in X} \{\mu(x_k),\mu(x_l)\} \tag{7.37}$$

$$NR(x_l,x_k) = \sup_{x_l \leq x_k} \min_{x_k,x_l \in X} \{\mu(x_k),\mu(x_l)\} \tag{7.38}$$

If the indicator has a maximization character, similar to Eqs. (5.29) and (5.30), the correlations Eqs. (7.37) and (7.38) are to be written for $x_k \geq x_l$ and $x_l \geq x_k$, respectively. Thus, the availability of fuzzy estimates for all $x_k \in X$ is essential to realize an automatic construction of $NR(x_k, x_l)$.

To continue the consideration of the stages indicated before, let us suppose (Ramalho 2017; Ramalho et al. 2019) that a group of experts $E = \{e_1, e_2, e_3, e_4\}$ expresses their preferences relative to a set of alternatives $X = \{x_1, x_2, x_3, x_4\}$ from the perspective of the criterion C (for example, "Innovation Level").

The expert e_1 may express his/her preferences by the order of alternatives

$$OA_{e_1}(x_k) = \{x_1, x_4, x_2, x_3\} \tag{7.39}$$

The expert e_2 may express his/her preferences using the multiplicative preference relation, whose construction is based on Saaty (1980):

$$MR_{e_2}(x_k, x_l) = \begin{bmatrix} 1 & 7 & 5 & 3 \\ 1/7 & 1 & 1/5 & 1/3 \\ 1/5 & 5 & 1 & 3 \\ 1/3 & 3 & 1/3 & 1 \end{bmatrix} \tag{7.40}$$

The expert e_3 may present his/her preferences applying the fuzzy estimates shown in Figure 7.1 (these estimates have been borrowed from Queiroz 2009; however, it is possible to apply other types of fuzzy estimates), indicating the following: x_1 – VH; x_2 – VL; x_3 – H; x_4 – M. The use of Eqs. (7.37) and (7.38), modified for $x_k \geq x_l$ and $x_l \geq x_k$, respectively, for these estimates permits one to construct the nonreciprocal fuzzy preference relation

$$NR_{e_3}(x_k, x_l) = \begin{bmatrix} 1 & 1 & 1 & 1 \\ 0 & 1 & 0 & 0.25 \\ 0.75 & 1 & 1 & 1 \\ 0.5 & 1 & 0.5 & 1 \end{bmatrix} \tag{7.41}$$

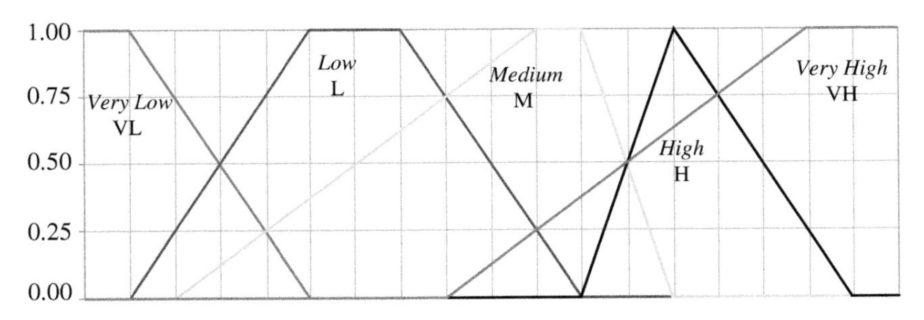

Figure 7.1 Fuzzy set-based qualitative scales utilized for objectives based on qualitative information.

The expert e_4 may also present his/her preferences applying the fuzzy estimates of Figure 7.1, indicating the following: x_1 – H; x_2 – M; x_3 – M; x_4 – L. It permits one, using Eqs. (7.33) and (7.34), also modified for $x_k \geq x_l$ and $x_l \geq x_k$, to build

$$NR_{e_4}(x_k, x_l) = \begin{bmatrix} 1 & 1 & 1 & 1 \\ 0.5 & 1 & 1 & 1 \\ 0.5 & 1 & 1 & 1 \\ 0 & 0.75 & 0.75 & 1 \end{bmatrix} \tag{7.42}$$

7.3.2 Representation of Preferences Within Multiplicative Preference Relations

As shown in Saaty et al. (2003), the components of the vector of preferences generated by the application of the AHP can be used as coefficients for forming the corresponding objective functions. The application of the AHP is based on processing the multiplicative preference relations. Thus, it is necessary to use transformation functions that permit one to convert different preference structures or formats (for example, ordering of alternatives, additive reciprocal fuzzy preference relations, and nonreciprocal fuzzy preference relations, etc.) to the multiplicative preference relations (Ramalho et al. 2019).

Taking this into account, it is necessary to indicate that there is not always a direct transformation of any preference format to the multiplicative preference relation. For instance, among preference formats discussed in Chapter 5, there is no direct conversion from the nonreciprocal fuzzy preference relations to the multiplicative fuzzy preference relations. It requires the preliminary conversion of the nonreciprocal fuzzy preference relations to the additive reciprocal fuzzy preference relations (Ramalho 2017; Ramalho et al. 2019).

Let us consider the transformation of the ordering of alternatives to the multiplicative preference relations. Applying the transformation function Eq. (5.40), we can obtain the additive reciprocal fuzzy preference relation from the ordering of alternatives. At the same time, the transformation function Eq. (5.50) permits one to generate the additive reciprocal fuzzy preference relation from the multiplicative preference relation. Substituting Eq. (5.50) in Eq. (5.40) and carrying out the corresponding transformations, one has

$$MR(x_k, x_l) = m^{\frac{OA(x_l) - OA(x_k)}{n-1}} \tag{7.43}$$

Accepting (as discussed in Chapter 5) the scale by Saaty (2008) with $m = 9$, it is possible to transform Eq. (7.43) as follows (Ramalho 2017; Ramalho et al. 2019):

$$MR(x_k, x_l) = 9^{\frac{OA(x_l) - OA(x_k)}{n-1}} \tag{7.44}$$

Applying Eq. (7.44) to process Eq. (7.39), we can obtain

$$MR_{e_1}(x_k,x_l) = \begin{bmatrix} 1.000 & 4.327 & 9.000 & 2.080 \\ 0.231 & 1.000 & 2.080 & 0.481 \\ 0.111 & 0.481 & 1.000 & 0.231 \\ 0.481 & 2.080 & 4.327 & 1.000 \end{bmatrix}. \tag{7.45}$$

The application of the transformation function Eq. (5.53) permits one to obtain the following additive reciprocal fuzzy preference relations from Eqs. (7.41) and (7.42):

$$RR_{e_3}(x_k,x_l) = \begin{bmatrix} 0.500 & 1.000 & 0.625 & 0.750 \\ 0.000 & 0.500 & 0.000 & 0.125 \\ 0.375 & 1.000 & 0.500 & 0.750 \\ 0.250 & 0.875 & 0.250 & 0.500 \end{bmatrix} \tag{7.46}$$

and

$$RR_{e_4}(x_k,x_l) = \begin{bmatrix} 0.500 & 0.750 & 0.750 & 1.000 \\ 0.250 & 0.500 & 0.500 & 0.625 \\ 0.250 & 0.750 & 0.500 & 0.625 \\ 0.000 & 0.375 & 0.375 & 0.500 \end{bmatrix} \tag{7.47}$$

respectively.

The transformation function for constructing the multiplicative preference relations on the basis of the additive reciprocal fuzzy preference relations can be obtained from Eq. (5.51) and has the following form (Ramalho et al. 2019):

$$MR(x_k,x_l) = 9^{(2RR(x_k,x_l)-1)} \tag{7.48}$$

The use of Eq. (7.48) permits one to transform Eqs. (7.46) and (7.47) to

$$MR_{e_3}(x_k,x_l) = \begin{bmatrix} 1.000 & 9.000 & 1.732 & 3.000 \\ 0.111 & 1.000 & 0.111 & 0.192 \\ 0.577 & 9.000 & 1.000 & 3.000 \\ 0.333 & 5.196 & 0.333 & 1.000 \end{bmatrix} \tag{7.49}$$

and

$$MR_{e_4}(x_k,x_l) = \begin{bmatrix} 1.000 & 3.000 & 3.000 & 9.000 \\ 0.333 & 1.000 & 1.000 & 1.732 \\ 0.333 & 3.000 & 1.000 & 1.732 \\ 0.111 & 0.577 & 0.577 & 1.000 \end{bmatrix} \tag{7.50}$$

respectively.

7.3.3 Definition of Preference Vectors on the Basis of Applying the AHP

All the multiplicative preference relations in Eqs. (7.40), (7.45), (7.49), and (7.50) have high levels of consistency: for all these multiplicative preference relations, the maximal eigenvalues $\lambda_{\max}$ are close to the dimension $n = 4$ (Saaty 1980). Taking this into account, the eigenvectors corresponding to Eqs. (7.45), (7.40), (7.49), and (7.50), given as

$$E_{e_1} = [0.549 \quad 0.127 \quad 0.061 \quad 0.264], \tag{7.51}$$

$$E_{e_2} = [0.584 \quad 0.053 \quad 0.234 \quad 0.130], \tag{7.52}$$

$$E_{e_3} = [0.459 \quad 0.039 \quad 0.349 \quad 0.081], \tag{7.53}$$

and

$$E_{e_4} = [0.537 \quad 0.156 \quad 0.226 \quad 0.081] \tag{7.54}$$

can serve as the preference vectors.

The vectors in Eqs. (7.51)–(7.54) are to be aggregated and used as representative combinations of initial data, states of nature, or scenarios.

7.3.4 Aggregation of Preferences and Generation of Representative Combinations of Initial Data, States of Nature, or Scenarios

The central idea of this stage is to utilize the level of "orness", offered by the OWA operator as a result of changing the set of the corresponding weights w_i, $i = 1, 2, \ldots, n$ (as discussed in Chapter 4) and the generation of representative combinations of initial data, on the basis of applying the LP_τ-sequences, whose application was discussed in Chapter 6. This generation is to provide representative combinations of initial data, states of nature, or scenarios balanced from the point of view of a justified mixture of pessimistic and optimistic situations (Ramalho et al. 2019).

To implement this idea, it is first necessary to aggregate the opinions of the experts by extracting the limits of pessimism/optimism accepted by the DM. For instance, applying OWA $(E_{e_i}, W), i = 1,2,3,4$ for the vectors $W^p = [1 \quad 0 \quad 0 \quad 0]$ (pessimistic) and $W^o = [0 \quad 0 \quad 0 \quad 1]$ (optimistic), respectively, it is possible (Ramalho et al. 2019) to obtain

$$\text{OWA}(E_{e_i}, W^p) = [0.459 \quad 0.039 \quad 0.061 \quad 0.081] \tag{7.55}$$

and

$$\text{OWA}(E_{e_i}, W^o) = [0.584 \quad 0.156 \quad 0.349 \quad 0.264] \tag{7.56}$$

The application of LP_τ-sequences is carried out taking into account the lower Eq. (7.55) and upper Eq. (7.56) limits for the generating, for example, $S = 3$

representative combinations of initial data, states of nature, or scenarios. It helps one to analyze the pth objective function with different coefficients

$$F_p(X) = 0.520x_1 + 0.100x_2 + 0.200x_3 + 0.170x_4 \tag{7.57}$$

$$F_p(X) = 0.490x_1 + 0.130x_2 + 0.130x_3 + 0.220x_4 \tag{7.58}$$

and

$$F_p(X) = 0.550x_1 + 0.070x_2 + 0.280x_3 + 0.130x_4 \tag{7.59}$$

7.4 General Scheme of Multicriteria Decision-Making under Conditions of Uncertainty

Taking into account the techniques presented previously, it is possible to suggest the *general scheme* of multicriteria decision-making under conditions of information uncertainty, which is associated with the following stages:

- The *first stage* is related to constructing the objective functions. The procedure for building objective functions, based on quantitative indices and the direct application of LP_τ-sequences, is described Chapter 6. The objective functions formed on the basis of qualitative information are constructed with the application of results discussed in Section 7.3.
- The *second stage* consists in constructing q payoff matrices (in accordance with the number of objective functions considered) for all combinations of the given representative combination of initial data, solution alternatives, or scenarios X_k, $k = 1, 2, \ldots, K$ and the given representative states of nature Y_s, $s = 1, 2, \ldots, S$. To construct the payoff matrices, it is necessary to analyze S problems formalized within the framework of $\langle X, F \rangle$ models. The results of this analysis are the distinct solution alternatives X_k, $k = 1, 2, \ldots, K$ $(K \leq S)$. Thereafter, X_k, $k = 1, 2, \ldots, K$ are substituted into $F_p(X)$, $p = 1, 2, \ldots, q$ for Y_s, $s = 1, 2, \ldots, S$. These substitutions generate q payoff matrices.
- The *third stage* is related to the analysis of the obtained payoff matrices. Its execution is based on the results presented in Sections 7.1 and 7.2. However, the insufficient resolving capacity of the present stage may lead to non-unique solutions and the necessity of applying the *fourth stage.*
- The *fourth stage* is associated with constructing and analyzing $\langle X, R \rangle$ models for the subsequent contraction of decision uncertainty regions. In this stage, the remaining rational solutions should be analyzed by applying additional criteria, including criteria based on knowledge, experience, and intuition of involved experts.

The flow chart in Figure 7.2 illustrates the general scheme described previously. This scheme can be used to construct solutions, including robust

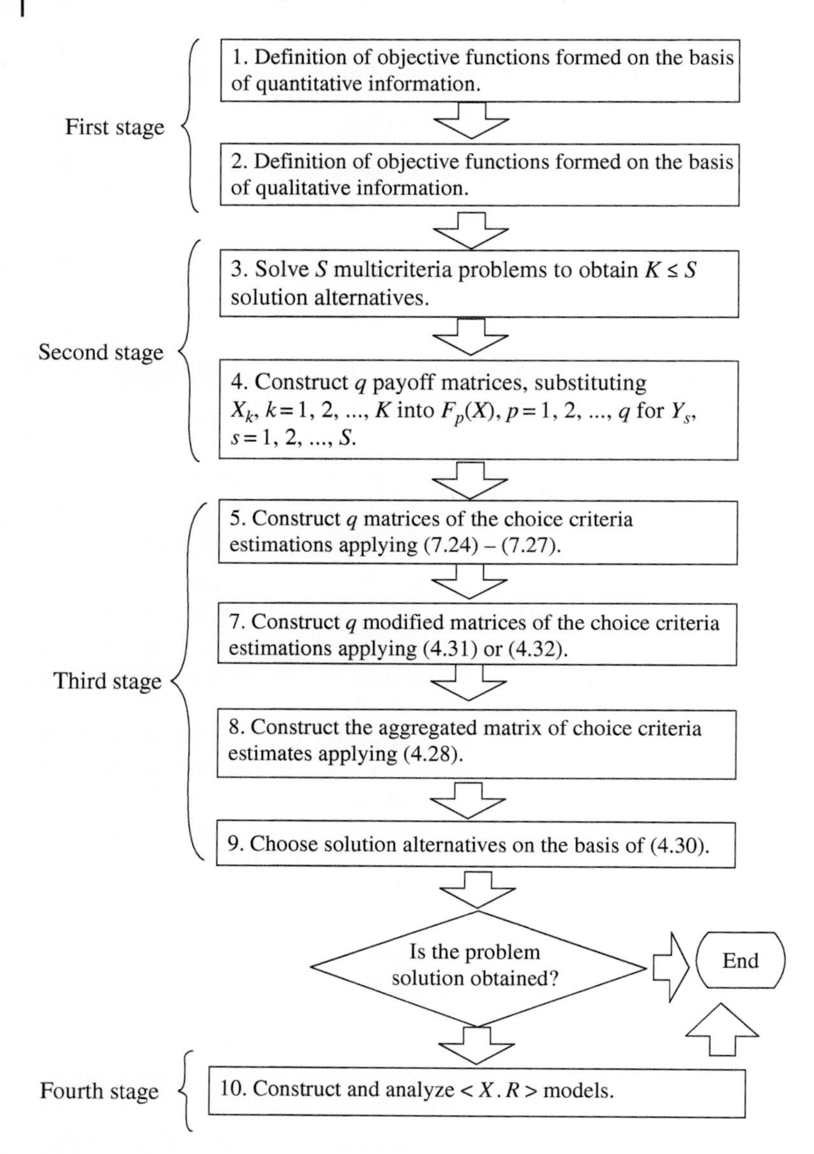

Figure 7.2 Flow chart of the general scheme of multiobjective decision-making under information uncertainty.

solutions, in multiobjective analysis under conditions of uncertainty (Ramalho 2017; Ramalho et al. 2019).

7.5 Application Studies

As examples illustrating the results of the present chapter, we consider the problems of multiobjective allocation of a shortage of financial resources in strategic planning formed on the basis of quantitative information as well as of quantitative and qualitative information (Ramalho et al. 2019). These problems are resolved within model 3 of allocating resources or their shortages, discussed in Section 4.7.

Example 7.5 Let us consider the problem of multiobjective allocating a shortage of financial resources among four strategic projects. The initial data for the problem are presented in Table 7.31.

Taking into account the demands D_i, $i = 1, 2, \ldots, 4$ as well as the minimally acceptable demands $D_i^m, i = 1, 2, \ldots, 4$ of the projects and the total available resource R = 52 000.00 kU$, it is not difficult to note that $A = \sum_{i=1}^{4} D_i - R =$ 58 927.00 – 52 000.00 = 6927.00.

Thus, we have to consider the following constraint:

$$\Delta x_1 + \Delta x_2 + \Delta x_3 + \Delta x_4 = 6{,}927.00 \tag{7.60}$$

Besides, we have to take into account that

$$\Delta x_1 \leq 2{,}512.00 \tag{7.61}$$

$$\Delta x_2 \leq 1{,}398.00 \tag{7.62}$$

$$\Delta x_3 \leq 1{,}976.00 \tag{7.63}$$

and

$$\Delta x_4 \leq 2{,}910.00 \tag{7.64}$$

Table 7.31 Example 7.5: Initial information for shortage allocation.

Project	D_i, kU$	D_i^m, kU$	A_i, kU$
1	12 532.00	10 020.00	2512.00
2	17 528.00	16 130.00	1398.00
3	9744.00	7768.00	1976.00
4	19 123.00	16 230.00	2910.00

The objectives are the following:

1) Prevailing financial constraint of projects generating a lower level of product supply abroad;
2) Prevailing financial constraint of projects generating a lower level of profit for each invested U$ 1 000 000.00.

The initial information needed for constructing objective functions corresponding to these objectives is presented in Table 7.32.

The application of LP_τ-sequences, considering the lower and upper limits given in Table 7.32 for generating, for example, $S = 7$ representative combinations of initial data, states of nature, or scenarios, gives rise to the following seven multiobjective problems:

$$F_{11}(\Delta X) = 26.000\Delta x_1 + 46.000\Delta x_2 + 29.500\Delta x_3 + 12.500\Delta x_4 \rightarrow \min \tag{7.65}$$

$$F_{12}(\Delta X) = 2,530.000\Delta x_1 + 2,100.000\Delta x_2 + 1,825.000\Delta x_3 + 2,645.000\Delta x_4 \rightarrow \min \tag{7.66}$$

$$F_{21}(\Delta X) = 23.250\Delta x_1 + 47.750\Delta x_2 + 27.500\Delta x_3 + 13.500\Delta x_4 \rightarrow \min \tag{7.67}$$

$$F_{22}(\Delta X) = 2,370.000\Delta x_1 + 2,225.000\Delta x_2 + 1,762.500\Delta x_3 + 2,797.500\Delta x_4 \rightarrow \min \tag{7.68}$$

$$F_{31}(\Delta X) = 28.750\Delta x_1 + 44.250\Delta x_2 + 31.500\Delta x_3 + 11.500\Delta x_4 \rightarrow \min \tag{7.69}$$

$$F_{32}(\Delta X) = 2,690.000\Delta x_1 + 1,975.000\Delta x_2 + 1,887.500\Delta x_3 + 2,492.500\Delta x_4 \rightarrow \min \tag{7.70}$$

$$F_{41}(\Delta X) = 21.875\Delta x_1 + 46.875\Delta x_2 + 32.500\Delta x_3 + 14.000\Delta x_4 \rightarrow \min \tag{7.71}$$

$$F_{42}(\Delta X) = 2,610.000\Delta x_1 + 1,912.500\Delta x_2 + 1,793.750\Delta x_3 + 2,568.750\Delta x_4 \rightarrow \min \tag{7.72}$$

$$F_{51}(\Delta X) = 27.375\Delta x_1 + 43.375\Delta x_2 + 28.500\Delta x_3 + 12.000\Delta x_4 \rightarrow \min \tag{7.73}$$

Table 7.32 Example 7.5: Initial information for constructing objective functions.

Project	$[c'_{1i}, c''_{1i}]$, %	$[c'_{2i}, c''_{2i}]$, kU$
1	[20.50, 31.50]	[2210.00, 2850.00]
2	[42.50, 49.50]	[1850.00, 2350.00]
3	[25.50, 33.50]	[1700.00, 1950.00]
4	[10.50, 14.50]	[2340.00, 2950.00]

$$F_{52}(\Delta X) = 2{,}290.000\Delta x_1 + 2{,}162.500\Delta x_2 + 1{,}918.750\Delta x_3 + 2{,}873.750\Delta x_4 \rightarrow \min \tag{7.74}$$

$$F_{61}(\Delta X) = 24.625\Delta x_1 + 45.125\Delta x_2 + 30.500\Delta x_3 + 11.000\Delta x_4 \rightarrow \min \tag{7.75}$$

$$F_{62}(\Delta X) = 2{,}770.000\Delta x_1 + 2{,}287.500\Delta x_2 + 1{,}731.250\Delta x_3 + 2{,}721.250\Delta x_4 \rightarrow \min \tag{7.76}$$

$$F_{71}(\Delta X) = 30.125\Delta x_1 + 48.625\Delta x_2 + 26.500\Delta x_3 + 13.000\Delta x_4 \rightarrow \min \tag{7.77}$$

$$F_{72}(\Delta X) = 2{,}450.000\Delta x_1 + 2{,}037.500\Delta x_2 + 1{,}856.250\Delta x_3 + 2{,}416.250\Delta x_4 \rightarrow \min \tag{7.78}$$

which are subject to the same constraints in Eqs. (7.61)–(7.64).

Using the results, discussed in Chapter 4, for analyzing these problems one can obtain the solutions given in Table 7.33. Substituting these solutions into Eqs. (7.65), (7.67), (7.69), (7.71), (7.73), (7.75), and (7.77), we can construct the payoff matrix for the first objective function (Table 7.34). Substituting these solutions into Eqs. (7.66), (7.68), (7.70), (7.72), (7.74), (7.76), and (7.78), we can construct the payoff matrix for the second objective function (Table 7.35).

The results presented in Tables 7.34 and 7.35 allow us to construct the characteristic estimates given in Tables 7.36 and 7.37. Then, we can apply Eqs. (7.9)–(7.11) and (7.13) to construct the matrices with the choice criteria estimates shown in Tables 7.38 and 7.39.

Applying Eq. (4.31) to the choice criteria estimates, we construct the corresponding modified matrices presented in Tables 7.40 and 7.41. The application of Eq. (4.28) to the modified matrices with the choice criteria estimates results in the construction of the matrix with the aggregated levels of the fuzzy choice criteria given in Table 7.42.

Table 7.33 Example 7.5: Solution alternatives for S = 7 scenarios.

Solution	Δx_1	Δx_2	Δx_3	Δx_4
1	1981.214	642.887	1976.000	2626.899
2	1971.689	360.733	1976.000	2618.578
3	2083.692	340.168	1976.000	2527.139
4	2221.119	69.114	1976.000	2660.768
5	2244.887	185.166	1976.000	2520.947
6	2440.874	59.381	1976.000	2450.745
7	2512.000	183.686	1976.000	2255.314

Table 7.34 Example 7.5: Payoff matrix for the first objective function.

Solution	Y_1	Y_2	Y_3	Y_4	Y_5	Y_6	Y_7
ΔX_1	164 412.62	159 589.23	169 236.00	167 908.48	161 747.26	159 574.07	168 420.65
ΔX_2	158 881.86	152 757.58	165 006.14	160 920.15	157 360.72	153 903.28	163 343.29
ΔX_3	159 704.99	153 145.27	166 264.71	161 126.11	158 437.55	154 727.55	164 528.73
ΔX_4	152 479.92	145 201.56	159 758.28	153 297.43	152 046.15	147 350.25	157 225.84
ΔX_5	156 688.54	149 408.09	163 968.99	157 299.82	156 052.73	151 634.38	161 767.24
ΔX_6	155 120.56	147 010.82	163 230.30	154 708.03	155 119.52	150 012.29	160 642.42
ΔX_7	160 244.97	151 961.74	168 528.21	159 354.67	160 113.14	155 223.28	166 288.81

Table 7.35 Example 7.5: Payoff matrix for the second objective function.

Solution	Y_1	Y_2	Y_3	Y_4	Y_5	Y_6	Y_7
ΔX_1	16 157 881.78	16 246 350.54	16 069 413.02	15 909 786.48	16 580 723.10	16 696 965.53	15 444 051.12
ΔX_2	16 278 251.23	16 283 705.90	16 272 796.56	16 106 932.25	16 611 841.59	16 833 510.54	15 560 720.54
ΔX_3	16 276 578.63	16 247 597.69	16 305 559.57	16 125 048.06	16 561 086.18	16 847 890.71	15 572 289.59
ΔX_4	16 408 499.27	16 344 026.60	16 472 971.94	16 308 596.36	16 673 651.02	16 972 160.10	15 679 589.59
ΔX_5	16 342 517.42	16 267 425.69	16 417 609.16	16 233 417.51	16 577 234.04	16 922 981.15	15 636 436.98
ΔX_6	16 388 531.83	16 255 653.26	16 521 410.40	16 324 048.48	16 552 291.34	16 987 094.82	15 690 692.67
ΔX_7	16 312 606.25	16 154 082.39	16 471 130.11	16 245 407.46	16 422 359.74	16 936 645.04	15 646 012.76

Table 7.36 Example 7.5: Matrix with the characteristic estimates for the first objective function.

Solution	$F^{max}(\Delta X_k)$	$F^{min}(\Delta X_k)$	$\bar{F}(\Delta X_k)$	$r^{max}(\Delta X_k)$
ΔX_1	164 412.62	159 589.23	169 236.00	167 908.48
ΔX_2	158 881.86	152 757.58	165 006.14	160 920.15
ΔX_3	159 704.99	153 145.27	166 264.71	161 126.11
ΔX_4	152 479.92	145 201.56	159 758.28	153 297.43
ΔX_5	156 688.54	149 408.09	163 968.99	157 299.82
ΔX_6	155 120.56	147 010.82	163 230.30	154 708.03
ΔX_7	160 244.97	151 961.74	168 528.21	159 354.67

Table 7.37 Example 7.5: Matrix with the characteristic estimates for the second objective function.

Solution	$F^{max}(\Delta X_k)$	$F^{min}(\Delta X_k)$	$\bar{F}(\Delta X_k)$	$r^{max}(\Delta X_k)$
ΔX_1	16 696 965.53	15 444 051.12	16 157 881.78	158 364.26
ΔX_2	16 833 510.54	15 560 720.54	16 278 251.23	203 383.53
ΔX_3	16 847 890.71	15 572 289.59	16 276 578.63	236 146.55
ΔX_4	16 972 160.10	15 679 589.59	16 408 499.27	403 558.91
ΔX_5	16 922 981.15	15 636 436.98	16 342 517.42	348 196.13
ΔX_6	16 987 094.82	15 690 692.67	16 388 531.83	451 997.38
ΔX_7	16 936 645.04	15 646 012.76	16 312 606.25	401 717.09

Table 7.38 Example 7.5: Matrix with the choice criteria estimates for the first objective function.

Solution	$F_1^W(X_k)$	$F_1^L(X_k)$	$F_1^S(X_k)$	$F_1^H(X_k)$
ΔX_1	169 236.00	164 412.62	14 611.05	161 989.55
ΔX_2	165 006.14	158 881.86	7622.72	155 819.72
ΔX_3	166 264.70	159 704.99	7943.71	156 425.13
ΔX_4	159 758.28	152 479.92	0.000	148 840.74
ΔX_5	163 968.99	156 688.54	4541.40	153 048.32
ΔX_6	163 230.30	155 120.56	3472.03	151 065.69
ΔX_7	168 528.21	160 244.97	9062.97	156 103.36
$\min_{1 \le k \le K} F_1(X_k)$	159 758.28	152 479.92	0.000	148 840.74
$\max_{1 \le k \le K} F_1(X_k)$	169 236.00	164 412.62	14 611.05	161 989.55

Table 7.39 Example 7.5: Matrix with the choice criteria estimates for the second objective.

Solution	$F_2^W(X_k)$	$F_2^L(X_k)$	$F_2^S(X_k)$	$F_2^H(X_k)$
ΔX_1	16 696 965.53	16 157 881.78	158 364.26	15 757 279.76
ΔX_2	16 833 510.54	16 278 251.23	203 383.53	15 878 918.04
ΔX_3	16 847 890.71	16 276 578.63	236 146.55	15 891 189.87
ΔX_4	16 972 160.10	16 408 499.27	403 558.91	16 002 732.25
ΔX_5	16 922 981.15	16 342 517.42	348 196.13	15 958 073.02
ΔX_6	16 987 094.82	16 388 531.83	451 997.38	16 014 793.24
ΔX_7	16 936 645.04	16 312 606.25	401 717.09	15 968 670.83
$\min\limits_{1\le k\le K} F_1(X_k)$	16 696 965.53	16 157 881.78	158 364.26	15 757 279.76
$\max\limits_{1\le k\le K} F_1(X_k)$	16 987 094.82	16 408 499.27	451 997.38	16 014 793.21

Table 7.40 Example 7.5: Modified matrix with the choice criteria estimates for the first objective function.

	$\mu_1^W(X_k)$	$\mu_1^L(X_k)$	$\mu_1^S(X_k)$	$\mu_1^H(X_k)$
ΔX_1	0.00	0.00	0.00	0.00
ΔX_2	0.45	0.46	0.48	0.47
ΔX_3	0.31	0.40	0.46	0.42
ΔX_4	1.00	1.00	1.00	1.00
ΔX_5	0.56	0.65	0.69	0.68
ΔX_6	0.63	0.78	0.76	0.83
ΔX_7	0.08	0.35	0.38	0.45

Table 7.41 Example 7.5: Modified matrix with the choice criteria estimates for the second objective function.

Solution	$\mu_2^W(X_k)$	$\mu_2^L(X_k)$	$\mu_2^S(X_k)$	$\mu_2^H(X_k)$
ΔX_1	1.00	1.00	1.00	1.00
ΔX_2	0.53	0.52	0.85	0.53
ΔX_3	0.48	0.53	0.74	0.48
ΔX_4	0.05	0.00	0.17	0.05
ΔX_5	0.22	0.26	0.35	0.22
ΔX_6	0.00	0.08	0.00	0.00
ΔX_7	0.17	0.38	0.17	0.18

Table 7.42 Example 7.5: Matrix with the aggregated levels of the fuzzy choice criteria.

	$\mu_D^W(X_k)$	$\mu_D^L(X_k)$	$\mu_D^S(X_k)$	$\mu_D^H(X_k)$
ΔX_1	0.00	0.00	0.00	0.00
ΔX_2	0.45	0.46	0.48	0.47
ΔX_3	0.31	0.40	0.46	0.42
ΔX_4	0.05	0.00	0.17	0.05
ΔX_5	0.22	0.26	0.35	0.22
ΔX_6	0.00	0.08	0.00	0.00
ΔX_7	0.08	0.35	0.17	0.18
$\max_{1 \le k \le K} \mu_D(X_k)$	0.45	0.46	0.48	0.47

The results presented in Table 7.42 convincingly demonstrate that problem solution is $\Delta X_2 = \{x_1^0 = 1{,}971.689, x_2^0 = 360.733, x_3^0 = 1{,}976.000, x_4^0 = 2{,}618.578\}$. However, if the aggregated fuzzy choice criteria levels indicate different solutions (the quantitative information does not lead to a unique solution), it is possible to apply the criteria of qualitative character at the final decision stage.

Example 7.6 Now let us resolve a modified problem of multiobjective financial shortage allocation in strategic planning using the data provided in Example 7.5. The modified problem is to be resolved for four projects, considering the construction of objective functions based on qualitative information.

The objectives are the following:

1) Prevailing financial constraint of projects generating a lower level of product supply abroad.
2) Prevailing financial constraint of projects generating a lower level of profit for each invested U$ 1 000 000.00.
3) Prevailing financial constraint of projects generating a lower level of innovation.

The constraints of the modified problem are the same as the previous problem, defined in Eqs. (7.61)–(7.64). The first two objective functions are the same as in the previous example. The construction of the third objective function is based on the use of qualitative information. This information is provided by a group of experts $E = \{e_1, e_2, e_3\}$, responsible for generating the necessary estimates related to a set of projects $X = \{x_1, x_2, x_3, x_4\}$ from the perspective of the indicator "Level of Innovation."

In particular, the expert e_1 has presented his/her preferences applying the fuzzy estimates of Figure 7.1, indicating the following: x_1 – VH; x_2 – L; x_3 – H; x_4 – M. The expert e_2 has also used the fuzzy estimates of Figure 7.1 and indicated x_1 – H; x_2 – M; x_3 – H; x_4 – L. Finally, the expert e_3 has ordered the alternatives as follows:

$$OA_{e_3}(x_k) = \{x_1, x_3, x_2, x_4\} \tag{7.79}$$

The first step in the problem solution is the processing of qualitative information for the construction of estimates for the third objective function. From the fuzzy estimates given by the experts e_1 and e_2, on the basis of Figure 7.1, and applying Eqs. (7.37) and (7.38), one can obtain the following nonreciprocal fuzzy preference relations:

$$NR_{e_1}(x_k, x_l) = \begin{bmatrix} 1.00 & 1.00 & 1.00 & 1.00 \\ 0.25 & 1.00 & 1.00 & 0.75 \\ 0.75 & 1.00 & 1.00 & 1.00 \\ 0.50 & 1.00 & 0.50 & 1.00 \end{bmatrix} \tag{7.80}$$

and

$$NR_{e_2}(x_k, x_l) = \begin{bmatrix} 1.00 & 1.00 & 1.00 & 1.00 \\ 0.50 & 1.00 & 0.50 & 1.00 \\ 1.00 & 1.00 & 1.00 & 1.00 \\ 0.00 & 0.75 & 0.00 & 1.00 \end{bmatrix} \tag{7.81}$$

Then, by applying Eq. (5.48), Eqs. (7.80) and (7.81) are converted into the reciprocal fuzzy preference relations

$$RR_{e_1}(x_k, x_l) = \begin{bmatrix} 0.500 & 0.875 & 0.625 & 0.750 \\ 0.125 & 0.500 & 0.000 & 0.375 \\ 0.375 & 1.000 & 0.500 & 0.750 \\ 0.250 & 0.625 & 0.250 & 0.500 \end{bmatrix} \tag{7.82}$$

and

$$RR_{e_2}(x_k, x_l) = \begin{bmatrix} 0.500 & 0.750 & 0.500 & 1.000 \\ 0.250 & 0.500 & 0.250 & 0.625 \\ 0.500 & 0.750 & 0.500 & 1.000 \\ 0.000 & 0.375 & 0.000 & 0.500 \end{bmatrix} \tag{7.83}$$

respectively.

Applying Eqs. (7.48) to Eqs. (7.82) and (7.83), we construct the multiplicative preference relations

$$MR_{e_1}(x_k, x_l) = \begin{bmatrix} 1.000 & 5.196 & 1.732 & 3.000 \\ 0.193 & 1.000 & 0.111 & 0.577 \\ 0.577 & 9.000 & 1.000 & 3.000 \\ 0.333 & 1.732 & 0.333 & 1.000 \end{bmatrix} \tag{7.84}$$

and

$$MR_{e_2}(x_k, x_l) = \begin{bmatrix} 1.000 & 3.000 & 1.000 & 9.000 \\ 0.333 & 1.000 & 0.333 & 1.732 \\ 1.000 & 3.000 & 1.000 & 9.000 \\ 0.111 & 0.577 & 0.111 & 1.000 \end{bmatrix} \tag{7.85}$$

respectively.

As the expert e_3 has expressed his/her preferences by Eq. (7.79), applying Eq. (7.44) to Eq. (7.79), we construct the following multiplicative preference relation:

$$MR_{e_3}(x_k, x_l) = \begin{bmatrix} 1.000 & 4.327 & 2.080 & 9.000 \\ 0.231 & 1.000 & 0.481 & 2.080 \\ 0.481 & 2.080 & 1.000 & 4.327 \\ 0.111 & 0.481 & 0.231 & 1.000 \end{bmatrix} \tag{7.86}$$

All the multiplicative preference relations Eqs. (7.84)–(7.86) have high levels of consistency: for all of them the maximal eigenvalues $\lambda_{\max}$ are close to the dimensionality of the relations ($n = 4$). Considering this, the eigenvectors corresponding to Eqs. (7.84)–(7.86)

$$E_{e_1} = [0.434 \;\; 0.063 \;\; 0.380 \;\; 0.122], \tag{7.87}$$

$$E_{e_2} = [0.413 \;\; 0.121 \;\; 0.413 \;\; 0.053] \tag{7.88}$$

and

$$E_{e_3} = [0.549 \;\; 0.127 \;\; 0.264 \;\; 0.061] \tag{7.89}$$

can serve as vectors of preferences.

On the basis of Eqs. (7.87)–(7.89) one can construct

$$\text{OWA}(E_{e_i}, W^p) = [0.413 \;\; 0.063 \;\; 0.264 \;\; 0.053] \tag{7.90}$$

and

$$\text{OWA}(E_{e_i}, W^O) = [0.549 \;\; 0.127 \;\; 0.413 \;\; 0.122] \tag{7.91}$$

Considering the lower Eq. (7.90) and upper Eq. (7.91) limits for the objective, based on the use of qualitative information, and the other two objectives, based on the application of quantitative information, we can now resume the second stage of the general scheme of multicriteria decision-making under conditions of information uncertainty. Therefore, we can use the LPτ-sequences for generating, for example, $S = 7$ representative combinations of initial data, states of nature, or scenarios. It gives rise to the following seven multiobjective problems:

$$F_{11}(\Delta X) = 26.000\Delta x_1 + 46.000\Delta x_2 + 29.500\Delta x_3 + 12.500\Delta x_4 \rightarrow \min \quad (7.92)$$

$$F_{12}(\Delta X) = 2,530.000\Delta x_1 + 2,100.000\Delta x_2 + 1,825.000\Delta x_3 + 2,645.000\Delta x_4 \rightarrow \min \quad (7.93)$$

$$F_{13}(\Delta X) = 0.481\Delta x_1 + 0.095\Delta x_2 + 0.338\Delta x_3 + 0.088\Delta x_4 \rightarrow \min \quad (7.94)$$

$$F_{21}(\Delta X) = 23.250\Delta x_1 + 47.750\Delta x_2 + 27.500\Delta x_3 + 13.500\Delta x_4 \rightarrow \min \quad (7.95)$$

$$F_{22}(\Delta X) = 2,370.000\Delta x_1 + 2,225.000\Delta x_2 + 1,762.500\Delta x_3 + 2,797.500\Delta x_4 \rightarrow \min \quad (7.96)$$

$$F_{23}(\Delta X) = 0.515\Delta x_1 + 0.079\Delta x_2 + 0.376\Delta x_3 + 0.070\Delta x_4 \rightarrow \min \quad (7.97)$$

$$F_{31}(\Delta X) = 28.750\Delta x_1 + 44.250\Delta x_2 + 31.500\Delta x_3 + 11.500\Delta x_4 \rightarrow \min \quad (7.98)$$

$$F_{32}(\Delta X) = 2,690.000\Delta x_1 + 1,975.000\Delta x_2 + 1,887.500\Delta x_3 + 2,492.500\Delta x_4 \rightarrow \min \quad (7.99)$$

$$F_{33}(\Delta X) = 0.447\Delta x_1 + 0.111\Delta x_2 + 0.301\Delta x_3 + 0.105\Delta x_4 \rightarrow \min \quad (7.100)$$

$$F_{41}(\Delta X) = 21.875\Delta x_1 + 46.875\Delta x_2 + 32.500\Delta x_3 + 14.000\Delta x_4 \rightarrow \min \quad (7.101)$$

$$F_{42}(\Delta X) = 2,610.000\Delta x_1 + 1,912.500\Delta x_2 + 1,793.750\Delta x_3 + 2,568.750\Delta x_4 \rightarrow \min \quad (7.102)$$

$$F_{43}(\Delta X) = 0.532\Delta x_1 + 0.103\Delta x_2 + 0.357\Delta x_3 + 0.114\Delta x_4 \rightarrow \min \quad (7.103)$$

$$F_{51}(\Delta X) = 27.375\Delta x_1 + 43.375\Delta x_2 + 28.500\Delta x_3 + 12.000\Delta x_4 \rightarrow \min \quad (7.104)$$

$$F_{52}(\Delta X) = 2,290.000\Delta x_1 + 2,162.500\Delta x_2 + 1,918.750\Delta x_3 + 2,873.750\Delta x_4 \rightarrow \min \quad (7.105)$$

$$F_{53}(\Delta X) = 0.464\Delta x_1 + 0.071\Delta x_2 + 0.282\Delta x_3 + 0.079\Delta x_4 \rightarrow \min \quad (7.106)$$

$$F_{61}(\Delta X) = 24.625\Delta x_1 + 45.125\Delta x_2 + 30.500\Delta x_3 + 11.000\Delta x_4 \rightarrow \min \quad (7.107)$$

$$F_{62}(\Delta X) = 2,770.000\Delta x_1 + 2,287.500\Delta x_2 + 1,731.250\Delta x_3 + 2,721.250\Delta x_4 \rightarrow \min \quad (7.108)$$

$$F_{63}(\Delta X) = 0.430\Delta x_1 + 0.119\Delta x_2 + 0.320\Delta x_3 + 0.096\Delta x_4 \rightarrow \min \quad (7.109)$$

$$F_{71}(\Delta X) = 30.125\Delta x_1 + 48.625\Delta x_2 + 26.500\Delta x_3 + 13.000\Delta x_4 \to \min \quad (7.110)$$

$$F_{72}(\Delta X) = 2,450.000\Delta x_1 + 2,037.500\Delta x_2 + 1,856.250\Delta x_3 + 2,416.250\Delta x_4 \to \min \quad (7.111)$$

$$F_{73}(\Delta X) = 0.498\Delta x_1 + 0.087\Delta x_2 + 0.394\Delta x_3 + 0.062\Delta x_4 \to \min \quad (7.112)$$

which are subject to the constraints in Eqs. (7.61)–(7.64).

Using the Bellman–Zadeh approach presented in Chapter 4 for analyzing the problems in Eqs. (7.92)–(7.112), one can construct Table 7.43. Substituting these solutions into Eqs. (7.92), (7.95), (7.98), (7.101), (7.104), (7.107), and (7.110), we construct the payoff matrix for the first objective function (Table 7.44). Substituting these solutions into Eqs. (7.93), (7.96), (7.99), (7.102), (7.105), (7.108), and (7.111), the payoff matrix for the second objective function is constructed (Table 7.45). Finally, substituting the same solutions into Eqs (7.94), (7.97), (7.100), (7.103), (7.106), (7.109), and (7.112), we construct the payoff matrix for the third objective function (Table 7.46).

The results shown in Tables 7.44–7.46 produce matrices with the characteristic estimates presented in Tables 7.47–7.49 and the corresponding matrices with the choice criteria estimates are given in Tables 7.50–7.52. Applying Eq. (4.31) to these estimates, we construct the corresponding modified matrices, presented in Tables 7.53–7.55. The modified matrices with the choice criteria estimates result in the construction of the matrix with the aggregated levels of the fuzzy choice criteria, presented in Table 7.56.

Finally, the problem's solution is $\Delta X_6 = \{\Delta x_1^0 = 1,569.496,\ \Delta x_2^0 = 797.186,\ \Delta x_3^0 = 1,716.699,\ \Delta x_3^0 = 2,843.620\}$, which differs significantly from the solution obtained with considering only the objective functions formed on the basis of quantitative information, presented in Example 7.5. Once we found a single solution to the problem, there is no need to apply the fourth stage of the general scheme of multicriteria decision-making under conditions of information uncertainty.

Table 7.43 Example 7.6: Solution alternatives for $S = 7$ scenarios.

Solution	Δx_1	Δx_2	Δx_3	Δx_4
1	1183.09	1398.00	1435.91	2910.00
2	1193.59	858.84	1976.00	2898.58
3	1294.36	947.73	1830.68	2854.24
4	1390.00	1329.14	1297.86	2901.00
5	1560.76	888.64	1589.21	2888.39
6	1569.50	797.19	1716.70	2843.62
7	2272.74	304.83	1439.44	2910.00

Table 7.44 Example 7.6: Payoff matrix for the first objective function.

Solution	Y_1	Y_2	Y_3	Y_4	Y_5	Y_6	Y_7
ΔX_1	173 802.68	173 033.86	174 571.50	178 818.39	168 868.78	168 023.58	179 499.97
ΔX_2	165 063.94	162 231.11	167 896.77	171 167.73	161 025.39	160 299.41	167 763.22
ΔX_3	166 931.72	164 223.62	169 639.81	172 195.13	162 965.89	161 872.04	170 693.82
ΔX_4	171 942.32	170 760.10	173 124.55	175 630.15	167 611.72	165 800.94	178 726.48
ΔX_5	164 443.82	161 416.82	167 470.83	167 883.47	161 223.78	158 776.85	169 891.19
ΔX_6	163 665.30	160 154.50	167 176.11	167 304.19	160 592.23	158 260.97	168 503.79
ΔX_7	151 951.54	146 266.09	157 637.00	151 526.53	151 381.96	145 634.24	159 263.45

Table 7.45 Example 7.6: Payoff matrix for the second objective function.

Solution	Y_1	Y_2	Y_3	Y_4	Y_5	Y_6	Y_7
ΔX_1	16 246 505.72	16 585 991.54	15 907 019.91	15 812 268.65	16 850 216.95	16 879 844.40	15 443 692.89
ΔX_2	16 096 266.88	16 331 180.09	15 861 353.66	15 747 955.80	16 711 782.23	16 579 526.17	15 345 803.31
ΔX_3	16 155 396.78	16 387 618.28	15 923 175.29	15 806 402.91	16 728 515.46	16 689 751.31	15 396 917.45
ΔX_4	16 373 437.98	16 679 839.11	16 067 036.85	15 972 978.62	16 910 246.03	17 056 464.75	15 554 062.52
ΔX_5	16 354 964.30	16 557 476.42	16 152 452.18	16 043 302.35	16 845 629.68	16 967 417.49	15 563 507.68
ΔX_6	16 299 263.71	16 474 150.88	16 124 376.55	16 004 878.25	16 783 827.30	16 881 300.51	15 527 048.80
ΔX_7	16 714 082.22	16 742 357.56	16 685 806.89	16 571 876.55	16 988 286.35	17 403 634.48	15 892 531.50

Table 7.46 Example 7.6: Payoff matrix for the third objective function.

Solution	Y_1	Y_2	Y_3	Y_4	Y_5	Y_6	Y_7
ΔX_1	1442.49	1463.72	1421.25	1615.90	1283.59	1414.00	1456.47
ΔX_2	1578.05	1628.57	1527.53	1757.48	1401.73	1525.97	1627.01
ΔX_3	1581.89	1629.75	1534.02	1763.29	1410.25	1529.13	1624.87
ΔX_4	1488.72	1512.92	1464.52	1669.54	1335.68	1450.63	1499.03
ΔX_5	1625.65	1673.90	1577.40	1816.53	1464.12	1562.73	1659.23
ΔX_6	1660.37	1715.90	1604.84	1852.18	1494.13	1592.06	1703.12
ΔX_7	1863.73	1939.47	1787.99	2083.95	1712.26	1753.61	1905.10

Table 7.47 Example 7.6: Matrix with the characteristic estimates for the first objective function.

Solution	$F^{max}(\Delta X_k)$	$F^{min}(\Delta X_k)$	$\bar{F}(\Delta X_k)$	$r^{max}(\Delta X_k)$
ΔX_1	179 499.97	168 023.58	173 802.68	27 291.86
ΔX_2	171 167.73	160 299.41	165 063.94	19 641.20
ΔX_3	172 195.13	161 872.04	166 931.72	20 668.59
ΔX_4	178 726.48	165 800.94	171 942.32	24 494.01
ΔX_5	169 891.19	158 776.85	164 443.82	16 356.94
ΔX_6	168 503.79	158 260.97	163 665.30	15 777.66
ΔX_7	159 263.45	145 634.24	151 951.54	0.00

Table 7.48 Example 7.6: Matrix with the characteristic estimates for the second objective function.

Solution	$F^{max}(\Delta X_k)$	$F^{min}(\Delta X_k)$	$\bar{F}(\Delta X_k)$	$r^{max}(\Delta X_k)$
ΔX_1	16 879 844.40	15 443 692.89	16 246 505.72	300 318.24
ΔX_2	16 711 782.23	15 345 803.31	16 096 266.88	0.00
ΔX_3	16 728 515.46	15 396 917.45	16 155 396.78	110 225.14
ΔX_4	17 056 464.75	15 554 062.52	16 373 437.98	476 938.59
ΔX_5	16 967 417.49	15 563 507.68	16 354 964.30	387 891.32
ΔX_6	16 881 300.51	15 527 048.80	16 299 263.71	301 774.34
ΔX_7	17 403 634.48	15 892 531.50	16 714 082.22	824 453.22

Table 7.49 Example 7.6: Matrix with the characteristic estimates for the third objective.

Solution	$F^{max}(\Delta X_k)$	$F^{min}(\Delta X_k)$	$\bar{F}(\Delta X_k)$	$r^{max}(\Delta X_k)$
ΔX_1	1615.899	1283.586	1442.487	0.000
ΔX_2	1757.480	1401.731	1578.050	170.548
ΔX_3	1763.294	1410.247	1581.885	168.402
ΔX_4	1669.536	1335.676	1488.718	53.637
ΔX_5	1816.528	1464.123	1625.653	210.182
ΔX_6	1852.183	1494.132	1660.373	252.181
ΔX_7	2083.953	1712.262	1863.730	475.750

Table 7.50 Example 7.6: Matrix with the choice criteria estimates for the first objective function.

	$F_1^W(X_k)$	$F_1^L(X_k)$	$F_1^S(X_k)$	$F_1^H(X_k)$
ΔX_1	179 499.97	173 802.68	27 291.86	170 892.68
ΔX_2	171 167.73	165 063.94	19 641.20	163 016.49
ΔX_3	172 195.13	166 931.72	20 668.59	164 452.81
ΔX_4	178 726.48	171 942.32	24 494.01	169 032.32
ΔX_5	169 891.19	164 443.82	16 356.94	161 555.44
ΔX_6	168 503.79	163 665.30	15 777.66	160 821.68
ΔX_7	159 263.45	151 951.54	0.00	149 041.55
$\min_{1 \le k \le K} F_1(X_k)$	159 263.45	151 951.54	0.00	149 041.55
$\max_{1 \le k \le K} F_1(X_k)$	179 499.97	173 802.68	27 291.86	170 892.68

Table 7.51 Example 7.6: Matrix with the choice criteria estimates for the second objective function.

	$F_2^W(X_k)$	$F_2^L(X_k)$	$F_2^S(X_k)$	$F_2^H(X_k)$
ΔX_1	16 879 844.40	16 246 505.72	300 318.24	15 802 730.77
ΔX_2	16 711 782.23	16 096 266.88	0.00	15 687 298.04
ΔX_3	16 728 515.46	16 155 396.78	110 225.14	15 729 816.95
ΔX_4	17 056 464.75	16 373 437.98	476 938.59	15 929 663.07
ΔX_5	16 967 417.49	16 354 964.30	387 891.32	15 914 485.13

Table 7.51 (Continued)

	$F_2^W(X_k)$	$F_2^L(X_k)$	$F_2^S(X_k)$	$F_2^H(X_k)$
ΔX_6	16 881 300.51	16 299 263.71	301 774.34	15 865 611.73
ΔX_7	17 403 634.48	16 714 082.22	824 453.22	16 270 307.24
$\min\limits_{1\leq k\leq K} F_2(X_k)$	16 711 782.23	16 096 266.88	0.00	15 687 298.04
$\max\limits_{1\leq k\leq K} F_2(X_k)$	17 403 634.48	16 714 082.22	824 453.22	16 270 307.24

Table 7.52 Example 7.6: Matrix with the choice criteria estimates for the third objective function.

	$F_3^W(X_k)$	$F_3^L(X_k)$	$F_3^S(X_k)$	$F_3^H(X_k)$
ΔX_1	1615.90	1442.49	0.00	1366.66
ΔX_2	1757.48	1578.05	170.55	1490.67
ΔX_3	1763.29	1581.89	168.40	1498.51
ΔX_4	1669.54	1488.72	53.64	1419.14
ΔX_5	1816.53	1625.65	210.18	1552.22
ΔX_6	1852.18	1660.37	252.18	1583.65
ΔX_7	2083.95	1863.73	475.75	1805.19
$\min\limits_{1\leq k\leq K} F_3(X_k)$	1615.90	1442.49	0.00	1366.66
$\max\limits_{1\leq k\leq K} F_3(X_k)$	2083.95	1863.73	475.75	1805.19

Table 7.53 Example 7.6: Modified matrix with the choice criteria estimates for the first objective function.

	$\mu_1^W(X_k)$	$\mu_1^L(X_k)$	$\mu_1^S(X_k)$	$\mu_1^H(X_k)$
ΔX_1	0.00	0.00	0.00	0.00
ΔX_2	0.41	0.40	0.28	0.36
ΔX_3	0.36	0.31	0.24	0.30
ΔX_4	0.04	0.09	0.10	0.09
ΔX_5	0.48	0.43	0.40	0.43
ΔX_6	0.54	0.46	0.42	0.46
ΔX_7	1.00	1.00	1.00	1.00

Table 7.54 Example 7.6: Modified matrix with the choice criteria estimates for the second objective function.

	$\mu_2^W(X_k)$	$\mu_2^L(X_k)$	$\mu_2^S(X_k)$	$\mu_2^H(X_k)$
ΔX_1	0.76	0.76	0.64	0.80
ΔX_2	1.00	1.00	1.00	1.00
ΔX_3	0.98	0.90	0.87	0.93
ΔX_4	0.50	0.55	0.42	0.58
ΔX_5	0.63	0.58	0.53	0.61
ΔX_6	0.76	0.67	0.63	0.69
ΔX_7	0.00	0.00	0.00	0.00

Table 7.55 Example 7.6: Modified matrix with the choice criteria estimates for the third objective function.

	$\mu_3^W(X_k)$	$\mu_3^L(X_k)$	$\mu_3^S(X_k)$	$\mu_3^H(X_k)$
ΔX_1	1.00	1.00	1.00	1.00
ΔX_2	0.70	0.68	0.64	0.72
ΔX_3	0.69	0.67	0.65	0.70
ΔX_4	0.89	0.89	0.89	0.88
ΔX_5	0.57	0.57	0.56	0.58
ΔX_6	0.50	0.48	0.47	0.51
ΔX_7	0.00	0.00	0.00	0.00

Table 7.56 Example 7.6: Aggregated payoff matrix of choice criteria estimates.

	$\mu_D^W(X_k)$	$\mu_D^L(X_k)$	$\mu_D^S(X_k)$	$\mu_D^H(X_k)$
ΔX_1	0.00	0.00	0.00	0.00
ΔX_2	0.41	0.40	0.28	0.36
ΔX_3	0.36	0.31	0.24	0.30
ΔX_4	0.04	0.09	0.10	0.09
ΔX_5	0.48	0.43	0.40	0.43
ΔX_6	0.50	0.46	0.42	0.46
ΔX_7	0.00	0.00	0.00	0.00
$\max \mu_D(X_k)$	0.50	0.46	0.42	0.46

7.6 Conclusions

We have considered the use of the choice criteria of the classic approach to handle information uncertainty in monocriteria decision-making as objective functions within the framework of multiobjective models. This consideration has permitted us to propose the general methodology for multicriteria decision-making under conditions of uncertainty. The proposed methodology is based on a possibilistic approach to produce solutions, including robust solutions, in multicriteria analysis. Its use allows one to use available quantitative information to the highest degree to reduce the decision uncertainty regions. If the solving capacity of quantitative information processing does not permit one to obtain unique solutions, the methodology supposes the utilization, at the final decision stage, of qualitative information within the framework of $<X, R>$ models. Thus, the general scheme is based on combining $<X, F>$ models, $<X, R>$ models, and the generalization of the classic approach to dealing with uncertainty of information.

However, in recent years, problems have appeared more often that require the consideration of the objectives formed with the use of qualitative information, at all stages, from the beginning of the decision-making process. Considering this, in this chapter an approach has been described that permits one to generate robust multiobjective solutions (within the framework of the possibilistic approach) by constructing representative combinations of initial data, states of nature, or scenarios with the direct use of qualitative information along with quantitative information, assuring convincing information fusion. The approach admits the possibility for experts to apply diverse preference formats with their processing on the basis of different transformation functions.

Exercises

7.1 Apply the approach described in Section 7.1, associated with constructing an aggregated payoff matrix, to analyze a bicriteria decision-making problem related to considering the solution alternatives X_k, $k = 1, 2, \ldots, 4$ with the presence of the representative combinations of initial data,

Table 7.57 Problem 7.1: Payoff matrix for the first criterion.

	Y_1	Y_2	Y_3	Y_4
X_1	2	3	3	5
X_2	7	1	2	3
X_3	3	8	2	6
X_4	1	5	8	4

Table 7.58 Problem 7.1: Payoff matrix for the second criterion.

	Y_1	Y_2	Y_3	Y_4
X_1	9	3	4	8
X_2	8	5	3	8
X_3	6	6	9	4
X_4	7	2	4	7

states of nature, or scenarios Y_s, $s = 1, 2, \ldots, 4$. Both objective functions are to be minimized. The corresponding payoff matrices are presented in Tables 7.57 and 7.58. Also, take into consideration $\alpha = 0.75$ in applying the Hurwicz choice criterion.

7.2 Solve the problem given in Problem 7.1 by applying the approach to dealing with uncertainty of information based on considering the choice criteria as objective functions, discussed in Section 7.2.

7.3 Suppose that members of a group of experts $E = \{e_1, e_2, e_3\}$ presented their preferences relatively to a set of alternatives $X = \{x_1, x_2, x_3, x_4\}$. The expert e_1 expressed his/her preferences by the order of alternatives $O_{e_1}(x_k) = \{x_3, x_1, x_2, x_4\}$. The experts e_2 and e_3 presented their preferences applying the fuzzy estimates shown in Figure 7.1, indicating the following: x_1 – H; x_2 – L; x_3 – M; x_4 – VH and x_1 – M; x_2 – H; x_3 – VH; x_4 – M, respectively. Based on the preferences given here, construct the corresponding multiplicative preference relations for each expert.

7.4 Based on the multiplicative preference relations, obtained in solving Problem 7.3, find the eigenvectors to define the preference vectors for each expert. Then, applying OWA (E_{e_i}, W), $i = 1,2,3$ for the vectors $W^p = [1 \ 0 \ 0]$ (pessimistic) and $W^o = [0 \ 0 \ 1]$ (optimistic), construct the vector indicating the limits of pessimism/optimism expressed by the preferences of the experts.

7.5 Solve the following multiobjective optimization problem:

$$F_1(x) = [2,5]x_1 + [3,6]x_2 + [4,7]x_3 + [4,6]x_4 \rightarrow \min$$
$$F_2(x) = [1,5]x_1 + [4,7]x_2 + [2,6]x_3 + [3,8]x_4 \rightarrow \min$$

constructing $S = 7$ representative combinations of initial data, states of nature, or scenarios and considering $\alpha = 0.75$ in the use of the Hurwicz choice criterion (Table 7.22 includes the necessary points of the LP_τ-sequences):

The constraints to be taken into account are the following:

$$0 \le x_1 \le 90;$$
$$0 \le x_2 \le 120;$$
$$0 \le x_3 \le 75;$$
$$0 \le x_4 \le 100;$$
$$x_1 + x_2 + x_3 + x_4 = 250$$

7.6 Consider the following preferences expressed by a group of experts:

- The expert e_1 ordered alternatives as follows: $O_{e_1}(x_k) = \{x_3, x_4, x_2, x_1\}$;
- The expert e_2 defined the following fuzzy estimates (using the fuzzy set-based qualitative scales in Figure 7.1): x_1 – H; x_2 – M; x_3 – VH; x_4 – L;
- The expert e_3 also defined fuzzy estimates (using the same fuzzy set-based qualitative scales): x_1 – L; x_2 – VH; x_3 – H; x_4 – M.

Construct $S = 7$ representative combinations of initial data, states of nature, or scenarios and on the basis of these given preferences.

7.7 Solve the multiobjective problem described in Problem 7.5 adding the results from solving Problem 7.6.

References

Bellman, R.E. and Zadeh, L.A. (1970). Decision-making in a fuzzy environment. *Management Science* 17 (4): B-141.

Damodaran, A. (2008). *Strategic Risk Taking: A Framework for Risk Management.* Upper Saddle River, NJ: Pearson Prentice Hall.

Ekel, P., Kokshenev, I., Parreiras, R. et al. (2016). Multiobjective and multiattribute decision making in a fuzzy environment and their power engineering applications. *Information Sciences* 360-361 (1): 100–119.

Ekel, P.Y., Kokshenev, I., Palhares, R. et al. (2011). Multicriteria analysis based on constructing payoff matrices and applying methods of decision making in fuzzy environment. *Optimization and Engineering* 12 (1–2): 5–29.

Ekel, P.Y., Martini, J.S.C., and Palhares, R.M. (2008). Multicriteria analysis in decision making under information uncertainty. *Applied Mathematics and Computation* 200 (2): 501–516.

Hodges, J.L. and Lehmann, E.L. (1952). The use of previous experience in reaching statistical decisions. *The Annals of Mathematical Statistics* 23 (3): 396–407.

Kokshenev, I., Parreiras, R., Ekel, P. et al. (2015). A web-based decision support center for electrical energy companies. *IEEE Transactions on Fuzzy Systems* 23 (1): 16–28.

Luce, R.D. and Raiffa, H. (1957). *Games and Decisions*. New York: Wiley.

Palomares, I., Liu, J., Xu, Y., and Martinez, L. (2012). Modelling experts' attitudes in group decision making. *Soft Computing* 16 (10): 1755–1766.

Pareto, V. (1886). *Cours d"Economie Politique, Lousanne Rouge*. Lousanne.

Pedrycz, W., Ekel, P.Y., and Parreiras, R. (2011). *Fuzzy Multicriteria Decision-Making: Methods and Applications*. Chichester: Wiley.

Pereira, J.G. Jr., Ekel, P.Y., Palhares, R.M., and Parreiras, R.O. (2015). On multicriteria decision making under conditions of uncertainty. *Information Sciences* 324 (1): 44–59.

Queiroz, J.C.B. (2009) Models and Methods of Decision Making to Support Strategic Management in Electric Power Companies: PhD thesis, Federal University of Minas Gerais, Belo Horizonte (in Portuguese).

Raiffa, H. (1968). *Decision Analysis: Introductory Lectures on Choices und Uncertainty*. Addison-Wesley.

Ramalho, F.D. (2017) *Utilization of Qualitative Information in the Process of Decision Making: M.Sc. Dissertation*, Pontifical Catholic University of Minas Gerais, Belo Horizonte (in Portuguese).

Ramalho, F.D., Ekel, P.Y., Pedrycz, W. et al. (2019). Multicriteria decision making under conditions of uncertainty in application to multiobjective allocation of resources. *Information Fusion* 49 (2): 249–261.

Saaty, T. (1980). *The Analytic Hierarchy Process*. New York: McGraw-Hill.

Saaty, T.L., Peniwati, K., and Shang, J.S. (2007). The analytic hierarchy process and human resource allocation: half the story. *Mathematical and Computer Modelling* 46 (7–8): 1041–1053.

Saaty, T.L., Vargas, T.L., and Dellman, K. (2003). The allocation of intangible resources: the analytic hierarchy process and linear programming. *Socio-Economic Planning Sciences* 37 (3): 169–184.

Sobol', I.M. (1966). On the distribution of points in a cube and integration grids. *Achievements of Mathematical Sciences* 21 (5): 271–272. in Russian.

Sobol', I.M. (1979). On the systematic search in a hypercube. *SIAM Journal on Numerical Analysis* 16 (5): 790–793.

Trukhaev, R.I. (1981). *Models of Decision Making in Conditions of Uncertainty*. Moscow (in Russian): Nauka.

Webster, T.J. (2003). *Managerial Economics: Theory and Practice*. London: Academic Press.

Yager, R.R. (1978). Fuzzy decision making including unequal objectives. *Fuzzy Sets and Systems* 1 (1): 87–95.

Index

Multicriteria Decision-Making under Conditions of Uncertainty: A Fuzzy Set Perspective,
First Edition. Petr Ekel, Witold Pedrycz, and Joel Pereira, Jr.

W

X